PURINE METABOLISM
IN MAN—III
Biochemical, Immunological, and
Cancer Research

ADVANCES IN EXPERIMENTAL MEDICINE AND BIOLOGY

Recent Volumes in this Series

PURINE METABOLISM IN MAN—III
Clinical and Therapeutic Aspects

Edited by

Aurelio Rapado
Fundación Jiménez Díaz
Madrid, Spain

R.W.E. Watts
M.R.C. Clinical Research Centre
Harrow, England

and

Chris H.M.M. De Bruyn
Department of Human Genetics
University of Nijmegen Faculty of Medicine
Nijmegen, The Netherlands

Springer Science+Business Media, LLC

Library of Congress Cataloging in Publication Data

International Symposium on Purine Metabolism in Man, 3d, Madrid, 1979.
  Purine metabolism in man, III.

  (Advances in experimental medicine and biology; v. 122A-122B)
  Includes index.
  CONTENTS: [1] Clinical and therapeutic aspects. — [2] Biochemical, immu-
nological and cancer research.
  1. Purine metabolism — Congresses. 2. Hyperuricemia — Congresses. 3. Immu-
nopathology — Congresses. 4. Cancer — Congresses. I. Rapado, A. II. Watts, R. W.
E. III. De Bruyn, C. H. M. M. IV. Title. V. Series. [DNLM: 1. Purine-pyrimidine
metabolism, Inborn errors — Congresses. 2. Purines — Metabolism — Congresses.
W3 IN922NM 3d 1979p/WD205.5P8 I61 1979p]
QP801.P8I56 1979   612'.0157   79-22555

ISBN 978-1-4615-9142-9   ISBN 978-1-4615-9140-5 (eBook)
DOI 10.1007/ 978-1-4615-9140-5

Proceedings of the first half of the Third International Symposium on
Purine Metabolism in Man, held in Madrid, Spain, June 11–15, 1979

© 1980 Springer Science+Business Media New York
Originally published by Plenum Press, New York in 1980
Softcover reprint of the hardcover 1st edition 1980

Preface

These volumes contain the papers which were presented at the
Third International Symposium on Purine Metabolism in Man held in
Madrid (Spain) in June, 1979. The previous meetings in the series
were held in Tel Aviv (Israel) and in Baden (Austria) in 1973 and
1976, respectively. The proceedings were also published by
Plenum.

Knowledge of the pathophysiology of the purines has developed
greatly since the 1950's when it was mainly related to clinical
gout, and it is now relevant to many fields of Medicine and Biology.
These volumes include papers reporting new work on clinical gout
and urolithiasis as well as on some of the subjects which have
featured prominently in the previous volumes, including: regulatory
aspects of the intermediary metabolism of purines and related compounds, enzymology, methodology, and the results of mutations which
affect purine metabolism. However, there have been many new develop-
ments during the last three years and the scope of the communications
reflects not only increasing depth of knowledge, but also a widening
of the field. This publication has clinical and fundamental implications for internal medicine, pediatrics, urology, biochemistry,
immunology, genetics, and oncology.

It is interesting to compare the scope of this volume with that
of its predecessors. The main emphasis has shifted from the study
of gout and the dissection of metabolic pathways to encompass investigations in the fields of oncology, immunology, and lymphocyte
physiology. There are pointers to possible implications in relation
to cardiology and neuromuscular diseases, which may well prove to
be growing points for the future. In spite of considerable work
on the mechanism of urinary stone formation, the inter-relationship
between uric acid and calcium oxalate urolithiasis remains obscure.

It is no longer logical to discuss clinically related purine
research without including comparable work in the less studied field
of pyrimidine metabolism. Some such studies were reported at the
Madrid meeting, and this development will be formally encouraged in
the future.

The use of some animal and single cell models as tools with complexity intermediate between man and the single or multi-enzyme systems represents another new development in this area of clinical investigation.

We acknowledge the support which we received from the distinguished members of the scientific community who served on the Organizing and Scientific Committees, as well as their contributions to the high standards of the material presented.

We also thank the "Fundacion Jimenez Diaz" and the Autonomous University of Madrid, both of whom sponsored the meeting, the Department of Cultural Relations in the Ministry of Foreign Affairs, the Madrid City Council and the Wellcome Research Laboratories (England) for their financial support, and Plenum Publishing Corporation (U.S.A.) for their assistance in the publication of the proceedings. The meeting would not have been possible without the cheerful and spirited help of Maria Luisa San Roman and Mireya Usano, and our special thanks are due to them.

A. Rapado
R.W.E. Watts
C.H.M.M. de Bruyn

Contents of Part A

III.   CLINICAL AND PHYSIOLOGICAL ASPECTS OF PURINE METABOLISM

VI. MUTATIONS AFFECTING PURINE METABOLISM

A. Phosphoribosyltransferases

B. Nucleoside Phosphoribosylating Enzymes

Contents of Part B

IV. LYMPHOCYTE PURINE METABOLISM RESEARCH

THE NATURAL HISTORY OF HYPERURICEMIA AMONG ASYMPTOMATIC RELATIVES

OF PATIENTS WITH GOUT

Ts'ai-fan Yü and Clara Kaung

Department of Medicine, Mt Sinai School of Medicine

City University of New York, New York, New York, USA

Hyperuricemic trait is inborn, but it can be modified by environmental changes. Two hundred twenty asymptomatic relatives from 140 patients with gout were studied, 172 males and 48 females.

Normal serum urate concentrations up to 7.0 mg/dl were found in 106 of 172 male relatives of patients with gout (62%), and in 41 of 48 female relatives (85%). If upper limit of normal is set at 6.0 mg/dl for women, then the % with normouricemia was 71% only (Fig 1).

Wide fluctuations in serum urate were found in 96 relatives who had repeated determinations followed over a period of few years to more than 20 years. Four different patterns of fluctuations were observed (Table 1). Twenty-seven relatives maintained their normal levels. Twenty-nine fluctuated between normal and abnormal ranges, 17 had higher levels in later years, and 12 showed higher level in earlier years. Twenty-one relatives had fluctuations within the hyperuricemic range, mostly having upward trend. Multiple fluctuations occurred in 19 relatives. In all 19 except one, the initial serum urate levels were followed by higher values, but became

Table 1 Patterns of Serum Urate Fluctuations

		No.	%
1.	Within Normal Range	27	38
2.	Between Normal and Abnormal Ranges	29	30
3.	Within Abnormal Range	21	22
4.	Bidirectional Fluctuation	19	20

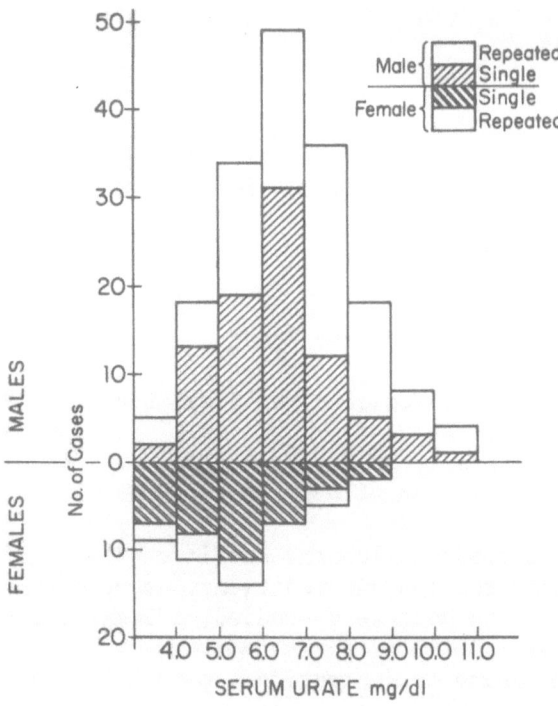

Figure 1

lower again. Differences of more than 2 mg/dl were frequently
observed. Satisfactory decrease in serum urate levels were often
related to careful dietary control, and steady but gradual weight
loss. Change in life style is another factor in serum urate
reduction.

Urinary uric acid, creatinine and total nitrogen were determined
in 100 subjects with 24 hour collection, and repeatedly in 33.
Using urinary uric acid nitrogen to total nitrogen ratios (UA-N/TN)
$1.6\pm0.2\%$ as the norm for normal subjects, (1) the distribution of
UA-N/TN for these subjects was skewed to the right (Fig 2). About
2/3 of them had a uric acid-N/TN>1.6%, 44%>2.0% ($1.6\pm2sd$). Relatively
more hyperuricemic subjects were over-excretors.

Urine pH was less than 6.0 in 75 of the 87 relatives. Each
had a normal creatinine clearance of more than 100 ml/min. There
were considerable differences in the output of TA, NH$_4^+$ and uric
acid among the different members. In comparing with 19 nongouty
control subjects of similar ages at same urine ph range, no difference in titratable acidity was found. There was, however, a trend
of decreased excretion of NH$_4^+$ and more of uric acid in relation to

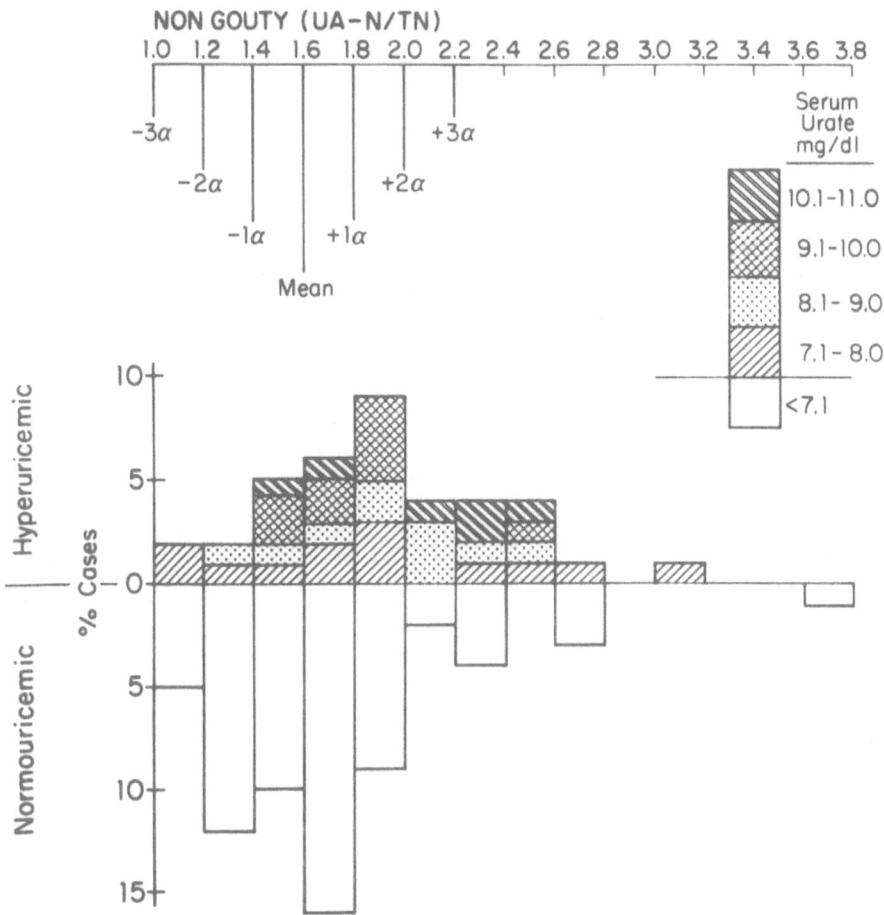

Figure 2

total nitrogen output, especially among those with hyperuricemia over 8 mg/dl (Fig 3). The decreased NH_4^+ and increased uric acid excretions were found to be more exaggerated by the group younger than 30 years old (2).

Plasma amino acids were determined in 37 relatives (Table 2). The mean plasma glutamic acid concentrations among the 21 normo-uricemic relatives and 9 with serum urate between 7.1 and 8.0 mg/dl, were not different from the 28 nongouty subjects. The hyperuricemic relatives with serum urate more than 8 mg/dl on the other hand, had plasma glutamic acid concentrations comparable to the 62 gouty men within similar serum urate concentration range. Plasma glutamine concentrations were not different from the gouty or nongouty sub-jects (3,4).

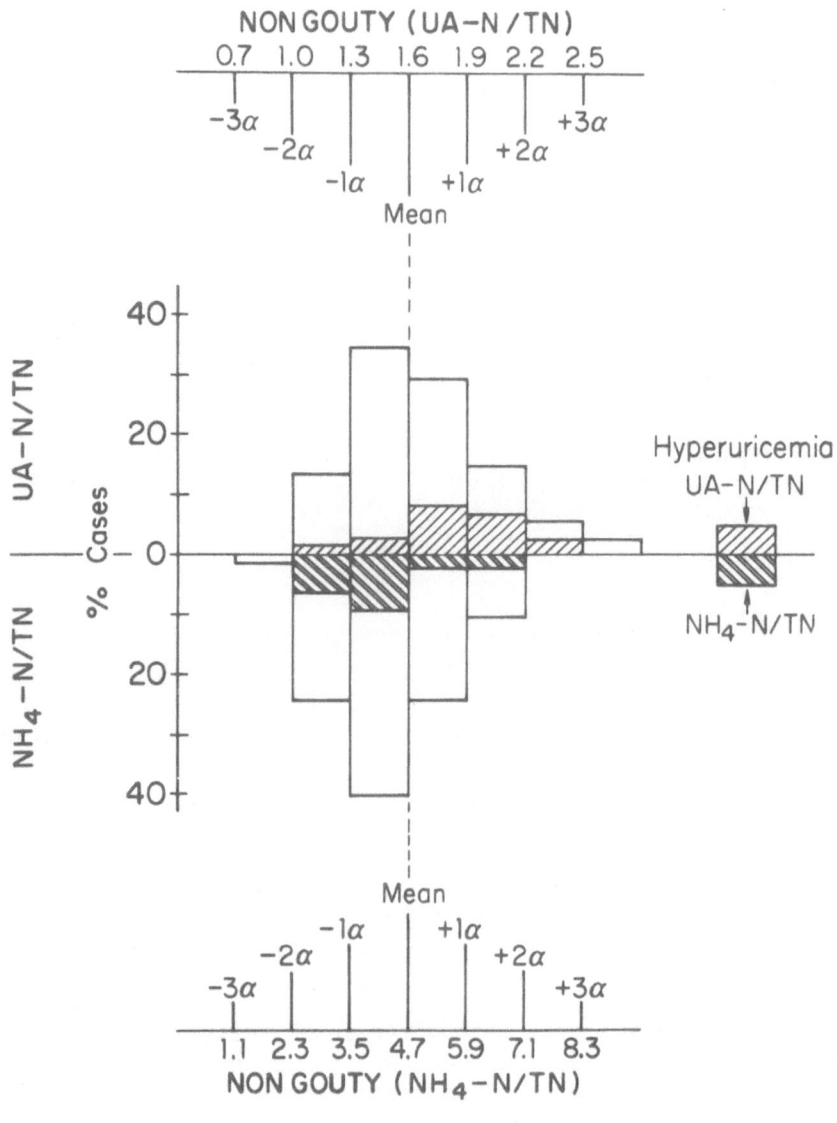

Figure 3

Table 2 Plasma Glutamic Acid (G.A.) and Glutamine

		No.	Purate (mg/dl)	Glutamic Ac (mM/L)	Glutamine (mM/L)
Relatives	a.	21	7.1	63$\pm$18	482$\pm$55
of Gouty	b.	9	7.1- 8.0	51$\pm$13	498$\pm$43
Subjects	c.	7	8.1-11.0	88$\pm$30	488$\pm$56
Gouty	a.	8	7.1- 8.0	67$\pm$31	480$\pm$80
Subjects	b.	62	8.1-11.0	80$\pm$27	495$\pm$75
Non-Gouty Subjects		28	7.1	53$\pm$16	500$\pm$78

Six relatives developed acute gouty arthritis. Three had early
onset before age 30, with serum urate repeatedly over 10 mg/dl,
excessive urinary uric acid more than 900 mg/day, and uric acid-N/TN
ratio exceeding 2%, and low NH_4^+ excretion. One was found to have
an accelerated rate of uric acid incorporation from ^{15}N-glycine (5).
Two others had serum urate < 9 mg/dl in early years. Gout developed
at ages 36 and 48 after being followed for 10 and 21 years respectively, with increasing serum urate > 9 mg/dl. One other elderly
man of 70 years of age developed gout after myocardial infarction,
but serum urate was never > 9 mg/dl.

Four relative had renal colic and passed uric acid gravels.
One child was known to stain his diaper with brownish crystals as
a baby; he developed painful uric acid crystalluria at an early
age of 8. He was known to have partial HGPRT deficiency (6). The
other three passed uric acid gravels at ages beween 33 and 35 years.
All had excessive urinary uric acid excretion and/or high plasma
glutamic acid. The immediate cause of acute renal colic in two
was acute dehydration after playing tennis and a third one from
severe diarrhea (7).

Summary: 1. Hyperuricemia is common among the gouty relatives
as reported by others (8-11). It is of interest to note that serum
urate fluctuates periodically. Hyperuricemia is not necessarily
maintained in a steady state throughout the years. Thus a single
determination of serum uric acid can be misleading.
              2. Development of gout from asymptomatic hyperuricemia
is often correlated with the degree of hyperuricemia as observed
from population or family studies (12-14). The data presented
indicate that unequivocal hyperuricemia is more often accompanied
by excessive excretion of uric acid, diminished excretion of ammonia
and abnormally high plasma glutamic acid. All are undoubtedly
important risk factors for gout.

3. The elevated glutamate could be due to a deficiency of glutamic dehydrogenase, as postulated by Pagliara and Goodman (15). In presence of intracellular accumulation of glutamate in glutamic dehydrogenase deficiency, renal production of ammonium may be reduced due to its inhibitory action on glutaminase 1. As a result of a renal block of ammonia formation, the glutamine in surplus may be diverted for uric acid synthesis.

4. Long-term studies indicate serum urate in most hyperuricemia relatives of gout can be modified by environmental factors, such as diet, weight and changes of life style. When hyperuricemia is under better control, the potential hazard of developing symptomatic gout may be circumvented.

1. A. B. Gutman, T. F. Yü, On the nature of the inborn metabolic error(s) of primary gout, Trans Assoc Amer Physicians 76:141 (1963).
2. A. B. Gutman and T. F. Yü, Urinary ammonium excretion in primary gout, J Clin Invest 44:1474-1481 (1965).
3. T. F. Yü, M. Adler, E. Bobrow, A. B. Gutman, Plasma and urinary amino acids in primary gout with special reference to glutamine, J Clin Invest 48:885 (1969).
4. A. B. Gutman and T. F. Yü, Hyperglutamatemia in primary gout, Am J Med 54:713 (1973).
5. E. J. Bien, T. F. Yü, J. D. Benedict, DeW Stetten, Jr, The relation of dietary nitrogen consumption to the rate of uric acid synthesis in normal and gouty man, J Clin Invest 32:778 (1953).
6. T. F. Yü, M. E. Balis, T. A. Krenitsky, J. Dancis, Silvers D. N., G. B. Elion, A. B. Gutman, Rarity of x-linked partial HPRT deficiency in a large gouty population, Ann Intern Med 76:255 (1972).
7. T. F. Yü, A. B. Gutman, Uric acid nephrolithiasis in gout. Predisposing factors, Ann Intern Med 67:1133 (1967).
8. J. H. Talbott, F. S. Coombs, Concentrations of serum uric acid in nonaffected members of gouty families, J Clin Invest 17:508 (1938).
9. C. J. Smyth, C. W. Cotterman, R. H. Freyberg, The genetics of gout and hyperuricemia, and analysis of nineteen families, J Clin Invest 27:749 (1948).
10. R. H. Stecher, H. H. Hersh, W. M. Solomon, The heredity of gout and its relationship to familial hyperuricemia, Ann Intern Med 31:595 (1949)
11. K. Mikanagi, K. Nishioka, Y. Ovi, Clinical aspects of gouty patients in Japan, XIII Int Congr Rheumatol, Kyoto (1973).
12. A. P. Hall, P. E. Barry, T. R. Dawber, M. McNamara, Epidemiology of gout and hyperuricemia, a long-term population study, Am J Med 42:27 (1967).
13. M. T. Rakic, H. A. Valkenburg, R. T. Davidson, J. P. Engel, W. M. Mikkelsen, J. R. Neel, I. F. Duff, Observations on the natural history of hyperuricemia and gout 1. An

eighteen year follow-up of nineteen gouty families, Am J Med 37:862 (1964).

14. W. J. Fessel, A. B. Siegdaub, E. S. Johnson, Correlates and consequences of asymptomatic hyperuricemia, Arch Int Med 132:44 (1973).

15. A. S. Pagliara, A. D. Goodman, Elevation of plasma glutamate in gout. Its possible role in the pathogenesis of hyperuricemia, N Engl J Med 281:767 (1969).

THE CLINICAL DIFFERENTIATION OF PRIMARY GOUT FROM PRIMARY RENAL DISEASE IN PATIENTS WITH BOTH GOUT AND RENAL DISEASE

B.T. Emmerson, P.J. Stride and Gail Williams

University of Queensland Medical School

Brisbane, Queensland, Australia

Renal disease is a well recognized complication of primary gout but it is less well recognized that patients with primary renal disease may also develop gout. Chronic lead nephropathy is the best recognized example of this, with 50% of these patients suffering from gout. However, gout also occurs in patients with primary renal disease not due to lead and, in the present study, patients with chronic lead nephropathy have been excluded.

In patients who present with both renal disease and gout, it can be difficult to determine whether they suffer from primary gout with secondary renal disease or primary renal disease with secondary gout and extensive investigation is usually necessary to elucidate this relationship. One point which might assist in the differentiation of these two groups of patients would be evidence which showed whether the gout antedated the renal disease (in which case the renal disease might be secondary to the gout) or whether the renal disease antedated the gout (in which case the gout might be secondary to the renal disease . Therefore, from a group of 120 well-documented patients attending a renal disease and gout clinic, 34 were chosen who had both gout and renal disease, but who did not suffer from chronic lead nephropathy. These patients were then separated into two groups using the single chronological criterion of whether the gout or the renal disease occurred first. Because of the importance of this criterion, special care was taken with a detailed clinical history over many years and medical records, which in many cases were extensive, were followed back as far as possible in order to assist in determining which came first. This gave one group of 17 patients in whom there was evidence that gout was present before any renal disease, and these patients were referred to as the gout-first group. The second group (referred to

as the renal-disease-first group) consisted of patients in whom there
was evidence of renal disease being present prior to the first attack
of gout.

These criteria were utilized in the knowledge that, because gout
is dramatically symptomatic and renal disease is often asymptomatic,
the finding that gout occurs first in time in the individual patient
may not indicate the absence of renal disease and that the presence
of minor degrees of renal disease may be difficult or impossible to
establish unless appropriate investigations have been undertaken.
This subdivision also depends upon the assumption that primary
renal gout, that is gout affecting the kidney before producing gouty
arthritis, is rare. Evidence for the rarity of primary renal gout
is provided by the very small number of published reports and by
the general experience of nephrologists. It is unlikely that any
one clinic could find 17 cases of primary renal gout or even
sufficient numbers of cases which might complicate the diagnosis of
primary renal disease with secondary gout. Using this criterion for
allocation between the gout-first group and the renal-disease-first
group, a number of variables was compared in the two groups.

TABLE I

Variable	Gout-First Group	Renal-Disease-First Group	Difference
Number	17	17	
Males	13	6	$p < .05$
Age	55	45	$p < .01$
Age at onset of renal disease	50 (23-69 years)	30 (15-45 years)	$p < .001$
Age at onset of gout (years)	39	40	N.S.
Presence of tophi	8	0	$p < .001$

As shown in Table I, there were significantly more males in
the gout-first group, and the age at the time of the survey was
significantly greater in this group (55 years versus 45 years).
Although the age of onset of gout was almost the same (approximately
40 years) in the two groups, the age of onset of the renal disease
was appreciably different, being 50 years in the gout-first group
and 30 years in the renal-disease-first group. Thus the latter
group tended to develop their renal disease at the age of 30 and
their gout at the age of 40 years, whereas the gout-first group
developed their gout at the age of 40 years and their renal disease
at the age of 50 years. Tophi were present only in the gout-first
group and had been present for an average of 6 years. The absence

of tophi from this renal-disease-first group may be contrasted with the presence of tophi in a significant number of patients with chronic lead nephropathy.

The criteria used for the diagnosis of gouty arthritis were graded into possible, probable and definite, largely on the clinical history. All had hyperuricaemia, but a case was classed as possible gout if the patient had suffered one or more typical attacks of acute gouty arthritis, whereas a case was classified as probable if the patient had suffered recurrent attacks of typical acute gouty arthritis with remissions. A definite case was either one in whom urate crystals had been demonstrated in joint fluid, or typical radiological changes were present in addition to the criteria needed for a diagnosis of probable gout. Using these criteria, the renal-disease-first group contained more cases of probable gout and fewer of definite gout, whereas the gout-first group contained many more cases of definite gout and no cases of possible gout.

TABLE II

NATURE OF GOUT

Variable	Gout-First Group	Renal-Disease-First Group	Difference
Age of onset of gout (yrs)	39	40	N.S.
Presence of tophi	8	0	$p < .001$
Duration of tophi (yrs)	6 (3-10)	-	$p < .001$
Number of attacks of gout	49	3	$p < .001$
Number of years of gout before renal disease known	10 (2-30)		
Number of attacks of gout before renal disease known	20-30 (<5 - >50)		
Duration of renal disease prior to the onset of gout		11 (4-22)	
Gout affecting hands	6	0	$p < .05$
Gout affecting wrists	7	1	$p < .05$
Podagra	16	12	N.S.
Moderate to heavy alcohol	7	1	$p < .05$
Family history of gout	8	4	N.S.
Family history of renal disease	2	4	N.S.
% Desirable weight	127	116	N.S.

This was also reflected (Table II) in the significant difference in the number of attacks of gout between the two groups, the mean for the renal-disease-first group being 3, while that for the gout-first group was 49. Looking at the gout-first group, the number of years of gout before the onset of renal disease was a mean of 10 years (with a range of 2-30 years) and the number of attacks of gout before the onset of renal disease showed a mean between 20 and 30 attacks, with a range between less than 5 and over 50 attacks. Despite the large ranges, the mean values suggest that recurrent attacks of gout have usually been occurring over a 10 year period before the development of renal disease secondary to the gout.

There was no significant difference in the incidence of podagra in the two groups, nor in the incidence of gout affecting other joints in the lower limbs. However, gout affecting the upper limbs was significantly more common in the gout-first group. Moderate to heavy alcohol consumption was significantly greater in the gout-first group but no significant differences were seen between the two groups in regard to a family history of gout or of renal disease. While it is well established that patients with primary gout tend to have an increase in the desirable weight for height, it was notable in this study that the mean per cent desirable weight was 127% in the renal-disease-first group and only 116% in the gout-first group. These figures are not significantly different. However, the mean per cent desirable weight in the gout-first group who developed renal disease was appreciably less than that in the whole group of patients with primary gout who did not develop renal disease.

TABLE III

COMPONENTS OF THE DIAGNOSIS OF THE RENAL DISEASE

Variable	Gout-First Group	Renal-Disease-First Group	Difference
Components of renal diagnosis:			
Glomerulonephritis	3	3	N.S.
Infective including chronic pyelonephritis without reflux	1	8	$p < .02$
Analgesic nephropathy	1	7	$p < .05$
Polycystic disease	0	1	
Essential hypertension	12	4	$p < .02$
Sulphinpyrazone therapy	5	0	$p < .02$
Calculi or crystalluria (prior to therapy	8	2	$p < .06$

In view of the frequent difficulties of making a single diagnosis of the nature of renal disease, a difficulty which is present even after renal biopsy has been undertaken, the various disease components which could be contributing to the renal disease were documented in the two groups of patients (Table III). The renal-disease-first group showed a significantly increased number of patients with an infective component to their renal disease, and in this were included patients diagnosed as having chronic pyelo-nephritis in whom no ureteric reflux was present. Analgesic nephropathy was also present in seven of the renal-disease-first group. As shown in the Table, essential hypertension was regarded as a contributory factor in the renal disease of many patients, most particularly in the gout-first group. In five of the gout-first group, the presence of renal disease was detected only after the patient had been given uricosuric therapy with sulphinpyrazone. Clearly, such a finding in a retrospective study does not establish an aetiological relationship between the two. However, in a hot climate, the problem of uric acid crystal obstruction to collecting ducts is a potential problem and the importance of this problem needs to be remembered, particularly in patients to whom sulphin-pyrazone is being administered to reduce platelet stickiness. In the renal-disease-first group, the duration of the renal disease prior to the onset of the gout showed a mean of 11 years, the range being between 4 and 22 years.

Many other variables compared showed no significant differences. Some of the more important ones in which no difference was demonstrated included the degree of proteinuria, of nitrogen retention, blood lipids, blood pressure, lead exposure, body weight and serum and urinary urate excretion.

In summary therefore, dividing patients with both gout and renal disease into those in whom the gout occurred first and those in whom the renal disease occurred first provides two distinct symptom complexes. In the gout-first group, the patients were older and the gout more severe as evidenced by the number of attacks, the more extensive joint involvement and the more frequent tophi. Their renal disease tended to be asymptomatic and not to develop until after 10 years and over 20 attacks of gout. The renal-disease-first group tended to have had renal disease for about 10 years before developing gout and the gout was less severe, both in the number of attacks and the number of joints involved. More of the renal disease group were suffering from analgesic nephropathy or chronic pyelonephritis, while more of the gout-first group had hypertensive nephrosclerosis. Knowledge of these differences may be applied in determining which is the underlying disease in patients in whom differentiation would otherwise be difficult.

RENAL FAILURE IN YOUNG SUBJECTS WITH FAMILIAL GOUT

H. A. Simmonds, J. S. Cameron, C. F. Potter, D. Warren,
T. Gibson and D. Farebrother
Purine Laboratory, Department of Medicine, Guy's Hospital;
St. Mary's Hospital, Portsmouth; Wellcome Laboratories,
Beckenham, United Kingdom

The classical patient with gout is usually an older male, and
gout is rare in premenopausal females[1]. Until recently, renal
failure was common in gout[2], but is now rare[3]. Renal involvement
however, as judged by low urine pH and proteinuria, remains
common[1,3,4], although concentrating ability and GFR are not usually
different from age-matched controls[3]. Conversely, gout is rarely
diagnosed in renal failure from other causes, despite mild hyperuricaemia as part of the general retention of nitrogenous waste[4,5].
If all patients entering terminal renal failure and offered dialysis
or transplantation are examined, less than 1% have gout recorded as
the cause (Table 1).

The combination of gout and renal failure is even more unusual
in young patients, yet an examination of the few individuals who
required dialysis and transplantation in our unit because of gout
showed that all were young (Table 1), and two of the four were

Table 1. Data of the European Dialysis and Transplant Association:
patients treated by dialysis or transplantation in Europe
1963-1977[15]

	No. dialysed for terminal renal failure	No. with gout causing t.r.f.	Women	No. of patients subsequently transplanted
Europe	69,400	454(0.7%)	36(8%)	40m, 2f
Guy's Hospital	472	4*(0.8%)	2(50%)*	2m, 1f

* Aged 32, 34 (see text), 40 and 41 at terminal renal failure (t.r.f.)

15

female. This led us to attempt a survey of the patients in South
East England who had both gout and renal failure, which revealed
several patients with a family history of gout and/or renal failure.
Studies of renal function and purine metabolism have been done in
five such families; one of these families is the subject of a more
detailed report to be published elsewhere. We know of two further
families not yet studied.

PATIENTS (Table II)

Family I (Fig. 1) included an affected brother and sister
(Table II (a)) who both developed gout and terminal renal failure
in their thirties. Investigation of this family was not possible
because all other members were abroad, and the proposita presented
in renal failure. Control of her blood pressure and plasma urate
concentration (allopurinol 200 mg/24h) failed to arrest the decline
in her renal function over the next three years; her creatinine
clearance fell from 35 ml/min to zero, and she was dialysed from 1972.
Two years later she received a cadaver allograft, but suffered
marrow depression from her azathioprine during the course of a
rejection episode, and died of infection and cerebral infarction.

Family II contained a pair of identical twins who, alone in
their family, developed gout, renal failure and hypertension
which was evident early in the course of their disease, and small
kidneys on IVU. Their mother was hypertensive (210/110) but
normouricemic. Both successfully completed pregnancies by Caesarian
section taking allopurinol. Despite treatment of both hypertension
and hyperuricemia, the first twin showed an initial decline in
renal function, but five years later had blood urea no higher than
at first measurement. The other twin is now in France.

Family III was identified when acute gout developed in a 9 year
old girl (a) whose nonidentical twin sister (b) was hyperuricemic,
and whose mother and her identical twin sister were being treated
for advanced renal failure, as was a third sister; all three died
of renal failure aged 36-46. The proposita and her twin had normal
blood pressure, even though renal disease was already evident
(Table II). A dominant mode of inheritance affecting nine females
in three generations was present. IVU demonstrated contracted
kidneys, and renal biopsy in three members showed patchy interstitial
nephritis with tubular lesions, dilatation and atrophy (Farebrother
et al, in preparation). Treatment initially with sulphinpyrazone
then with allopurinol in the proposita, her sister, and allopurinol
in two cousins (d,e), appears to have arrested the progress of the
disease over periods of up to 15 years.

In Family IV there was a history of gout on both sides of the
proposita's family when she presented with mild gout aged 28 and

Table II. Patients studied with familial gout/renal failure

	Sex	Age onset gout	First BP noted	GFR / Ccr	Max. Uosm/l	Pur mmol/l	HGPRT in RBC	Treatment: before study	After study	Present renal function
Fam 1.a	f	31	180.110	35	250	0.67	64*	–	All, hypo	Deceased
Fam 2.a	f}	22	170.110	Pcr180	339	0.60	127	All,hypo(3y)	As before	Purea 9.0
b	f}	22	170.120	Pcr172	313	0.59	136	All,hypo(3y)	As before	?
Fam 3.a	f}	9	110.70	12	324	0.78	70*	Sulph(7y)All(9y)	As before	Pcr 140-180
b	f}	9x	90.60	58	391	0.46	77*	Sulph(3y)All(9y)	As before	Pcr 170
c	f	31x	130.90	20	339	0.41	114	All(6y)hypo(5y)	As before	GFR 15
d	f	14x	100.65	68	492	0.42	101	–	All	GFR 66
Fam 4.a	f	28	200.105	24	303	0.45	107	–	All, hypo	RDT
Fam 5.a	m	33	140.105	118	707	0.87	99	–	All	GFR 72

x   – hyperuricemia only
*   – stored sample: within normal limits
RDT   – regular dialysis therapy
All   – allopurinol
Sulph – sulphinpyrazone
Hypo – hypotensive agents
Cr   – creatinine

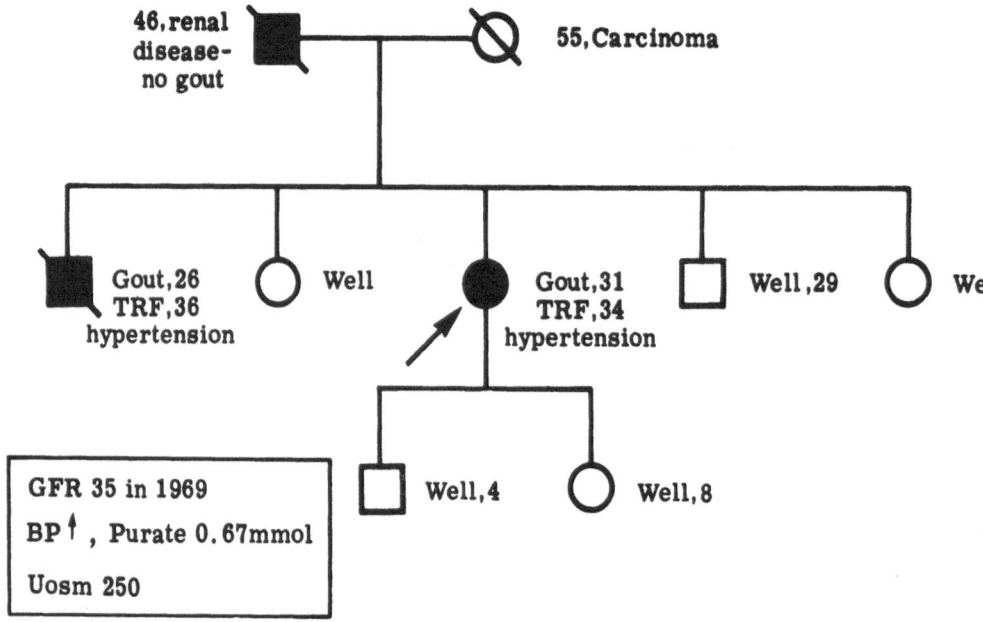

Fig. 1.   Family tree of family I.   Data within the box refer to the
          proposita (arrowed).

renal failure and hypertension at 32, the GFR already reduced to
24 ml/min. Renal biopsy revealed severe tubulointerstitial disease,
urate crystals and severe vessel changes. Control of hypertension,
and of hyperuricemia with allopurinol failed to arrest decline in
renal function and 23 months later she began dialysis.

      Family V is presented for comparison. The father and brother
of the propositus had gout, and he developed severe tophaceous gout
aged 33, when obese, hypertensive and a heavy drinker. His GFR
fell from 118 ml/min to 65 ml/min over 2½-years, but control of his
gout and hyperuricemia with allopurinol resulted in no further loss
of renal function over a further 18 months. The hypertension
disappeared with weight loss. Renal biopsy in this man showed mild
tubuloinstertitial disease but severe vascular changes.

      The biochemistry and haematology in all the patients studied
was compatible with the degree of renal function observed, with
the exception of the hyperuricemia. Urinary urate excretion was
not excessive in any patient. although Cur/Ccr ratios were raised
in those with severe renal disease,   as would be expected[4].
Urinary xanthine and hypoxanthine excretion were not raised except in
those taking allopurinol. Activities of HGPRT in erythrocyte
lysates were normal, as was APRT (Table II).

DISCUSSION

The patients described here fall neither into the group of
older, predominantly male patients with renal function relatively
normal for age, nor into the group of juvenile gouty patients with
identifiable enzymic abnormalities such as the Lesch-Nyhan syndrome.
The histories of these families do not suggest associated primary
renal disease, such as Alport's syndrome, Bartter's syndrome, or
nephronophthisis, and polycystic kidney disease was excluded by
the IVU examination. Analgesic abuse was not a feature, and
although lead excretion was not studied, the family histories of
several generations all affected does not suggest saturnine gout
which also leads to renal failure in young adults with a higher
proportion of females than classical gout[6]. The possibility of
primary tubulointerstitial nephritis ("chronic pyelonephritis")
remains, with gout secondary to this; but we, and others[5], rarely
see this combination, only two of our uremic patients having
suffered secondary gout of this type in a large clinic followed
over a decade.

A feature of particular interest was the relationship of
blood pressure to the renal failure in these patients. Whilst
the members of families I, II, IV and V were hypertensive, in
family III all five affected female members studied were normotensive at diagnosis (Table II). Of families previously reported[7-11],
this kindred resembles most closely that of Leumann and Wegmann[9] in
which three normotensive females, including two children, suffered
renal failure and hyperuricemia; the mother had gout from childhood.
Other reported families have been hypertensive, and both males and
females[8,10,11] or males alone[7] affected. The fifth family discussed
here, in contrast, resembles the classical severe gouty pattern of
earlier reports, which is now much less common and renders the
other families more evident.

The families described, together with those in the literature,
suggest that several patterns of familial gout and renal failure
may be seen in young adults, one of which affects females and is not
associated with hypertension. The others may be heterogenous and
comprise several groups associated in general with hypertension,
and affecting both sexes. These types of young gouty patients may
be more common than usually recognised. Grahame and Scott[1] noted
renal failure and severe hypertension more commonly in their patients
aged less than 40, and in Yü's series of female gouty patients[12], 9
of those aged less than 40 had chronic renal disease and almost two
thirds had a family history of gout.

In some of our patients, particularly in family III, treatment
with allopurinol appears to have arrested or ameliorated the disease,
which makes early identification of children at risk in these
families of importance. Equally, treatment failed in some of the

other patients, but these were already severely uremic and hypertensive when treatment began. Study of these and similar families may throw light on the controversial question of the relationship between gout, renal damage, hyperuricemia and vascular disease[13,14].

REFERENCES

1. R. Grahame and J. T. Scott, Clinical survey of 354 patients with gout, Ann. Rheum. Dis., 29:461 (1970).
2. J. H. Talbott and K. H. Terplan, The kidney in gout, Medicine (Baltimore), 39:405 (1960).
3. T. Gibson, H. A. Simmonds, C. F. Potter, N. Jeyarajah and J. Highton, Gout and renal function, Eur. J. Rheumatol. Inflamm., 1:79 (1978).
4. L. Berger and T-F. Yü, Renal function in gout. IV. An analysis of 524 gouty subjects including long-term follow-up studies, Amer. J. Med., 59:605 (1975).
5. G. Richet, F. Mignon and R. Ardaillou, Goutte secondaire des néphropathies chroniques, Presse Médicale, 73:633 (1965).
6. B. T. Emmerson, Chronic lead nephropathy, Kidney Int., 4:1 (1973).
7. F. M. Rosenbloom, W. N. Kelley, A. A. Carr and J. E. Seegmiller, Familial nephropathy and gout in a kindred, Clin. Res., 15:270 (abstract) (1967).
8. H. Duncan and A. StJ. Dixon, Gout, familial hyperuricaemia and renal disease, Quart. J. Med. (n.s.), 29:127 (1960).
9. W. Van Goor, C. J. Kooiker and E. J. Dourhout Mees, An unusual form of renal disease associated with gout and hypertension, J. Clin. Path., 24:354 (1971).
10. E. P. Leumann and W. Wegmann, Hereditary nephropathy with hyperuricaemia. Abstracts, 6th meeting of the European Society of Paediatric Nephrology, Dublin (1972).
11. P. U. Massari, C. H. Hsu, I. H. Fox, P. W. Gikas, R. V. Barnes and J. M. Weller, Hereditary nephropathy and hyperuricemia. Abstracts, 7th International Congress of the International Society of Nephrology, Montreal, 1978, p.P.3 (hereditary renal disorders).
12. T-F. Yü, Some unusual features of gouty arthritis in females, Sem. Arthr. Rheum., 6:247 (1977).
13. B. T. Emmerson and P. G. Row, An evaluation of the pathogenesis of the gouty kidney, Kidney Int., 8:65 (1975).
14. H. A. Simmonds, Crystal-induced nephropathy. A current view, Eur. J. Rheumatol. Inflamm., 1:86 (1975).
15. A. J. Wing, Personal communication (1979).

FAMILY STUDY OF LIPID AND PURINE LEVELS IN GOUT PATIENTS AND

ANALYSIS OF MORTALITY

L. G. Darlington, J. Slack, J. T. Scott

Epsom Rheumatology Unit, Epsom, Surrey.
Institute of Child Health, London, W.C.1
Kennedy Institute of Rheumatology, London, W.6

Many authors have noted an association between hyperlipidaemia
and hyperuricaemia and Berkowitz (1964), Feldman and Wallace (1964)
and Darlington and Scott (1972) all found raised triglyceride levels
in gout.

Hypertriglyceridaemia occurs frequently and was found in 52% of
gout patients by Darlington and Scott (1972) and Frank (1974) and in
75% of gouty patients by Berkowitz (1964). This association was not
seen in patients with symptomless hyperuricaemia (Frank 1974).

In 1973, Mielants et al. found an increase in pre-β lipoproteins
and a reduction in α- and β-lipoproteins.

In 1971, Emmerson and Knowles demonstrated hypertriglyceridaemia
in gout patients persisting after correction for body weight and,
in 1972, Darlington and Scott also described gout and
hypertriglyceridaemia independent of obesity.

In 1974, Gibson and Grahame suggested that obesity, alcohol or
both were the main causes of hypertriglyceridaemia in gout.

In 1979, Gibson et al. showed significant reductions in
triglycerides in gout patients by reducing either alcohol intake
or weight and, in the same year, Darlington and Scott, also found
reductions in triglycerides on alcohol abstention although these
did not achieve statistical significance.

In 1977, Elkeles and Chalmers raised triglyceride concentrations
by fat infusion but did not find any subsequent effect on plasma

uric acid.

In 1979, Gibson et al. also used a fat tolerance test but did not demonstrate any effect of triglyceride on uric acid or of hyperuricaemia on triglyceride removal.

Unaffected relatives of gout patients may have an increased serum level of uric acid which is thought to be genetically determined (Smyth et al., 1948; Talbott, 1940; Hauge and Harvald, 1955).

In 1970, Rondier et al. concluded that gout does not have special features when associated with hypertriglyceridaemia but a family history is more frequent.

To determine whether the hypertriglyceridaemia of gout patients occurs in their families or is simply the result of the life-style of these usually obese, alcohol-drinking patients who frequently originate from higher social classes a family study was designed to measure lipid and uric acid levels in the blood of gout patients and their first-degree relatives and to compare them with those found in normal controls of the same age and sex.

A pedigree was constructed for 135 male and 9 female gout patients.

Index patients and relatives who agreed to help were weighed and fasting blood samples were taken.

Due to the regional and ethnic (Bronte-Stewart, 1955) differences in lipid concentrations control data was needed from subjects from the same geographical area and ethnic background as the gout group. A suitable control sample was available from a working population in N.W. London (Slack et al., 1977) and this was used to calculate standard deviation scores for lipid and uric acid levels.

The means and standard deviations for index patients and relatives are shown in Table 1 as standard deviation scores and significance was assessed. Triglyceride levels in gout patients were significantly higher than in relatives or controls ($p < 0.001$).

These changes were, predictably, reflected by changes in the proportion of pre-β lipoprotein which was significantly higher in the gout patients than in relatives or controls ($p < 0.001$).

There was a corresponding, significant reduction in the proportion of β-lipoprotein in gout patients when compared with controls ($p < 0.001$) and relatives ($0.01 > p > 0.001$) resulting in a raised $\alpha : \beta$ ratio. There was no significant difference between the

Table 1   <u>Means and Standard deviations compared with controls</u>

Variable	Index patients	No of Subjects	1st degree relatives	No of Subjects
cholesterol	+0.20 $\pm$ 1.15	143	+0.25 $\pm$ 1.12*	165
triglycerides	+0.60 $\pm$ 1.28***	143	-0.16 $\pm$ 1.18	165
% β-lipoprotein	-0.78 $\pm$ 1.49***	142	-0.34 $\pm$ 1.18**	162
% pre-β lipoprotein	+0.71 $\pm$ 1.55***	142	-0.04 $\pm$ 1.07	161
% α-lipoprotein	-0.07 $\pm$ 1.32	141	+0.27 $\pm$ 1.24*	162
Uric acid	+0.61 $\pm$ 1.44***	140	+0.05 $\pm$ 1.28	159

* = 0.05 > p > 0.01   ** = 0.01 > p > 0.001   *** = p < 0.001

α -lipoprotein levels in gout patients and controls.

There were no significant differences between cholesterol levels in gout patients and controls or between gout patients and relatives. The marginal elevation of cholesterol in relatives when compared with controls could be explained by the presence among the relatives. of some patients with familial hypercholesterolaemia - probably by chance and unrelated to gout.

Uric acid levels in male, gout patients were, predictably, significantly higher than in male controls or relatives (p < 0.001) and there was no evidence of familial hyperuricaemia since uric acid levels were not significantly increased in relatives. Female uric acid data also did not show familial hyperuricaemia but numbers of female index patients were too small for valid conclusions.

The families of the 30 male and 2 female gout patients with lipid levels more than 1.69 standard deviations greater than the mean i.e. above the 95th centile were assessed to determine whether they constituted a particular sub-group in the family study.

There was some elevation of cholesterol levels in both male and female relatives of <u>hyperlipidaemic</u> gout patients (0.05 > p > 0.025 and 0.02 > p > 0.01 respectively). However, assessment of these families showed that 4 out of 5 families whose pedigrees included at least one other hyperlipidaemic subject had a relative with hypercholesterolaemia. In the fifth family, the gout patient

himself had hypercholesterolaemia. The slightly raised cholesterol
levels in these relatives may possibly, therefore, be explained by
a few families with familial hypercholesterolaemia. This would be
expected in a few families by chance and unrelated to gout.

Weights were measured in gout patients, relatives and controls.
There were too few female gout patients for analysis but male gout
patients were significantly heavier than male relatives and controls
$(p < 0.001)$.

These data show that the male, gout patients were significantly
heavier than male relatives and controls ($p = <0.001$). The mean
weight of the hyperlipidaemic male gout patients was not, however,
significantly heavier than the mean of the whole male gout group.

This obesity may well be associated with the hypertriglyceridaemia
in certain male subjects and contribute to the raised triglyceride
levels in the whole male gout population.

The alcohol intake of the hyperlipidaemic group varied widely,
emphasizing that alcohol is only one factor in inducing
hyperlipidaemia but is doubtless significant in some cases.

The clinical implications of hypertriglyceridaemia remain
controversial.

A mortality study was undertaken in these families to determine
whether either the gout patients or their relatives died more
frequently from coronary and/or cerebrovascular pathology than non-
gouty subjects of the same age and sex. (Full details will be
described in a later publication but a short description is given
below).

Cause of death was sought for all dead gout patients and
relatives.

Life tables were constructed for patients and relatives and the
Registrar General's tables were used to give risks of dying from
coronary and cerebrovascular disease for the general population.

Observed and expected deaths were compared in each age group.

No significantly increased mortality was found from coronary
or cerebrovascular disease in male gout patients or their relatives.
There were too few female gout patients for assessment.

A further analysis was performed to determine whether there was
an increased incidence of coronary or cerebrovascular deaths in
families of hyperlipidaemic gout patients. This revealed a slight

increase in male coronary deaths in the hyperlipidaemic group although this may have resulted from hypercholesterolaemia occurring by chance rather than from gout.

Two possible reasons for the absence of a definite increase in vascular mortality may, firstly, be the relatively mild risk from hypertriglyceridaemia and, secondly, the fact that the increased proportion of pre-β lipoprotein is associated with a raised $\alpha : \beta$ ratio which may reduce the overall risk of vascular disease.

Summary

A family study was performed to determine whether the hypertriglyceridaemia associated with gout is present in families of gout patients or simply due to their life-style.

The study revealed hypertriglyceridaemia in gout patients, reflected by hyperprebetalipoproteinaemia and with reciprocal reduction in the proportion of β-lipoprotein. These abnormalities were not seen in first-degree relatives.

No definite increase in mortality was found from coronary or cerebrovascular disease in male gout patients after presentation to hospital or in their relatives.

Families of hyperlipidaemic gout patients did reveal a slight increase in male coronary deaths although the significance of this finding was doubtful since some hypercholesterolaemia was found in these hyperlipidaemic families.

References

Berkowitz, D., 1964, Blood Lipid and Uric Acid Inter-relationships, J.Amer.med.Ass., 190: 856-8.
Bronte-Stewart, B., Keys, A., and Brock, J.F., 1955, Serum-Cholesterol, Diet, and Coronary Heart-Disease, Lancet, ii, 1103-7.
Darlington, L.G., and Scott, J.T., 1972, Plasma lipid levels in gout Ann.rheum.Dis., 31, 487-9.
Darlington, L.G., and Scott, J.T., 1979, Ann.rheum.Dis., (Submitted for publication).
Elkeles, R.S., and Chalmers, R.A., 1977, Hypertriglyceridaemia And Hyperuricaemia, Lancet, ii, 252.
Emmerson, B.T., and Knowles, B.R., 1971, Triglyceride Concentrations in Primary Gout and Gout of Chronic Lead Nephropathy, Metabolism, 20, 721-9.
Feldman, E.B., and Wallace, S.L., 1964, Hypertriglyceridaemia in Gout, Circulation, 29, 508-13.
Frank, O., 1974, Observations concerning the incidence of disturbance of lipid and carbohydrate metabolism in gout, Adv.Exp.Med.Biol.,

41, 495-98.

Fredrickson, D.S., and Levy, R.I., 1972, Familial hyperlipoproteinaemia In The Metabolic basis of inherited disease.
3rd ed. eds. J.B. Stanbury, J.B. Wyngaarden, and D.S. Fredrickson,
New York, McGraw-Hill.

Gibson, T., and Grahame, R., 1974, Gout and hyperlipidaemia,
Ann.rheum.Dis., 33, 298-303.

Gibson, T., Kilbourn, K., Horner, I., and Simmonds, H.A., 1979,
Mechanism and treatment of hypertriglyceridaemia in gout,
Ann.rheum.Dis., 38,31-35.

Hauge, M., and Harvald, B., 1955, Heredity in gout and hyperuricaemia,
Acta.Med.Scand., 152, 247-57.

Hutchinson, J., 1889, Case report, Lancet, i, 789.

Mertz, D.P., Schwoerer, P., and Batucke, G., 1972, Zur
Klassifizierumg der Hyperlipoproteinämie bei primärer Gucht,
Dtsch.med.Wschr., 97, 600-604.

Mielants, H., Veys, E.M., and De Weerdt, A., 1973, Gout and its
relation to lipid metabolism, Ann.rheum.Dis., 32, 501-5.

Patterson, D., and Slack, J., 1972, Lipid Abnormalities in Male and
Female Survivors of Myocardial Infarction And Their First-degree
Relatives, Lancet, i, 393-9.

Rondier, J., Truffert, J., LeGo, A., Brouilhet, H., Saporta, L.,
de Gennes, J.L., and Delbarre, F., 1970, Goutte et hyperlipidémies
(etude) portant sur 50 goutteux et sur 50 sujets non goutteux,
Rev.Europ.etud.clin.biol., 15, 959-968.

Slack, J., Noble, N., Meade, T.W., and North, W.R.S., 1977,
Lipid and Lipoprotein concentrations in 1604 men and women in
working populations in north-west London, Brit.med.J., 2, 353-7.

Smyth, C.J., Cotterman, C.W., and Freyberg, R.H., 1948, The
Genetics Of Gout and Hyperuricaemia An Analysis of Nineteen Families,
J.clin.Invest., 27, 749-59.

Talbott, J.H., 1940, Serum urate in relatives of gouty patients,
J.clin.Invest., 19, 645-48.

URIC ACID TURNOVER IN NORMALS, IN GOUT AND IN CHRONIC RENAL FAILURE USING ^{14}C-URIC ACID

C. Vitali, G. Pasero, A. Clerico, L. Riente, N. Molea,
A. Pilo, G. Mariani, R. Bianchi

Service of Rheumatology and Metabolic Disease Unit,

University of Pisa, and C.N.R. Clinical Physiology

Laboratory, Via Roma 2, 56100 Pisa, Italy

Turnover studies using ^{14}C-uric acid (UA) have clarified several important aspects of UA metabolism in man under normal and pathological conditions[1]. However, the metabolic characterization of gouty patients in normoexcretors and overexcretors (renal and metabolic gout) is obtained, at present, by measuring basal urate excretion and/or incorporation of labeled glycine into urinary UA[2]. Recently, we have reinvestigated some aspects of UA turnover in man by using a chromatographic purification of 2-^{14}C-UA in plasma and urine samples[3] and different computational techniques for calculating metabolic parameters[4,5]. This chromatographic technique shows obvious advantages of accuracy and practicability respect to the classical crystallization procedure[4,6]. The present paper reports the results obtained in 6 control subjects, in 17 gouty patients and in 4 patients with hyperuricemia due to chronic renal failure.

MATERIALS AND METHODS

We have investigated a total of 27 subjects: 6 normal males, 17 male gouty patients, and 4 patients with chronic renal failure (3 males and one female). All the subjects were hospitalized and maintained on purine-free diet before and during the study. None of the gouty patients had clinical or radiological evidence of tophi, nor significant renal impairment Furthermore, none was receiving allopurinol, diuretics and any other drugs known to interfere

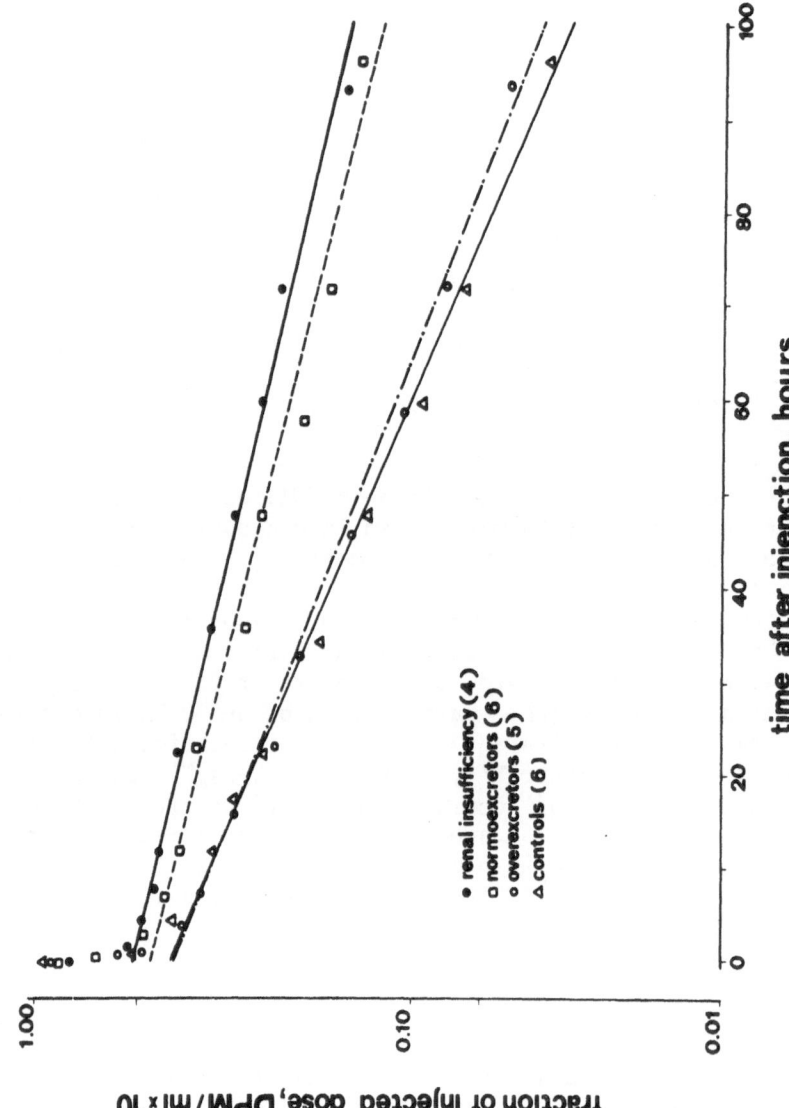

Figure 1: Mean plasma disappearance curves of injected ^{14}C–UA

with UA metabolism since at least 3 months. According to their urate excretion on purine free diet[7], the gouty patients were subdi-
vided into three groups: 1) 5 overexcretors (>600 mg/day), 2) 6
normoexcretors (<460 mg/day), 3) 6 borderlines (between 460 and 600
mg/day). The patients with chronic renal failure had a glomerular
filtration rate lower than 20 ml/min.

The laboratory techniques and computational procedures used
were previously reported[4,5]. Briefly, about 35 μCi of 2-^{14}C-UA were
impulsively injected after overnight fasting. Heparinized venous
blood samples were taken frequently up to 72 or 96 hours and urines
were also collected, in plastic containers, at 6 and 12 hours intervals. Plasma and urine samples were chromatographed through Bio-
gel P2 resin columns (0.9 x 28 cm in size), which specifically ad-
sorbs UA[3]. Fractions containing UA were pooled and counted in an
automatic scintillation beta-counter. Besides, all plasma and urine
samples were counted without chromatography. An enzymatic spectrophotometric method was employed for the determination of the UA
levels in plasma and urine[8].

The experimental data of plasma and urine radioactivities were analyzed by different computational approaches[4,5]. The exchange-
able pool (M), the turnover rate (TR), the metabolic clearance rate
(MCR), the fractional catabolic rate (FCR), the initial (IDV) and
total distribution volumes (TDV) were calculated from the plasma
disappearance curves by non-compartmental analysis, while renal
(TRr) and extra-renal (TRe) removal were calculated by the urine/
plasma ratio method[4,5].

RESULTS

The mean plasma disappearance curves of injected ^{14}C-UA of
control subjects, overexcretor and normoexcretor gouty patients
and patients with chronic renal failure are reported in Figure 1.
Normoexcretors and patients with renal failure showed a slower removal than overexcretors and controls. The mean values with the
respective standard deviations of plasma urate (PU), daily urate
excretion (UU), IDV, TDV, M, TR, MCR, FCR and TRr are reported in
Table 1.

The values of IDV and TDV were non significantly different a-
mong all the groups. The M of UA was significantly increased in
gouty and renal failure patients (P<0.01). Overexcretor patients
showed TR values significantly increased (P<0.05) respect to controls, normoexcretors and patients with renal failure. MCR and FCR
in this group were non different from normal subjects. Normoexcretor gouty patients were characterized by significantly lower
(P<0.001) MCR and FCR values than controls and overexcretors, but
normal TR values.

Table 1

Turnover studies with ^{14}C-UA in normal controls (C), gouty patients as a whole (G), overexcretor (OE) and normoexcretor (NE) gouty patients, and in patients with chronic renal failure (RF). The metabolic data have been normalized to the body surface area (m^2).

	C	G	OE	NE	RF
No. of subjects	6	17	5	6	4
PU (mg/dl)	4.5±0.5	7.4±1.3	6.9±1.2	7.7±1.4	7.5±1.4
UU (mg/day)	464±95	532±142	709±71	393±68	272±21
IDV ($1/m^2$)	6.9±0.7	7.2±1.4	6.7±0.8	7.3±2.2	7.5±1.2
TDV ($1/m^2$)	12.7±1.1	12.6±2.3	12.8±0.7	11.7±2.3	12.3±2.5
M (mg/m^2)	563±37	915±161	886±202	881±152	898±54
TR ($mg/dayxm^2$)	381±78	437±107	545±79	348±75	288±77
MCR ($1/dayxm^2$)	8.5±1.7	6.1±1.7	8.0±0.6	4.7±1.3	4.0±1.5
FCR (%/day)	67.3±11.5	48.4±11.8	62.5±7.6	39.8±7.1	31.2±7.1
TRr (% of TR)	65.6±7.0	65.4±8.9	68.8±7.4	62.4±9.5	50.6±5.1

Of the 6 patients defined as borderline, 5 showed a metabolic pattern similar to normoexcretors and one comparable to overexcretors, as previously reported[5]. When considering the gouty patients as a whole, MCR and FCR were significantly reduced respect to the normal subjects ($P<0.01$).

In the patients affected by renal failure, the MCR and FCR values were significantly lower than in controls ($P<0.05$ and $P<0.001$, respectively) and in overexcretors ($P<0.001$), but similar to the normoexcretors. Besides, only this group of patients was characterized by a significant reduction of percent TRr respect both to the controls ($P<0.01$) and gouty patients as a whole ($P<0.05$).

The mean difference between total radioactivity in plasma and urine samples and that due to UA purified by chromatography was about 3% in all groups.

DISCUSSION

Our method for metabolic investigation of ^{14}C-UA turnover in man shows obvious advantages of accuracy and technical practicability respect to classical crystallization procedures[4]. In addition, the metabolic parameters obtained allow to clarify several important aspects of UA removal in normal and gouty patients[5]. In particular, it was possible to confirm the presence of two differ-

ent metabolic patterns in gouty patients. In fact, overexcretors
have increased TR and normal MCR, while normoexcretors show decreased MCR and normal TR. These findings indicate that in overexcretors hyperproduction of UA is responsable for expanded M,
whereas in normoexcretors there appears to exist a primitive defect
in excretion of urate, without a significant increase of TRe. On
the contrary, the significant decrease of percent TRr in renal
failure suggests a compensatory, but not sufficient increase of TRe
of UA in this pathological condition, as previously reported[9].
The allantoin and urea as degradation products of ^{14}C-UA in plasma
and urine samples, and their role in the UA metabolism in humans
appear unremarkable.

REFERENCES

1. L.B. Sorensen, The elimination of uric acid in man studied by
     means of ^{14}C-labeled uric acid, Scand. J. Clin. Lab. Invest.
     12, suppl. 54 (1960)
2. A.B. Gutman, T.F. Yu, H. Black, R.S. Yalow, S.A. Berson, In-
     corporation of glycine-1-C^{14}, glycine-2-C^{14}, and glycine-N^{15}
     into uric acid in normal and gouty patients. Am. J. Med.
     25:917 (1958)
3. P.A. Simkin, Uric acid metabolism in Cebus monkeys, Am. J.
     Physiol. 221:1105 (1971)
4. R. Bianchi, C. Vitali, A. Clerico, A. Pilo, L. Fusani, L. Riente,
     G. Mariani, A chromatographic method for the determination of
     uric acid turnover in man, J. Nucl. Med. All. Sci. 22:37 (1978)
5. R. Bianchi, C. Vitali, A. Clerico, A. Pilo, L. Fusani, L. Riente,
     G. Mariani, Uric acid metabolism in normal subjects and in
     gouty patients by chromatographic measurement of ^{14}C-uric acid
     in plasma and urine, Metabolism (in press)
6. L.A. Johnson, B.T. Emmerson, Isolation of crystalline uric acid
     from urine, for urate pool or turnover measurements, Clin.
     Chim. Acta, 41:389 (1972)
7. N. Zöllner, Moderne Gichtprobleme: Aetiologie, Pathogenese,
     Klinik, Ergbn. inn. Med. Kinderheilk. 14:321.(1960)
8. E. Praetorius, An enzymatic method for the determination of uric
     acid by ultraviolet spectrophotometry, Scand. J. Clin. Lab.
     Invest. 1:222 (1949)
9. L.B. Sorensen, The pathogenesis of gout, Arch. intern. Med.
     109:55 (1962)

ERYTHROCYTE ADENOSINE-DEAMINASE ACTIVITY IN GOUT AND HYPERURICEMIA

A. Carcassi, P. Macrì, G. Chiaroni, S. Boschi

Institute of Clinical Medicine, University of Siena

Piazza Selva 7   53.I00   Siena   Italy

Adenosine-Deaminase (adenosine-aminohydrolase EC 3.5.4.4.) (ADA) catalyses the irreversible hydrolytic deamination of adenosine to produce inosine and ammonia. Inosine is then cleaved by Purine-Nucleoside-Phosphorylase (PNP) to hypoxanthine and oxidized to xanthine and uric acid by xanthine-oxidase.

ADA is an enzyme present in most mammalian tissues the activity being highest in organs containing many lymphoid cells (I,2). In erythrocytes ADA activity is present but lower of that observed in lymphoid cells.

ADA deficiency has been described in a group of patients with the autosomal recessive form of severe combined immunodeficiency (SCID), a disorder of infancy characterized by defects of both cellular and humoral immunity, with severe impairment of T and B lymphocytes (3).

Although the main involvement of this enzyme in man appears to be related to immune system, ADA activity can also be related to the catabolism of AMP, via the 5'-Nucleotidase pathway, alternatively with AMP-Deaminase, that is its main catabolic pathway (4).

The importance of this catabolic pathway in the production of uric acid in primary gout has not been proved, even if increased ADA activity in erythrocytes has already been reported in patients with primary gout (5).

To verify these data we have studied erythrocyte ADA activity
in a group of patients with primary gout, in subjects with primary
asymptomatic hyperuricemia and in patients with hyperuricemia secon
dary to chronic renal failure (on hemodialysis treatment).

MATERIALS AND METHODS

Studies were carried out on : 20 control subjects (I5 males
and 5 females), I6 patients with primary gout, 8 subjects with
primary asymptomatic hyperuricemia and IO patients with hyperurice
mia secondary to chronic renal failure (on hemodialysis treatment)
(6 males and 4 females).

Fructose load (0,5 g / Kg b.w. / IO min') was administered
intravenously to 5 control subjects and to 4 patients with primary
hyperuricemia (all males). Erythrocyte ADA activity was studied
before and 30, 60 and I20 min' following intravenous load.

All subjects were fasting for I2 hours. All gouty patients
did not receive any drug affecting uric acid metabolism for at
least I5 days.

Serum uric acid was determined by the method of Archibald (6).

ADA activity was determined by the method of Seligson and
Seligson (7), and ammonia was evaluated with the colorimetric
method of Chaney and Marbach (8).

RESULTS

The results obtained are reported in Tables I and II and in
Figures I and 2.

Gouty patients and the subjects with the primary asymptomatic
hyperuricemia showed slightly higher values in erythrocyte ADA
activity than control subjects. The difference was not significant
($p > 0,05$).

No definite changes in ADA activity were observed in patients
with hyperuricemia secondary to chronic renal failure (on hemodialysis treatment).

ADA activity in subjects with primary hyperuricemia and in controls did not increase after intravenous infusion of fructose.

TABLE I

ERYTHROCYTE ADENOSINE–DEAMINASE ACTIVITY

		μmoles NH_3 / 60' / ml packed cells	
		Mean	SD
Control	(20)	I3,6 *	4,4
Primary Gout	(I6)	16,6 *	4,3
Primary hyperuricemia	(8)	I4,5	3,4
Renal hyperuricemia	(IO)	I2,6	3,2

* (p > 0,05)

TABLE II

ERYTHROCYTE ADENOSINE–DEAMINASE ACTIVITY
AFTER INTRAVENOUS FRUCTOSE LOAD

		μmoles NH_3 / 60' / ml packed cells				
		0'	30'	60'	I20'	
Control	(5)	I2,3	I2,5	I2,8	I3,I	mean
		3,9	4,3	5,2	4,7	SD
Primary hyperuricemia	(4)	I5,0	I5,2	I5,2	I5,4	mean
		5,4	6,5	5,7	6,3	SD

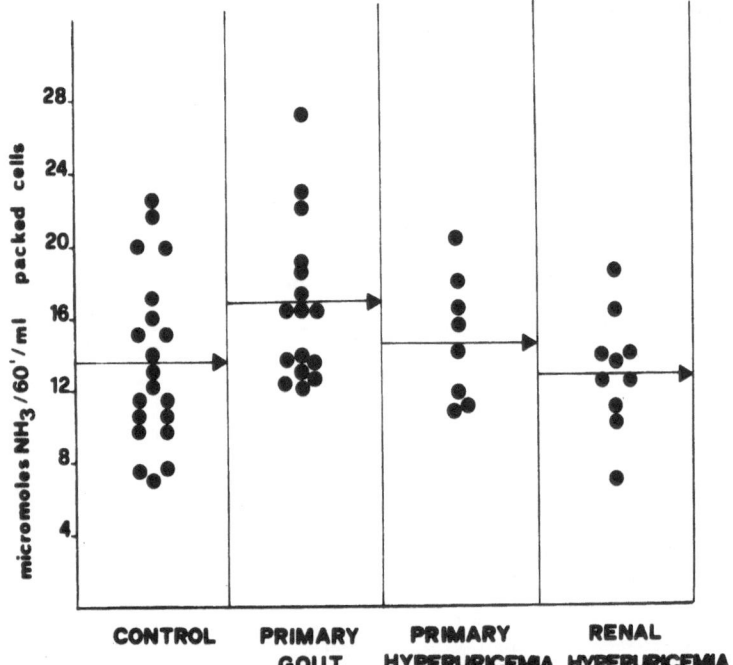

Fig. I – Erythrocyte Adenosine –Deaminase activity

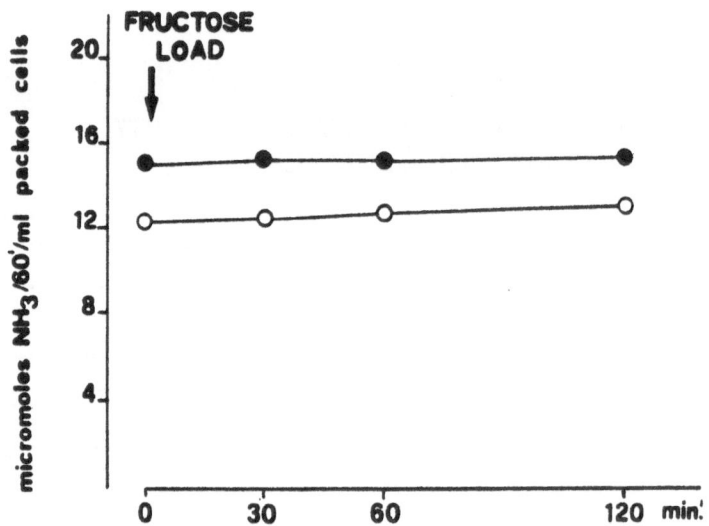

Fig. 2 – Erythrocyte Adenosine–Deaminase activity
        after intravenous fructose load in controls (O)
        and in patients with primary hyperuricemia (●).

DISCUSSION

The results obtained in our gouty patients and in subjects with primary asymptomatic or renal hyperuricemia did not demonstrate a significant increase in erythrocyte ADA activity in comparison with controls.

These findings are not in agreement with those obtained by Nishizawa et al. (5) and are not consistent with the hypothesis that hyperuricemia in primary gout may be due to an increase in catabolism of purine ribonucleotides.

Erythrocytes are not the best cells to study this catabolic pathway, since they lack any 5'-Nucleotidase activity (9). 5'-Phosphomonoesterase activity may be supplied by nonspecific alkaline phosphatase, but the hydrolysis of nucleoside monophosphate preferentially occur by means of 5'-Nucleotidase than alkaline phosphatase (I0).

The lack of any significant increase in erythrocyte ADA activity in primary hyperuricemia and after fructose load support this view.

On the contrary studies in rat liver cells have demonstrated the activation of both 5'-Nucleotidase and ADA following intravenous administration of fructose at high doses, but ADA activity in human liver is very low (2) and its activation, not yet demonstrated, seems not to be involved in the pathogenesis of primary hyperuricemia.

REFERENCES

I. Conway E.J. and Cooke R.
   Biochem. J. 333 : 479, (I939)
2. Van der Weyden M.B. and Kelley W.N.
   J. Biol. Chem. 25I : 5448, (I976)
3. Giblett E.R., Anderson J.E., Cohen F., Pollara B., and
   Meuwissen H.J.
   Lancet 2 : I067, (I972)
4. Woods H.F. and Alberti K.G.M.M.
   Lancet 2 : I354, (I972)
5. Nishizawa T., Nishida Y., Akaoka I. and Yoshimura T.
   Clin. Chim. Acta 58 : 277, (I975)

6. Archibald R.M.
   Clin. Chem. 3 : I02,(I957)
7. Seligson D. and Seligson H.
   J. Lab. Clin. Med. 38 : 394, (I961)
8. Chaney A.L. and Marbach E.P.
   Clin. Chem. 8 : I30, (I962)
9. Shenoy T.S. and Clifford A.J.
   Biochem Biophys. Acta 4II : I33, (1975)

I0. Schmidt G. in "The Nucleic Acid" Chargoff E. and Davidson J.N.
    eds. Academic Press, New York, (I955)
II. Fox I.H. and Marchant P.
    Advances Exp. Med. Biol. 76A : 249, (I977)
I2. Schwerzmeier J.D. Müller M.M. and Marktl W.
    Advances Exp. Med. Biol. 76A : 542, (I977)

CLINICAL VARIABILITY OF THE GOUTY DIATHESIS

William J. Arnold, M.D.
Robert A. Simmons, M.D.

Lutheran General Hospital
Park Ridge, Illinois
Brooke Army Medical Center
San Antonio, Texas

Gouty arthritis with or without tophaceous deposits and uric acid nephrolithiasis are the clinical manifestations of hyperuricemia known collectively as the gouty diathesis. While the pathogenesis of hyperuricemia may involve either a primary (genetic) or secondary abnormality of uric acid metabolism, the incidence and severity of the clinical manifestations have been related to two other factors, the magnitude and the duration of hyperuricemia[1]. However, recent investigations into the pathogenesis of gouty arthritis have demonstrated that both soluble and cellular mediators of the inflammatory process are important in the pathogenesis of the articular inflammation associated with monosodium urate (MSU) crystals[2]. The relationship of these factors to those previously shown to be of importance in the pathogenesis of gouty arthritis has not been clarified. However, they introduce new variables which potentially are of significance in determining the eventual clinical manifestations of hyperuricemia. This report concerns an 83 year-old patient who noted the gradual appearance of multiple tophaceous deposits over a period of at least 20 years, yet had never experienced an attack of gouty arthritis. Therefore, while no signs of acute inflammation have been noted, this patient clearly had significant hyperuricemia which resulted in the accumulation of MSU crystals. Stimulated by this unusual clinical syndrome, we have investigated this patient's uric acid metabolism and certain components of the acute inflammatory response for an explanation of the striking absence of acute gouty arthritis.

39

Case Report

The patient is an 83 year-old white female who was admitted to
the hospital with the chief complaint of dyspnea on exertion for three
months prior to admission. The present and past history was otherwise
unremarkable except for a 15 to 20 year history of firm nodules which
first appeared on a fingertip. The nodules had gradually enlarged
and progressed to involve the majority of her fingers, the right
second toe and the left elbow. The nodules on her fingers had
enlarged to such a degree that she was forced to abandon her avoca-
tion as a concert pianist. The patient steadfastly denied any
episodes of articular or periarticular inflammation, renal stones or
gross hematuria. The patient had no children. There was no family
history of a similar illness, renal stones or arthritis. Physical
examination at the time of admission revealed a tachypneic white
female otherwise in no acute distress. Vital signs were normal save
for an irregularly irregular pulse at a rate of 120/minute and a
respiratory rate of 22/minute. Examination of the eyes revealed
bilateral lenticular opacities. Cardiopulmonary examination was
consistent with congestive heart failure. The remainder of the
physical examination was otherwise normal except for the joints.
There were firm nodular, nontender, pale yellow subcutaneous deposits
on the extensor surfaces of the DIP joints of the index, long and ring
fingers of both hands, the PIP joints of both index fingers and the
MCP joint of the left index finger (Fig. 1A). Also a firm lobulated,
subcutaneous mass was noted on the flexor surface of the distal right
long finger. There was a single 7x7 nodule present in the left
olecronon bursa. Examination of the feet revealed bilaterally
enlarged first metatarsophalangeal joints with minimal tenderness
and a fusiform enlargement of the second right toe from which a
chalky white material was draining. No evidence of articular or
periarticular inflammation was present.

Routine laboratory data was within normal limits. On two
separate occasions the blood urea nitrogen was 14 and 17 mg%
(normal up to 22 mg%). Serum creatinine was 1.3 and 1.5 mg%
(normal up to 1.5 mg%). Chest x-ray on admission revealed cardio-
megaly, pulmonary vascular congestion and scattered calcifications.
The EKG initially showed atrial fibrillation. Appropriate therapy
resulted in rapid subjective and objective improvement of the signs
of congestive heart failure. X-rays of the hands revealed periarticular soft tissue swelling with calcification in a pattern corresponding to the subcutaneous nodules (Fig. 1B). Also joint space
narrowing and sclerosis with subchondral and bony cysts were noted
at several DIPs. X-rays of the feet revealed almost total dissolution
of the distal phalanx of the right second toe. Needle aspiration of
the nodule overlying the DIP of the left index finger was performed.
The chalky white material obtained was examined by compensated,
polarized light microscopy and revealed sheets of needle-shaped,

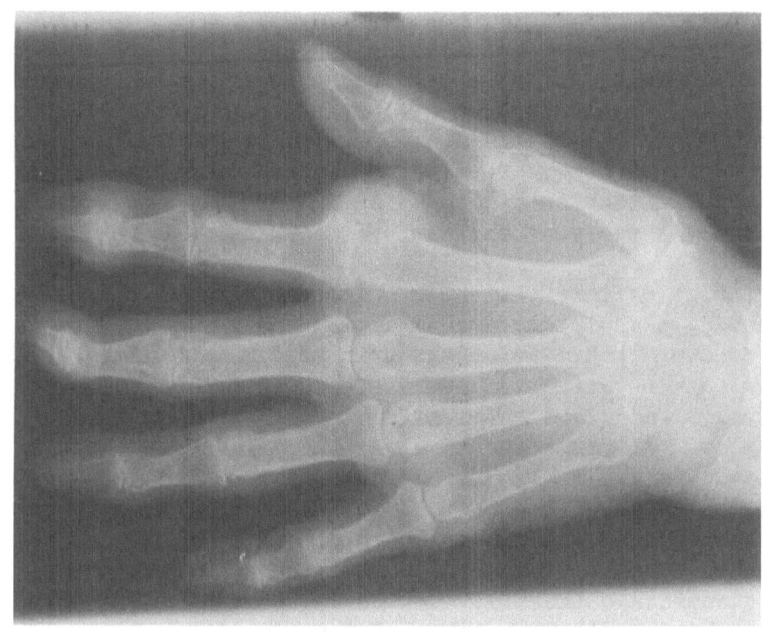

Fig. 1B Roentgenogram of the patient's left hand. See text for description.

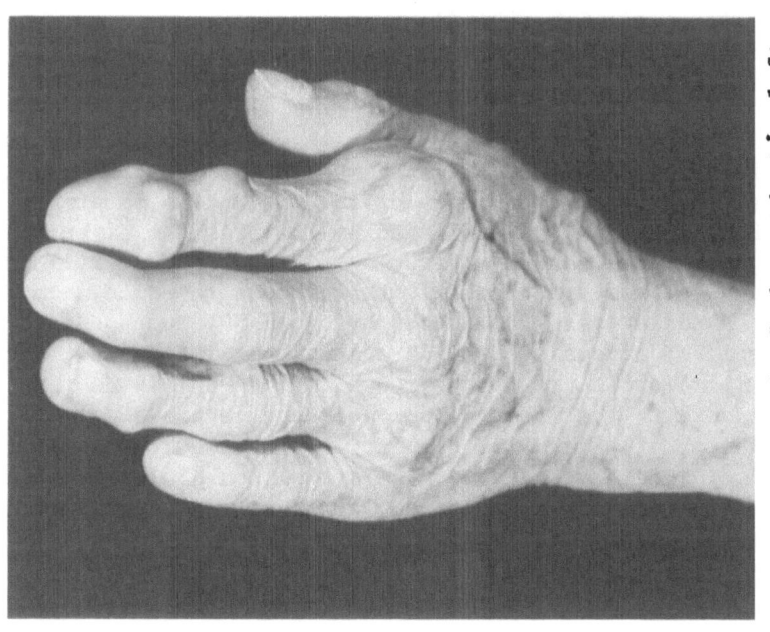

Fig. 1A Photograph of the patient's left hand. Needle aspiration of the distal interphalangeal joint of the index finger revealed sheets of MSU crystals.

strongly negatively-birefringent crystals of MSU.

The patient was followed outside the hospital for three years after this initial admission. She remained hyperuricemic throughout this time as she was unable to take allopurinol or a uricosuric due to gastrointestinal intolerance. Two months after her last visit sh passed away suddenly at home. Permission for an autopsy was denied. No articular or periarticular inflammation was ever reported or observed during our three year follow-up.

Further studies of this patient's uric acid metabolism were performed following circulatory stabilization and while the patient was receiving no drugs known to effect uric acid metabolism. Serum salicylate level was negative. Serum urate level by the colorimetri method was 9.5, 9.3 and 9.8 mg% (normal up to 8.5 mg%) on three separate occasions. Plasma urate level by the uricase method was 9.1 mg% (normal up to 7.5 mg%). Two 24-hour urine collections were judged to be incomplete but the ratio or uric acid (mg%) to creatini (mg%) in the two different specimens was 0.27 and 0.67 (normal less than 0.75). Activity of the erythrocyte enzymes Hypoxanthine-Guanin Phosphoribosyltransferase (HGPRT), Adenine Phosphoribosyltransferase (APRT), and Phosphoribosylpyrophosphate Synthetase (PRPP-Synthetase) were determined by previously described radiochemical techniques[3,4,] (Table 1) The borderline low level of PRPP-Synthetase was attribute to instability during shipment and was not repeated.

The plasma level of Hageman factor was 83% (normal 40-90%). Th whole blood clotting time and bleeding time were normal. The levels of the complement component C_3 and C_4 as well as total hemolytic complement were normal. Serum protein electrophoresis was normal.

The levels of circulating immunoglobulins were low normal with IgG of 688 mg% (normal 800-1800), IgA 57 mg% (normal 90-450) and IgM 202 mg% (normal 70-280). The ability of the patient to mount a specific antibody response was reflected by the presence of antibodi to S. parathyphi (1:160), and by a rubella titre of 1:10. Evaluatio of the patient's delayed hypersensitivity response revealed negative skin tests (1:100 dilutions) for Trichophyton, SKSD, Mumps, Histo-plasmosis, Coccidioidomycosis, Monilia and Intermediate-strength PPD Lymphocyte stimulation in vitro revealed normal H^3-thymidine incorp-oration in response to the mitogens pokeweek and phytohemagglutinin (stimulation idices of 46 and 15, respectively) but no response to Monilia or SKSD. Attempted cutaneous sensitization of the patient to DNCB revealed no primary response and a markedly delayed second-ary response after twenty-one days.

Table I.   ACTIVITY OF ERYTHROCYTE ENZYMES OF PURINE METABOLISM

ENZYME	PATIENT	NORMAL
	(nMoles/mg Protein/hr)	
HGPRT	84	98+14
APRT	30.6	31+ 6
PRPP-Synthetase	14	20-50

Normal polymorphonuclear cell function was strongly suggested by the patient's negative past history for recurrent bacterial infections and the presence in the peripheral blood of 7500 WBC's with 75% PMNs and 20% lymphocytes. The nitroblue tetrazolium test revealed that 20% of the PMNs were positive for dye reduction when unstimulated and 95% were positive following stimulation with latex particles.

Synthetic MSU crystals were prepared by the procedure of Seegmiller, Howell and Malawista to specifically evaluate the ability of the patient's PMNs to phagocytize these crystals[6]. Prior to use the crystals were heated to 200° for three hours. Leukocytes were obtained from the patient and two controls by sedimentation of heparinized fresh whole blood in a 4% clinical Dextran solution. Synthetic MSU crystals (3mg) were added to 5x10[5] WBC's obtained from each subject. The mixture was brought to a final volume of 10 mg with 0.9% NaCl. The samples were then centrifuged at 1,000 g for 90 seconds and incubated at 37° C for 45 minutes. Following this, the percentage of PMNs containing MSU crystals was determined by two observers in two specimens from each sample. For the patient and two controls the average values were 55%, 60% and 55%, respectively from one observer and 40%, 32% and 42%, respectively from the second observer. The ability of the patient's PMNs to respond to a specific chemotactic agent, C_{5a}, was evaluated by using a modified Boyden chamber[7]. Samples were run in triplicate and revealed a value of 13.9 WBC/high-powered field for the patient and 17.2 WBC/high-powered field for a control. These differences are not felt to be significant.

Discussion

The data presented here cannot explain the striking absence of acute gouty arthritis in this patient with hyperuricemia and extensive tophaceous deposits of at least 20 years duration. Although it is most likely that the hyperuricemia in this patient was related to relative renal underexcretion of uric acid, complete data is lacking. The normal values for HGPRT and APRT eliminate the possibility that the patient is a heterozygote for partial HGPRT deficiency. Such patients may have hyperuricemia due to an over-production of uric acid [8].

While the exact role of the complement system in the inflammation of acute gouty arthritis has not yet been defined, recent studies
indicate that MSU crystals can activate complement by the classical
pathway [9, 10]. Complement activation may occur in the presence or
absence of IgG bound to the surface of the crystal since crystals
lacking surface bound IgG can bind C1Q and therefore initiate the
complement cascade. Studies by McCarty and Phelps using a complement-
depleted dog model of gouty arthritis have shown that a normal
articular inflammatory response to MSU crystals can be generated
despite very low levels of serum complement [11]. While we did not
directly study the ability of MSU crystals to activate the complement
system in this patient's serum, the patient was shown to have normal
levels of C_3, C_4 and total hemolytic complement.

In vivo activation of Hageman factor by MSU crystals could
activate the kinin system with subsequent generation of bradykinin[12].
The increased capillary permeability, dilatation of blood vessels
and local increase in bloodflow due to this vasoactive peptide would
mimic the clinical acute gouty attack and assist the influx of PMNs.
However, in animals with an absence of Hageman factor or in animals
depleted of bradykinin with carboxypeptidase B, the articular inflammatory response to MSU crystals is undiminished [13,14]. The
Hageman factor level was normal in this patient and could be activat-
ed normally as demonstrated by the normal whole blood clotting time.

The pivotal importance of the PMN in the pathogenesis of the
acute gouty attack has been demonstrated in vivo. Depletion of
circulatory PMNs with vinblastine or anti-PMN serum clearly blunts
the articular inflammatory response to MSU crystals in the dog
model [15,16]. This critical role for the PMN in the acute inflammation associated with MSU crystals may derive from the PMN production
of a MSU crystal-induced chemotactic substance. Phelps first demonstrated the in vitro release of a potent chemotactic substance
released from PMNs following phagocytosis of MSU crystals [17]. The
study of this MSU-induced chemotactic substance has been extended by
Spilberg [18,19]. These studies demonstrate that highly-purified preparations of the chemotactic substance can produce inflammation
when injected in vivo in the absence of MSU cyrstals· Colchicine at
concentrations which do not effect PMN chemotaxis, does inhibit the
production and release of the PMN-derived MSU crystal-induced chemotactic factor. This is associated with an inhibition of the inflammatory response to MSU crystals. Normal PMN function has been documented in this patient. The patient's PMNs reduced NBT dye and
phagoctyzed synthetic MSU crystals normally. Also the PMNs responded
normally to a potent chemotactic stimulus, C_{5a}. An absence or impairment in the production of the PMN-derived, MSU crystal-induced
chemotactic factor would be an attractive hypothesis to explain the
absence of acute gouty arthritis. Unfortunately, on two separate
occasions prior to the patient's demise we were unable to demonstrate

the generation of PMN-derived, MSU crystal-induced chemotaccic factor from either the patient's or normal PMNs.

The role of humoral and cell-mediated immunity in the pathogenesis of gouty arthritis is uncertain. IgG as well as other proteins may bind to the surface of MSU crystals and mediate complement activation [20]. While our patient did have low IgG levels it is unlikely that this minor reduction would produce an inability to activate complement.

Lymphocytes are consistently found at the periphery of tophaceous deposits and in the synovium of patients with chronic gout [21,22]. Their role, if any, in the pathogenesis of acute gouty arthritis is unknown. The skin test anergy and markedly impaired response to DNCB present in our patient suggest that a deficiency in cell-mediated immunity was present. Since the patient's chest x-ray showed multiple calcified granulomas it is likely that the patient had an acquired deficiency of cell-mediated immunity possibly related to aging.

In summary, we have reported a patient with extensive tophaceous deposits in the absence of gouty arthritis. While our investigations failed to reveal an explanation for the absence of an inflammatory response to MSU crystals they illustrate that the presence of crystals alone is not sufficient to cause acute gouty arthritis. Recently, Agudelo et al reported the finding of MSU crystals in aspirates from the first MTP joint of 9 of 10 asymptomatic patients with a previous history of podagra [22]. MSU cyrstals were found in two patients who had been asymptomatic for one year. While recent experimental evidence suggests that other factors participate in MSU crystal-induced inflammation, further studies of patients similar to ours may reveal which of these factors is (are) absolutely required for a clinically obvious acute inflammatory response to MSU crystals and therefore help to explain the clinical variability of the gouty diathesis.

References

1.   A.P. Hall, P.E. Barry, T.R. Dawber, et al. Epidemiology of Gout and Hyperuricemia: A Long-Term Population Study. Am J Med 42:27, 1967.
2.   I. Spilberg. Current Concepts of the Mechanism of Acute Inflammation in Gouty Arthritis. Arthritis Rheum. 18;129, 1975.
3.   W.J.Arnold and W.N.Kelley. Human Hypoxanthine-Guanine Phosphoribosyltransferase. J. Biol. Chem. 246:7398, 1971.
4.   C.B. Thomas, W.J. Arnold, and W.N. Kelley. Human Adenine Phosphoribosyltransferase. J. Biol. Chem. 248:2529, 1973.
5.   I.H. Fox and W.N. Kelley. Human Phosphoribosylpyrophosphate Synthetase. J. Biol. Chem. 247:2126, 1972.

6.   J.E. Seegmiller, R.R. Howell, and S.E. Malawista. Inflammatory
Reaction to Sodium Urate: Its Possible Relationship to Genesis of
Acute Gouty Arthritis. JAMA 180:469, 1962
7.   R. Snyderman, H. Gerwuz, and S.E. Mergenhagen. Interactions of
the Complement System with Endotoxic Lipopolysaccharide. J.Exp.Med.
127:259, 1968.
8.   B.T.Emerson. Urate Metabolism in Heterozygotes for HGPRTase
Deficiency. In Adv.Exp.Med.Biol., O.Sperling, A.deVries, and
J.B. Wyngaarden, eds. Plenum, N.Y. 41A:287, 1974.
9.   G.B.Naff and P.H. Byers. Complement as a Mediator of Inflammation in Acute Gouty Arthritis. J.Lab.Clin.Med. 81:747, 1973.
10.   P.Hasselbacher. Activation of C3 by MSU, Potassium Urate and
Steroid Crystals. Arhtritis Rheum. 21:565, 1978.
11.   P.Phelps, and D.J. McCarty. Crystal-Induced Arthritis. Postgrad
Med 45:87, 1969.
12.   R.W.Kellermeyer. Hageman Factor and Acute Gouty Arthritis.
Arthritis Rheum 11:452, 1968.
13.   I.Spilberg. Urate Crystal Arthritis in Animals Lacking Hageman
Factor. Arthritis Rheum 17:143, 1974.
14.   P. Phelps, D.J. Prockop, and D.J.McCarty. Crystal-induced
Inflammation in Canine Joints. J.Lab.Clin.Med. 68:433,1966.
15.   P.Phelps, and D.J.McCarty. Crystal-induced Inflammation in
Canine Joints. II Importance of Polymorphonuclear Leukocytes.
J.Exp.Med. 124:115, 1966.
16.   Y.H. Chang, and E.J.Gralla. Suppression of Urate Crystal-induced
Canine Joint Inflammation by Heterologous Anti-Polymorphonuclear
Leukocyte Serum. Arhtritis Rheum. 11:145, 1968.
17.   P.Phelps. Appearance of Chemotactic Activity Following Intra-
Articular Injection of MSU Crystals: Effect of Colchicine. J.Lab.
Clin.Med. 76:622, 1970.
18.   I.Spilberg, B.Mandell, and R.D.Wachner. Studies on Crystal-
Induced Chemotactic Factor. J.Lab.Clin.Med. 83:56, 1974.
19.   I.Spilberg, D.Rosenberg and B. Mandell.Induction of Arthritis by
Purified Cell-Derived Chemotactic Factor. J.C.I. 59:582, 1977.
20.   F.Kozin and D.J. McCarty. Protein Absorption of Monosodium Urate,
Calcium Pyrophosphate Dihydrate and Silica Crystals. Arthritis
Rheum. 19:433, 1976.
21.   J.B. Wyngaarden and W.N.Kelley, 1976. The Pathology of Gaut.
in Gout and Hyperuricemia. J.B. Wyngaarden and W.N.Kelley, eds.
Grune and Stratton, New York.
22.   H.R.Schumacher. Pathogenesis of Crystal-Induced Sunovitis.
Clin. Rheum Dis 3(1):105, 1977.
23.   C.A.Agudelo, A.Wienberger,H.R.Schumacher, et al. Definitive
Diagnosis of Gout by Identification of Urate Crystals in Asymptomatic
Metatarsophalangeal Joints. Arhthritis Rheum. 22:559, 1979.

CLINICAL FEATURES OF 4,000 GOUTY SUBJECTS IN JAPAN

N. Nishioka, M.D.
K. Mikanagi, M.D.

Rheumatology Division, Dept. of Medicine
School of Medicine, Mie University
Medical Center, Mie Pref. Japan, 514

Rheumatology Division, Dept. of Orthopaedics
Jichi Medical School, Tochigi, Japan 329-04

INTRODUCTION

Before World War II, gout had been an extremely rare disease in Japan. The post war years have seen an increasing number of cases, especially among middle-aged males. Since 1965, cases have become more and more prevalent and now, gout is frequently observed at rheumatology and metabolic clinics.

The number of gouty patients has been increasing in Japan year by year. From 1958 to 1964, the number of patients was less than 100, from 1965 to 1969, less than 300. However, Since 1970, more than 300 additional patients have entered our clinic every year.

In this paper we describe the general profile of patients with gout in Japan.

MATERIAL AND METHOD

Four thousand one hundred and ten gouty subjects were examined at Toranomon General Hospital in Tokyo, between the years 1958 and 1977.

The 2,455 cases seen between 1958 and 1973 were described based on a retrospective computer assited analysis. Most of the serum and urine urate values had been estimated with the use of Technicon Auto-Analyzer. The upper limit of normal for the serum uric acid

was taken to be 7.9 mg/100 ml for males and 6.0 mg/100 ml for females.
Gout was diagnosed as the episodes of reccurrent acute arthritis
attacks. The presence of urate crystals in synovial fluid or tophaceous changes were taken as confirming the diagnosis of the cases
that did not have typical clinical features of Gout,[1].

The following information was analyzed: age, sex distribution,
genetic implications, the site of acute gouty attack, Tophii, laboratory data including serum and urine uric acid, complications and
prognosis.

RESULTS

Age Distribution

The age distribution of the patients at the first visit to the
clinic had its peak at the fifth decade (29.1%), followed by the
sixth (23.5%), seventh (20.7%) and fourth (14.1%). Even the age
group above eighty years had an incidence of 0.9 %. The mean age
of the first attack was 49.5 years in males and 52.7 years in females.
However, the distribution of the age of the patients among second
and third decade has increased. (Fig. 1).

Sex Distribution

The female with gout was less than one per cent of the subjects,
(23 cases). The mean age of the initial attack was 52.7 years and
uric acid value was 93 mg/dl for this female group. Four cases had
initial gouty attack before menopause, but two of the patients had
secondary gout associated with renal damage. Primary gout in females
before menopause was extremely rare in Japanese cases compared
with the Caucasians.

Genetic Implications

Patients with more than one gouty patient among their relatives
amounted to 186 cases and the total number of these family occurences

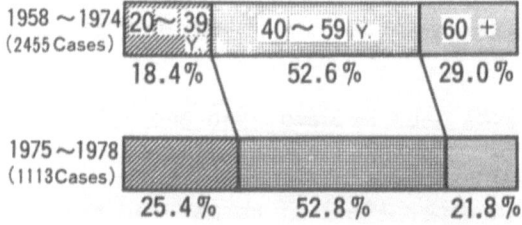

Fig. 1. Age distribution of gouty patients.

was 202, which were classified according to their relation with the
proposit. Male siblings were the most prevalent (48.0%), male children (5.5%), nephews (4.0%), mothers (3.5%), first male cousins
(3.5%), uncles (3.0%) and grandfathers (2.0%).

Gouty Attack

The initial attack of gouty arthritis occurred predominantly
in the MP joint of the great toe (68.4%), and the following sites
of involvement were in the instep and ankle (18.0%), heel cord (3.5%)
and knee (2.9%). In the upper extremes, the first attack occurred
less frequently; e.g. phalangeal joint of finger 1.1% and the first
wrist joint 0.2% (Table 1).

Tophii

9.2% of the subjects had tophaceous gout. Tophii predominantly
involved the helix and antihelix (56.5%). The next most common site
was the great toe (32.8%), followed by the ankle joint (8.9%), ten-
dons and phalangeal joints of fingers (6.2%). Tophii were also
formed in the knee joint (3.1%), ankle joint (8.8%) and heel cord
(4.4%).

Table 1. First Manifestation of Gout.

Joint involved	Right	Left	Total	%
Big Toe	879	801	1,680	(68.4)
Other Toe				
Other Toe	42	31	73	(3.5)
Ankle and Foot joint	230	212	442	(18.0)
Heel	42	43	85	(3.5)
Knee	41	29	70	(2.9)
Finger	14	14	28	(1.1)
Wrist	3	1	4	(0.2)
Other	-	-	73	(3.0)
Total			2,455	

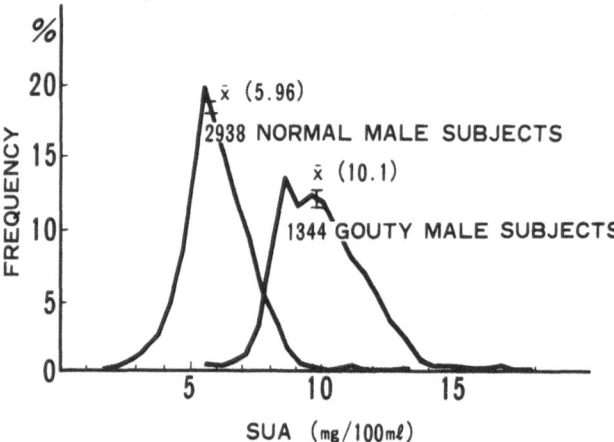

Fig 2. Comparative study of serum uric acid in normal male subjects
        and gouty patients.

Laboratory Data

Serum Uric Acid: Figure 2 shows the age distribution of serum
uric acid in 1,344 subjects compared with the normal subjects who
were living in Tokyo, under similar environment as the patients.
The mean value of the measurements was 10.1 mg/dl (+1.70), 93.7% of
the cases fulfilled the definition of hyperuricemia at our laboratory.

Excretion of Uric Acid

The daily excretion of uric acid showed a peak at 800-899 mg/
24 hrs. (12.2%). There were 11.3% of the patients in the range of
900-999 mg/24 hrs. and 10.6% excreted 1000-1099 mg/ 24 hrs. In this
study, we define overexcretion of uric acid as above 1200 mg/24 hrs.
and there were 19.3% in this group, because the patients were not
under the conditioning of purine free diet.

Other Laboratory Findings

In our laboratory, the mean value of RBC, Hb, thrombocyte and
WBC increased more than those of the age-matched standard average.
ESR also had increased to 67.2%. The value of the blood urea was 18.5
mg/dl (+ 5.5) and was also increased compared with the age-matched
control group. Thirty-one per cent of the patients showed high values for GOT and 18.5% showed high values for GPT. There were 28.1%
patients who were CRP positive and 8.1 % were RA positive. 10.3%
showed proteinuria and 3.1% glucosuria.

Complications

In the cases with more than two complications, such as hypertension or chronic renal failure, the one which seemed to be primary was chosen. 770 cases (31.0%) had complications.

Hypertension and renal failure had a high incidence, the former being 14.1% and the latter 11.2%. Pulmonary tuberculosis was seen in 62 cases (2.5%) and diabetes in 53 cases (2.2%). Less common were polycythemia (3 cases), Sarcoidosis (3 cases) and psoriasis (1 case). Among the renal complications (275 cases), nephrolithiasis was the most prevalent (49.5%), followed by acute glomerulonephritis (22.5%), chronic renal failure (11.2%), renal tuberculosis (9.5%) and nephrotic syndrome (4.4%). The nephrolithiasis comprised about half the renal complications and its occurrence in all gouty subjects was 5.5%.

Prognosis

In 74 cases the cause of death was confirmed. Heart failure was the most common (29.7%), followed by malignant tumors (24.3%) apoplexy (23.0%) and chronic renal failure (6.8%). The mean age of death from cardiac involvement was 66.2 years, from malignant tumors 68.0 years, from apoplexy 66.6 years and from uremia 54.6 years. Other causes of death were; accidents 4.1% and ageing 4.1%.

SUMMARY

The clinical aspects of gout were described based on the 4,112 patients observed in the same clinic. In this paper mainly, 2,455 cases seen between 1958 and 1973 were described based on a pretrospective computer assisted analysis. The age distribution of the patient had its peak at 50 years of age and the average age at the time of onset was 49.5 years for males and 54.2 years for females. Recently, in Japan, the incident of gout among second and third decades has increased.

The sex distribution, showed a predominance in males (99%). Primary gout in females before the neopause was very rare, only 0.08 % of all subjects.

Cases with a definite hereditary incidence comprised 7.5% and male sibling involvement was the most frequent.

68.4% of the patients had their initial gouty attack in the MTP joint of the great toe. 47.4% of the patients had a history of gouty attacks in two or more joints.

Cases with tophaceous gout composed 9.2%, however, recently it has been decreasing in number every year.

Table 2. Distribution of Serum Uric Acid Level in Gouty Patients by Age Group.

Age group mg/day	-29	30-39	40-49	50-59	60-69	70-	Total
-299		4 (3.2)	5 (1.8)	11 (3.2)	6 (2.3)	3 (2.6)	29 (2.5)
300-399		3 (2.4)	8 (2.8)	9 (.26)	8 (3.1)	3 (2.6)	31 (2.7)
400-499		4 (3.2)	8 (2.8)	11 (3.2)	16 (6.2)	10 (8.7)	49 (4.2)
500-599	3 (8.3)	9 (7.2)	20 (7.1)	30 (8.6)	23 (8.8)	13(11.3)	98 (8.4)
600-699	3 (8.3)	4 (3.2)	26 (9.3)	25 (7.2)	24 (9.2)	15(13.0)	97 (8.3)
700-799	5(13.9)	11 (8.8)	26 (9.3)	31 (8.9)	30(11.5)	11 (9.6)	114 (9.8)
800-899	3 (8.3)	22(17.6)	37(13.2)	41(11.8)	33(12.7)	6 (5.2)	142(12.2)
900-999	6(16.7)	16(12.8)	29(10.3)	43(12.4)	25 (9.6)	12(10.4)	131(11.3)
1000-1099	2 (5.5)	20(16.0)	24 (8.5)	36(10.4)	23 (8.8)	18(15.6)	123(10.6)
1100-1199	3 (8.3)	15(12.0)	34(12.1)	33 (9.5)	31(11.9)	6.(5.2)	122(10.5)
1200-1299	5(13.9)	7 (5.6)	19 (6.8)	22 (6.3)	15 (5.8)	6 (5.2)	74 (6.4)
1300-1399	1 (2.8)	4 (3.2)	16 (5.7)	17 (4.9)	5 (1.9)	3 (2.6)	46 (4.0)
1400-1499		2 (1.6)	8 (2.8)	12 (3.5)	5 (1.9)	3 (2.6)	30 (2.6)
1500-1599	2 (5.5)	2 (1.6)	8 (2.8)	7 (2.0)	5 (1.9)	3 (2.6)	27 (2.3)
1600-1699	1 (2.8)		3 (1.1)	5 (1.4)			9 (0.8)
1700-1799		1 (0.8)	1 (0.4)	5 (1.4)	3 (1.2)	2 (1.7)	12 (1.0)
1800-1899		1 (0.8)	4 (1.4)	2 (0.6)	2 (0.8)		9 (0.8)
1900-1999			3 (1.1)	1 (0.3)	2 (0.8)		6 (0.5)
2000-2099			1 (0.4)	3 (0.9)	1 (0.4)		5 (0.4)
2100-2199					1 (0.4)	1 (0.9)	2 (0.2)
2300-2399				1 (0.3)			1 (0.1)
2400-2499			1 (0.4)				1 (0.1)
2500-2599					1 (0.4)		1 (0.1)
2600-2699							
2700-2799							
2800-2899							
2900-2999				1 (0.3)			1 (0.1)
3000-	2 (5.5)			1 (0.3)			3 (0.3)
Total	36	125	281	347	260	115	1164

Table 3. Cause of Death of Gouty Patients.

Cause of death	Nr. of Case	Average of death age
Cardiac involvement	22 (24.7%)	66.2
Malignant tumor	18 (24.3%)	68.0
Apoplexy	17 (23.0%)	66.6
Uremia	5 (6.8%)	58.2
Accident	3 (4.1%)	54.6
Senility	3 (4.1%)	81.6
Miscellaneous	6 (13.5%)	59.8
Total	74	65

93.7% of the patients had hyperuricemia. The mean value of the serum uric acid was 10.1 mg ($\pm$ 1.70). The overexcretion type was more common in the younger group, in those over the age of fifty. The cases with over excerator was up to 15 - 20% of all the subjects.

Complications and past illness were as follows; hypertension (14.4%), renal disease (11.2%), pulmonary tuberculosis (2.5%) and diabetes mellitus (2.2%). Among the gout with renal involvement, nephrolithiasis was the most common (49.5%). In 61 patients, heart disease including coronary ischemic disease was the most prevalent cause of death (27.6%), followed by malignant tumors (24.6%), apoplexy (18.0%) and uremia (6.6%).

REFERENCES

1. CIOMS Criteria. New York: in Bennett PH, Wood PHN (eds): Population Studies in the Rheumatic Disease Proc. IIIrd Internat. Symp,,
   Amsterdam, Exerpta. Medica, 1968, P457.
2. A.P. Hall, P.E. Barry, T.R. Dawber and P.W. McNamara: Epidemiology of gout hyperuricemia, A long term Population study, Am. J.
   Med., 42: 27-37, 1967.
3. W.M. Mikkelson, H.J. Dudge, H. Valkenburg: The distribution of
   Serum Uric Acid Values in a population unselected as to gout or
   hyperuricemia. Am. J. Med., 29: 242-251. 1965.
4. J.B. O'Sullivan: Gout in a New England Town. A prevalence Study
   in Sadbury, Massachusetts. Ann. Rheum. Dis., 31: 166-169, 1972.
5. J.T. Scott, R. Grahame: Clinical Survey of 354 patients with
   gout. Ann. Rheum. Dis., 29: 461-465, 1970.

FREQUENCY OF CHONDROCALCINOSIS OF THE KNEES AND AVASCULAR

NECROSIS OF THE FEMORAL HEADS IN GOUT, A CONTROLLED STUDY

A.Stockman*, L.G.Darlington**, J.T.Scott

*400 Albert Street, East Melbourne, 3002 Victoria
Australia
**Epsom District Hospital, Epsom, Surrey England
Charing Cross Hospital and Kennedy Institute of
Rheumatology, London W6, England

The pathogenesis of chondrocalcinosis remains obscure. An association has been reported with various metabolic disorders such as hyperparathyroidism, haemochromatosis and gout. The association between gout and chondrocalcinosis remains ill-defined because of the lack of appropriate controls, small numbers of patients or retrospective study. An association has also been claimed between gout and avascular necrosis of the femoral head.

In order to obtain further information on these points, we have conducted a radiological survey of a group of patients with gout and a group of non-gouty control subjects.

Patients and Methods

136 male and 2 female patients with gout and 142 male non-gouty controls were studied.

The gout patients were drawn without selection from the Charing Cross Hospital Gout Clinic. All of them had been diagnosed as gout by acceptable clinical criteria: all were hyperuricaemic (or had been before treatment was commenced) and had experienced acute episodes of arthritis which had taken the form of typical podagra in 110 of them. 23 had tophi and urate crystals had been identified from synovial fluid in seven.

The controls were volunteers, carefully matched for age, who were also attending Charing Cross Hospital. There were 130 out-patients, 40 of whom were having radiotherapy for malignant skin

lesions and 90 of whom were attending a fracture clinic for
treatment of traumatic fractures and abrasions. 12 in-patients
with acute medical conditions made up the rest of the control
group. These control subjects were questioned as to a history of
gouty arthritis: only one of them gave such a history: his X-rays
were normal.

To detect chondrocalcinosis and avascular necrosis standard
antero-posterior films of the knees and pelvis were taken,
radiographs of the knees alone being expected to detect 89 - 99%
of patients with articular chondrocalcinosis. The films from both
groups were mixed and read "blind" by all three authors. The films
were also graded for severity of osteoarthrosis.

Plasma uric acid, using an enzymatic method, was estimated in
all of the subjects. In those in whom chondrocalcinosis was found,
serum calcium, phosphate, alkaline phosphatase and iron were
estimated, and the wrists were X-rayed.

Results

The mean age of the gout patients was 55.2 years ($\pm$ 12.9),
that of the controls 54.9 years ($\pm$ 13.4). There was no difference
in the prevalence of osteoarthrosis.

Chondrocalcinosis of the knees was detected in eight patients
with gout, a prevalence of 5.8%. No cases were found in the
control group. This difference is significant (p <0.025, Yates
X^2 test; p <0.01, Fisher's exact probability test).

Six of the eight gout patients with chondrocalcinosis of the
knees gave a history of symptoms involving the knees, mainly in
the form of acute synovitis, sometimes recurrent. In only two of
them, however, had the opportunity arisen of aspirating synovial
fluid, urate crystals being found in one and no crystals in the
other. Thus the cause of the synovitis in the knees (urate or
pyrophosphate) in most of them remains unknown, although there is
little doubt that all had urate gout involving other joints.

In five of the eight patients the deposits in the knee joints
appeared linear on radiological examination, involving both knees
in four of them. One other patient showed both linear and
irregular deposits, and the other two patients showed irregular
deposits only. In all of them the appearance of calcification,
though definite, tended to be rather faint. Evidence of chondrocalcinosis in the symphysis pubis or hips was seen in four of the
patients, all of them with linear deposits in the knees. Three
patients showed radiological evidence of calcium deposition in the
wrists, mainly in relation to the triangular ligament. Six of the
eight patients thus showed evidence of chondrocalcinosis in sites

additional to the knees.

Tophaceous deposits were present in two of the eight patients and were not extensive.

Urinary urate on a low-purine diet was normal or only slightly elevated in all of the patients. Some evidence of renal functional impairment (creatinine clearance less than 60ml/min or plasma creatinine >140μmol/l) was present in two of the eight patients. There was no evidence of other metabolic disorders commonly associated with chondrocalcinosis.

Comparison of the patients with chondrocalcinosis with the rest of the gout patients shows no significant difference in age, prevalence of tophi or prevalence of renal impairment. The duration of gout appeared to be longer in the patients with chondrocalcinosis and osteoarthrosis of the knees was relatively more common.

No cases of avascular necrosis of the femoral head were found in any of the gout patients or control subjects.

Discussion

These results appear to confirm some form of association between gout and chondrocalcinosis. Perhaps rather surprisingly, no cases of chondrocalcinosis were found in 142 age-matched controls. The reported prevalence in older subjects ranges from 2.2 to 27.6%: the frequency of chondrocalcinosis increases with age. Special radiographic techniques will detect early and subtle calcification which would be missed on standard radiographs.

The mean age of the 142 control subjects − 54.9 ± 13.4 − was lower than that of the elderly subjects in whom relatively high prevalences of chondrocalcinosis have been reported, and this difference probably accounts for the lack of calcification in these subjects. Since only routine radiographs were used early calcification may have been missed, but this of course applied also to the gout patients. There is certainly no reason to believe that selection of controls was unrepresentative of the population as a whole.

The cause of episodes of acute synovitis of the knee which had occurred in six of the eight patients with gout and chondro-calcinosis remains largely unknown, since joint fluid had been examined in only two of them, urate crystals being found in one, and no crystals in the other. The linear pattern of calcification seen in six patients is strongly suggestive of calcium pyro-phosphate deposition. The irregular pattern of calcification seen in three patients could well be due to other types of calcium deposits. Calcification was not restricted to the knee joint, being found elsewhere (hips, pubic symphysis and wrists) in six of the eight patients.

There was no overall difference in the prevalence of radiological osteoarthrosis between the gouty and control groups, but most of the patients with chondrocalcinosis showed evidence of osteoarthrosis of the knees. This may have been secondary to chondrocalcinosis. An association with tophaceous deposition and renal impairment, claimed by other investigators, is not supported by the present study. Nor has any evidence been found of an association with other metabolic disorders.

It thus appears likely that there is some association between calcium pyrophosphate deposition and urate gouty arthritis. The nature of this association remains quite obscure, but it suggests some underlying common predisposition to crystal deposition in connective tissue or perhaps urate crystals may act as nucleating substances leading to deposition of pyrophosphate or apatite.

The failure to detect evidence of avascular necrosis of the femoral head in any of the gout patients (together with the fact that we have never to our knowledge encountered such an association in a larger experience of several hundred patients with gout) is at variance with the findings of other investigators, and probably reflects the heterogeneity of hyperuricaemia and gout in different populations and communities, with different influences of such factors as, for example, hyperlipidaemia and alcohol, which may perhaps predispose to osteonecrosis.

GAS-CHROMATOGRAPHIC EVALUATION OF URINARY 17-KETOSTEROIDS,

ETIOCHOLANOLONE, AND DEHYDROEPIANDROSTERONE IN PRIMARY

GOUT AND HYPERURICEMIA

A. Carcassi, F. Loré, G. Manasse, P. Macrì,

and M. Pisani

Institute of Clinical Medicine, University of Siena
Piazza Selva 7, 53100 Siena, Italy

Sex and age differences in the incidence of gout have suggested that hormonal influences play an important role in the pathogenesis of hyperuricemia and gout. Several possible modes of action have so far been suggested.

In "in vitro" studies it has been observed that testosterone accelerates the crystal lysis of lysosomes which leads to breakdown of leucocytes (1). The cosequent release of inflammatory substances in joint fluid is considered an important step in the onset of the acute arthritis of gout.

On the other hand a clear uricosuric action of estrogens has been demonstrated by uric acid clearance studies carried out in a group of transsexual men before and during estrogen therapy (2). In these subjects the administration of synthetic estrogens caused a highly significant increase in uric acid clearance with a fall in plasma uric acid.

As it concerns adrenal androgens some authors have described low urinary excretion of dehydroepiandrosterone (DHA) in subjects with gout (3,4). Other researchers have reported excretion not only of DHA, but also of androsterone and etiocholanolone (E) significantly below normal (5).

It has been proposed that a reduced production of DHA may be involved in the pathogenesis of hyperuricemia through a stimulatory effect on the biosynthesis of purines.

DHA is considered an inhibitor of glucose-6-phosphate dehydrogenase (G-6-PD) and therefore also a regulator for the pentose-phos-
phate cycle; a reduced production of DHA would activate G-6-PD
with a consequent increase in $NADPH_2$ and phosphoribosyl-I-pyrophos-
phate (PRPP) that would be responsible for an enhanced synthesis
of uric acid (6).

Most of the authors who described low levels of urinary DHA
and/or etiocholanolone in gouty patients had performed their studies by methods which are no longer used. For instance today it is
known that acid hydrolysis is responsible for a partial degradation of some steroids, in particular DHA (7). More accurate determinations are achieved by enzymatic hydrolysis.

To verify the above data by a modern gas-chromatographic method
we studied the urinary excretion of total 17-ketosteroids (17-ks),
E and DHA in gouty patients and in subjects with primary hyperuricemia.

MATERIALS AND METHODS

The study was carried out in: 20 patients with primary gout
(aged 31-80 years, mean 56.9), 20 subjects with asymptomatic hyperuricemia (aged 30-81 years, mean 57.1) and 20 normal subjects (aged
25-80 years, mean 56.7) as a control. All subjects of each group
were males.

An aliquot (1/1000) of a 24-hour urine collection was hydrolysed
with beta-glucuronidase and sulfatase (Hélicase, Industrie Biologique
Francaise) for 24 hours at 37° C at pH 5.2. The urine was then
extracted with ethyl ether, the extract was washed with NaOH 2 N and
with distilled water, dried over anidrous sodium sulfate and taken
to dryness in a water bath. The extracted steroids were then converted into trimethylsilyl ethers and injected into a Perkin Elmer
900 gas-chromatograph equipped with hydrogen flame detector and a
glass column measuring 200 X 0.4 (o.d.) cm. The stationary phase
was 3% OV 225 on Chromosorb W 80/100 mesh (Carlo Erba). The injec-
tor and detector temperature was set at 225° C and the oven at
245° C. The nitrogen carrier gas was maintained at a flow of 30
ml/min. Androstandiol was used as internal standard. It was added
prior to enzymatic hydrolysis (8).

Statistical analysis was performed by the Student's unpaired
t test.

RESULTS

Our results are shown in Table I and Figures 1-3.

TABLE I

Urinary excretion of total 17-ks, E and DHA in normal subjects, patients with primary gout and patients with primary asymptomatic hyperuricemia.

		Total 17-ks	E	DHA	
Control	20	7.43	2.24	0.35	mean
		3.74	1.39	0.46	SD
Primary Gout	20	7.28	2.02	0.18	mean
		3.80	1.13	0.15	SD
		0.12	0.54	1.58	t
Primary hyperuricemia	20	7.40	1.95	0.29	mean
		3.42	1.44	0.24	SD
		0.03	0.65	0.54	t

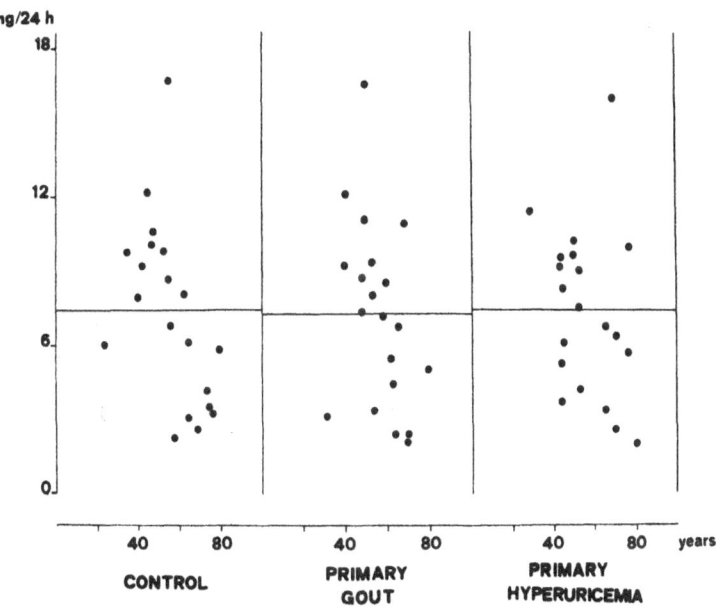

Fig. 1. Urinary excretion of total 17-KS.

ETIOCHOLANOLONE

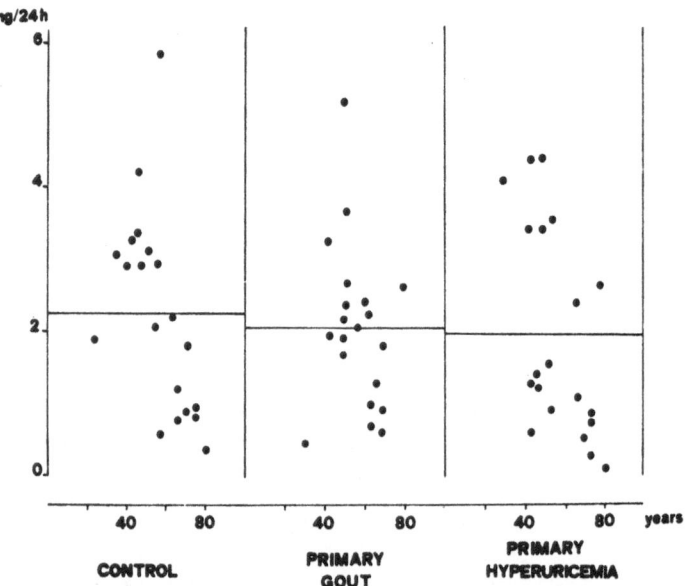

Fig. 2. Urinary excretion of etiocholanolone.

DEHYDROEPIANDROSTERONE

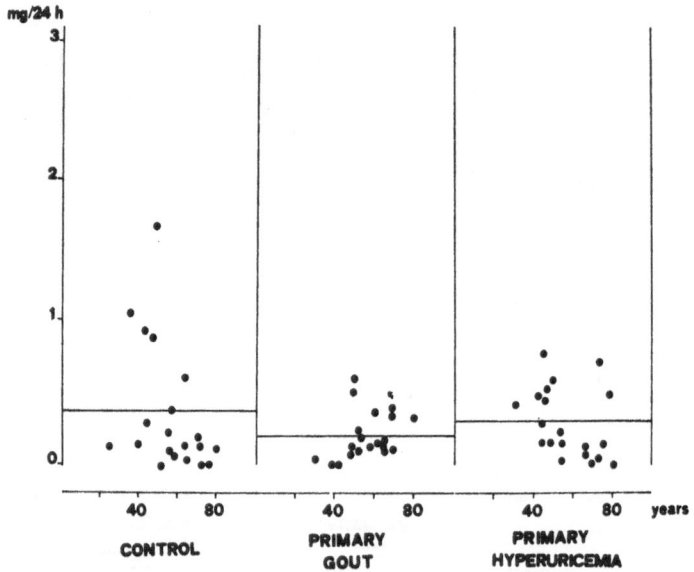

Fig. 3. Urinary excretion of dehydroepiandrosterone.

DISCUSSION

Our findings show that total 17-ks excretion is within the normal range in each group of patients. Differences in etiocholanolone excretion were very small and statistically not significant. In disagreement with other authors we found that DHA was absent in urine of only a few gouty patients. Subjects with primary asymptomatic hyperuricemia showed DHA urinary excretion slightly below the control subjects, but higher than gouty patients. None of these differences was significant.

It should be stressed that, as one can see in Figs. 1-3, there is a clear trend toward a decrease in total and individual 17-ks when age increases (9). As it is known that gout is most common in the age group between 40 and 50 years it could be suggested that if gouty patients are compared with normal subjects as control without paying attention to age, differences between these groups may become erroneously significant.

Our results seem to lessen the importance of DHA in the pathogenesis of hyperuricemia and gout.

REFERENCES

1. Rita, G. A. and Weissmann, G., Boll. Soc. It. Biol. Sper. 48: 142 (1973).
2. Nicholls, A., Snaith M. L. and Scott J. T., Brit. Med. J. 1:449 (1973).
3. Sonka, J., Gregorova, J. and Krizek, V., Steroids 4:843 (1964)
4. Pittman, J. A., Starnes, W. R., Beethan, W. P., and Luketic, G. C., Clin. Res. 14:66 (1966).
5. Sparagana, M. and Phillips, G., Steroids 19:477 (1972).
6. Marks, P. A. and Banks, J., Proc. Nat. Acad. Sci. Wash. 46:447 (1960).
7. Van Kampen, E. J., Clin. Chim. Acta 34:241 (1971).
8. Loré, F., Boll. Soc. It. Biol. Sper. 52:1736 (1976).
9. Loré, F., Min. Med. 69:2927 (1978).

HORMONAL ASPECTS OF GOUTY PATIENTS

U. Valentini[1], G. Riario-Sforza[2], R. Marcolongo[3]
and E. Marinello[4]

[1]Hormones Department, Hospital of Brescia
[2]Clinical Chemistry Laboratory, Hospital of Chieti
[3]Service of Rheumatology, University of Siena
[4]Department of Biochemistry, EULO Brescia, Italy

Attempts have been made from time to time to identify gout as
an endocrine disorder. The well known sex difference in the incidence of gout and the rarity of its occurrence in hypogonadal men
and premenopausal women have led to an interest in the role of the
sex hormones in this metabolic disease. Hormonal influences are
responsible for the known age and sex differences in plasma uric
acid values. Estrogen therapy falls plasma uric acid and rises
urinary uric acid (1), while the development of acute attacks of
gout has been observed in men and in women shortly after receiving
testosterone for a variety of conditions (2,3). Reduction in the
excretion of 17-ketosteroids has been advanced as evidence of en-
docrinological abnormality in gouty patients (4), but this finding
has not been confirmed (5). Sonka et al. (6, 7) claimed that the
absence of urinary dehydroepiandrosterone (DHA) was a constant finding in gouty patients, but others (5,8,9) found that DHA was significantly decreased but not always absent in the urine of gouty
subjects.

The present study was designed to examine some hormonal aspects
of gouty patients. Fifty-three male patients with primary metabolic
gout were studied. They were treated only with colchicine or non-
steroidal and antiinflammatory drugs and received no hypouricemic
treatment during the study period. Twenty four of the 53 patients
were overproductors (urinary uric acid in excess of 800 mg per day).
The age at time of study ranged from 32 to 73 years (mean: 51 years)
and the duration of the disease ranged from 1 to 23 years (mean: 11
years). Control subjects were out- or inpatients matched for sex
and age and not affected by gout or other metabolic disease.

Table. I. Hormonal Pattern of Gouty Patients and Control Subjects

HORMONE	GOUTY PATIENTS (N.53)	NORMAL CONTROL
F S H	6.12 ± 4.1 (■)	7.48 ± 3.5
L H	7.55 ± 5.5 (■)	10.50 ± 3.5
PROGESTERONE	268 ± 83 (□)	293 ± 56
ESTRADIOL 17	32 ± 10 ;(□)	36.5 ± 14
TESTOSTERONE	2.90 ± 1.1 (▲)	5.22 ± 0.88
17-KETOSTEROIDS (urine)	15.20 ± 8.1 (△)	15.05 ± 5.2
DEHYDROEPIANDROSTERONE (DHA) (urine)	0.16 ± 0.1 (△)	0.36 ± 0.1

(■) THE VALUES ARE EXPRESSED AS mUI/ml   (▲) THE VALUES ARE EXPRESSED AS ng/ml
(□) THE VALUES ARE EXPRESSED AS pg/ml   (△) THE VALUES ARE EXPRESSED AS mg/24 hr

Urinary excretion of DHA and 17-ketosteroids has been evaluated by chromatographic methods; serum luteinizing hormone (LH), follicle stimulating hormone (FSH), progesterone, testosterone and estradiol were measured by radioimmunological techniques, LH, FSH, testosterone and estradiol were also measured daily after the administration of clomiphene given orally for 5 days in a dose of 50 mg. daily. Clomiphene is a drug known to stimulate pituitary gonadotropin secretion probably mediated through the hypothalamus or higher centers (10).

The results obtained in gouty patients and in controls are reported in Table I. Urinary excretion of 17-ketosteroids showed no significant variations in gouty patients in comparison to control subjects, while urinary DHA was reduced in gouty patients (0.16 mg/24 hr) if compared to controls (0.36 mg/24 hr) of the same age. The difference is statistically significant (< 0.001). The serum levels of LH, FSH, progesterone and estradiol were normal in all the gouty patients tested. In some cases LH and FSH values were low to normal but the difference with controls is not statistically significant. The mean serum testosterone of the group of gouty patients was significantly lower than the mean serum testosterone of controls (2.9 ng/ml vs. 5.2 ng/ml, p < 0.001) (Fig.1).

Gouty patients despite significantly depressed serum testosterone had normal LH values and there was no correlation between LH levels and testosterone concentrations. There was no correlation between serum levels of LH, FSH or testosterone and either plasma uric acid levels or the duration of the disease. A slightly significant correlation seems on the contrary to exist between estradiol levels and serum uric acid values. In several cases the higher was the value of serum estradiol, the lower were uric acid levels (Fig. 2).

After clomiphene test, serum LH, FSH, estradiol and testosterone levels were increased sharply and significantly and high levels were

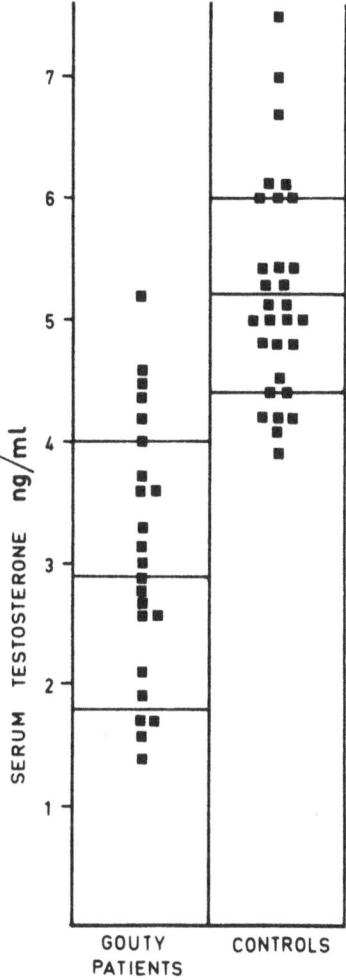

Fig. 1

maintained for several days. The increase in serum LH, FSH and estradiol showed no significant differences in gouty patients tested in
comparison to control subjects. Also the increase in serum testosterone after clomiphene was normal in gouty patients thus indicating
substantial testicular reserve (Fig. 3). Moreover, serum testosterone
remained significantly elevated over baseline values both in gouty
and control subjects also when measured 5 days after the last dose of
clomiphene. This effect may be due to the drug long half-life (11).

    Lower urinary DHA excretion in gouty patients may be due to
reduced production rate. Since exogenous DHA inhibits glcose-6-
phosphate dehydrogenase (G-6-P.D.) activity, Sonka et. al. (6) postulated that failure to produce DHA results in increased G-6-P.D.

activity which in turm causes an increase in precursors of purine
biosynthesis. However, the finding of normal concentrations of plasma DHA sulphate in some gouty patients who failed to excrete DHA
seems to indicate that there is a defect in the peripheral metabolism
of the steroid rather than a deficient production (5). On the other
hand, no correlation has been found between plasma or urinary uric
acid and the levels of urinary DHA (5,9). In addition, the oral
administration of DHA for 1 week had no effect on plasma or urinary
uric acid levels (9). Some authors suggested an enhanced transformation of DHA into androgens in gout. DHA is a precursor for testosterone which is known to facilitate the rupture of lysosomes by
urate crystals (12). Sonka et al. (13) showed an inverse correlation between plasma DHA and urinary androsterone and etiocholanolone,
both these steroids being excreted in larger quantities in gouty
patients if compared to controls.

The present study does not provide an explanation for the decreased serum testosterone levels found in gouty patients. LH, FSH,
estradiol and testosterone response to climiphene test suggests
that testicular function is normal in gout. The results obtained
indicate that the testes of gouty patients responded to stimulation
with an increase in testosterone output, so showing significant reser
capacity to secrete testosterone. On the other hand, gouty patients
had normal secondary sexual characteristics despite their reduced
serum testosterone levels. Many gouty patients are obese and it is
known that most obese males show a low or low normal serum testosterone levels and normal response to clomiphene (10,11). There are
several mechanisms by which obesity might result in low serum testosterone levels. Testosterone is normally bound in serum to a

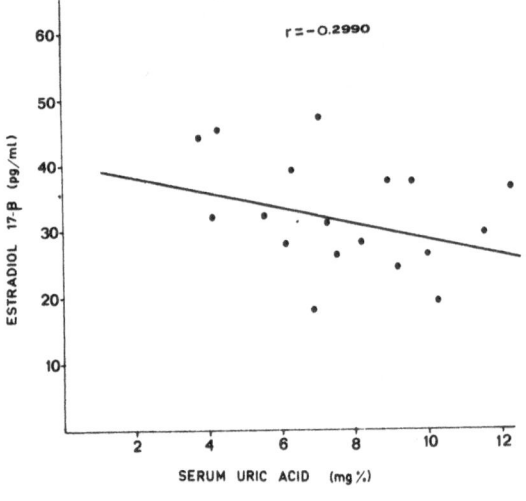

Fig. 2

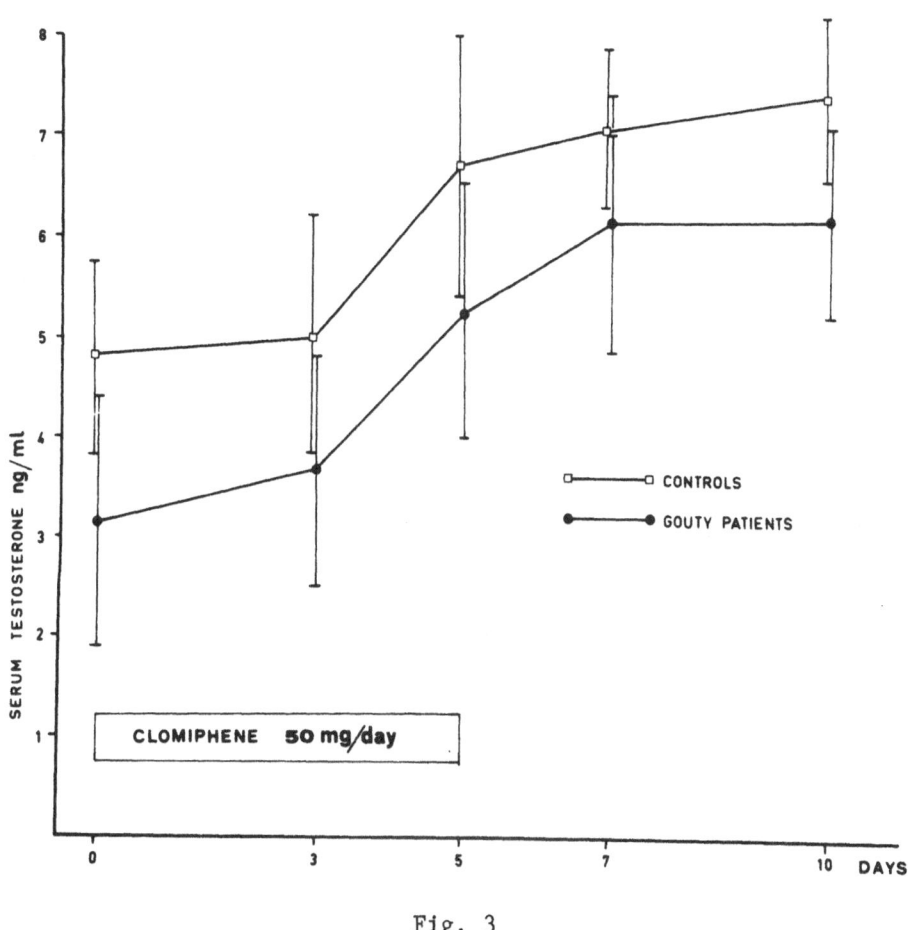

Fig. 3

sex-hormone-binding-globulin (SHBG) and the lower serum testosterone
in obesity was related to reduced levels of this globulin. On the
other hand, adipose tissue is able to metabolize androgens and estrogens in vivo and vitro (11). Some of these steroid metabolites might
be capable of displacing testosterone from SHBG. Adipose tissue may
also metabolize testosterone to rapidly cleared end products (11).
Obese subjects might have altered affinity of their SHBG for testosterone or possess a primary defect in the hypothalamic-pituitary-
gonadal axis too subtle to be detected by our tests (11).

For instance, it is possible that neuroendocrine axis is not sensitive to the decrease of serum testosterone or that some estrogen or androgen metabolite may be substituting for testosterone in negative feedback to the hypothalamus(11).

In conclusion, the significance of the decreased serum testosterone levels in gout remains unclear and further work is required to clarify if sex differences in gout incidence are related to differences in estrogen or androgen metabolism and if testosterone does or not play some role in the expression of gout.

References

1. A. Nicholls, M. L. Snaith, and J. T. Scott, Effect of oestrogen therapy on plasma and urinary levels of uric acid, Brit. Med. J. 1:449 (1973).

2. G. Maranon, Alcunos aspectos del problema de la gota, Rev. Iber. Endocr. 4:77 (1957).

3. J. Graber-Duvernay, and B. Graber-Duvernay; A propos de la goutte aigue féminine, Rhumatologie 9:261 (1957)

4. W. Q. Wolfson, H. S. Guterman, R. Levine, C. Cohn, H. D. Hunt, and E. F. Rosenberg, An endocrine finding apparently characteristic of gout: very low urinary 17-ketosteroid excretion with clinically normal androgen function, J. Clin. Endocr. 9:497 (1949)

5. J. H. Casey, M. M. Hoffman, and S. Solomon, The excretion of urinary dehydroepiandrosterone in gout, Arthr. Rheum. 11:444 (1968)

6. J. Sonka, I. Gregorova, and V. Krizek, Dehydroepiandrosterone in gout, Lancet 1:671 (1964)

7. J. Sonka, V. Krizek, I. Gregorova, H. Vrbova, J. Pikalova, M. Josifko, and J. Stas, Gout and dehydroepiandrosterone. I. DHEA excretion, Endokrynologia Polska 24:193 (1973)

8. J. A. Pittman, W. R. Starnes, W. P. Beetham, and G. C. Luketic, Dehydroepiandrosterone excretion in gout, Clin. Res. 14:66 (1966)

9. M. Sparagana, and G. Phillips, Dehydroepiandrosterone metabolism in gout, Steroids 19:477 (1972)

10. C. G. Heller, M. J. Rowley, and G. V. Heller, Clomiphene citrate: a correlation of its effect on sperm concentration and morphology, total gonadotropins, ICSH, estrogen and testosterone excretion, and testicular cytology in normal men, J. Clin. Endocr. 29:638 (1969)

11. A. R. Glass, R. S. Swerdloff, G. A. Bray, W. T. Dahms, and R. L. Atkinson, Low serum testosterone and sex-hormone-binding-globulin in massively obese men, J. Clin. Endocr. Metab. 45:1211 (1977)

12. G. Weissmann, and G. A. Rita, Molecular basis of gouty inflammation: interaction of monosodium urate crystals with lysosomes and liposomes, Nature 240:167 (1972)

13. J. Sonka, V. Krizek, I. Gregorova, Z. Tomsova, J. Pikalova, M. Strakova, M. Josifko, and J. Stas, Gout and dehydroepiandrosterone. II. Plasma and urinary dehydroepiandrosterone correlated to some clinical and biochemical data, Endokrynologia Polska 24:201 (1973)

DETERMINATION OF TUBULAR SECRETION OF URATE IN HEALTHY AND

GOUTY MEN

L.B. Sorensen and D.J. Levinson

Departments of Medicine, The University of Chicago
Hospitals and Michael Reese Medical Center, Pritzker
School of Medicine, University of Chicago, Chicago,
Illinois, U.S.A.

Recently, a number of clinical and pharmacological observations
have provided evidence for two reabsorptive sites for urate within
the nephron, one proximal to and one distal to the secretory
locus.[1-6] Our own data conform to a four-component system for the
renal handling of urate, involving complete filtration of uric acid
at the glomerulus, reabsorption of 99.3% of the filtered load,
secretion further distal in the nephron, and reabsorption of the
majority of secreted urate.[7] This reference should be consulted
for details of the studies on renal tubular secretion of urate.

The first piece of evidence for tubular secretion of uric acid
in man was provided by Praetorius and Kirk,[8] who reported a patient
with a urate clearance that was 46% higher than the simultaneously
determined glomerular filtration rate. This finding was interpreted
as indicating not only a defect in urate reabsorption, but also the
existence of tubular secretion of urate. Gutman and co-workers[9]
demonstrated net tubular secretion of urate in some patients with
a modest decrease in renal function by the administration of sulfin-
pyrazone during osmotic diuresis and urate loading. Under these
defined experimental conditions, urate excretion exceeded the
filtered load by approximately 20%. On the basis of these findings,
Gutman and Yu[10] proposed a three-component hypothesis for the renal
handling of uric acid. According to that formulation, urate is
completely filtered at the glomerulus and subsequently undergoes
both reabsorption and secretion. The authors speculated that all
of the uric acid excreted in the urine might represent uric acid
that had been secreted in the tubule.

In recent years, evidence has accumulated indicating that
uric acid secretion is far greater than heretofore believed and

that most of the secreted uric acid is reabsorbed at a post-secretory site further distal in the nephron.

Indirect evidence favoring the existence of reabsorption of secreted uric acid has been derived from studies of patients with Hodgkin's disease[1] and Wilson's disease[2] who have hypouricemia related to defective tubular reabsorption of uric acid. When such patients were given pyrazinamide, a compound known to block tubular secretion of urate, the urine became almost free of uric acid, indicating that their inappropriate handling of urate did not result from a defect in tubular reabsorption of filtered urate.

Analogous to these cases, we have reported a 26-year-old woman who was found to have a defect in reabsorption of urate localized solely at the post-secretory site, but who had no under-lying disease such as Wilson's disease or Fanconi's syndrome.[11]

Perhaps the most compelling argument for reabsorption of secreted uric acid has been gathered from pharmacologic studies using pyrazinamide and uricosuric drugs. We have found that the uricosuric response to benzbromarone is completely abolished when tubular secretion of urate is blocked by prior administration of pyrazinamide.[6] Similar but quantiatively less dramatic responses have been observed with probenecid,[3,4] uricosuric doses of chloro-thiazide,[3] and benziodarone[5] after inhibition of tubular secretion by either pyrazinamide or low-dose salicylate. The finding that the uricosuric response to benzbromarone can be completely elimi-nated by pyrazinamide indicates that uricosuric drugs act distally to the secretory site by inhibiting reabsorption of secreted uric acid. This means, in turn, that the magnitude of tubular secretion of urate can be determined if it is possible to block distal reab-sorption completely. To the extent that benzbromarone selectively inhibits post-secretory reabsorption of urate, it is possible to assess tubular secretion of urate by measuring the maximum rate of uric acid excretion following administration of a suitable dose of the drug. In other words, the maximum uricosuric response can be equated to the minimum secretory rate.

We have employed a dose of 80 mg micronized benzbromarone to study the uricosuric response in normal and gouty men. Due to a more complete reabsorption of the micronized preparation in the gastrointestinal tract, a dose of 80 mg micronized benzbromarone is equivalent to 150-200 mg of the commercial preparation that is available in Europe. Benzbromarone is especially suitable for this purpose, since it is not excreted via the renal organic anion transport system and, therefore, does not interfere with tubular secretion of uric acid.[12] The paradoxical effect of urate reten-tion seen when conventional uricosuric drugs are given in low dosage is not observed in the case of benzbromarone.

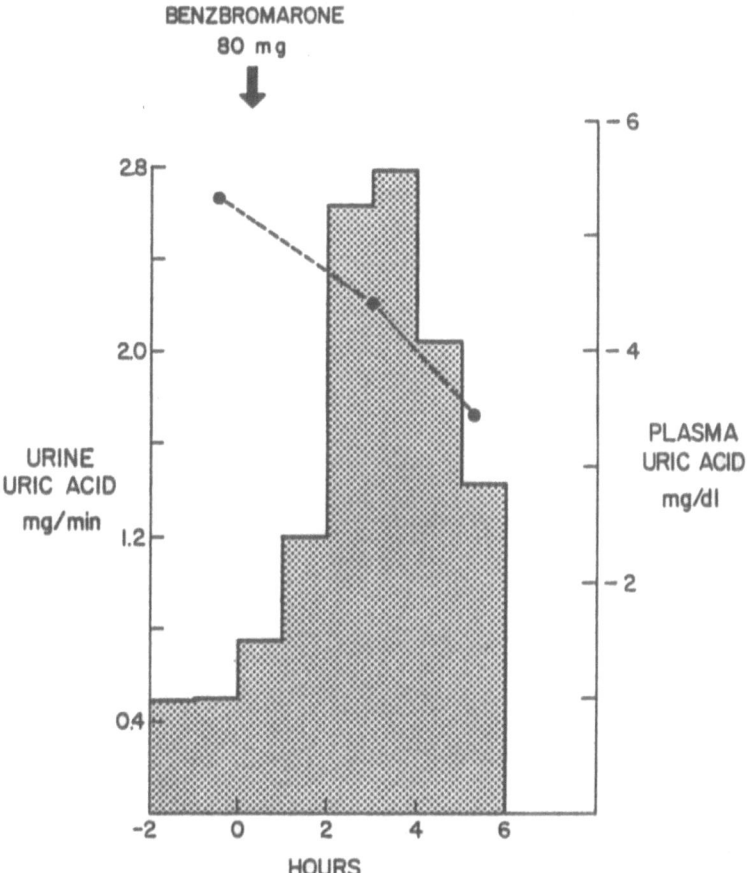

Fig. 1.   The effect of an oral dose of 80 mg micronized benzbro-
          marone upon plasma and urinary uric acid in a normal
          subject.

        A typical study in a healthy volunteer is presented in
Figure 1.   The urine is collected hourly prior to and after administration of 80 mg benzbromarone.   A urine output of 4-6 ml/min
is obtained by aggressive oral hydration before and during the
study.   In this particular study, the excretion of uric acid in
the control period is 495 µg/min, with a plasma urate concentration
of 5.32 mg/dl.   At the height of the benzbromarone response, uric
acid excretion has risen to 2,780 µg/min, and plasma urate has
fallen to 4.20 mg/dl.   The urate clearance rose from 9.3 ml/min in
the control period to 66.2 ml/min 3½ hours after benzbromarone
administration.

        Figure 2 depicts the baseline excretion and the uricosuric

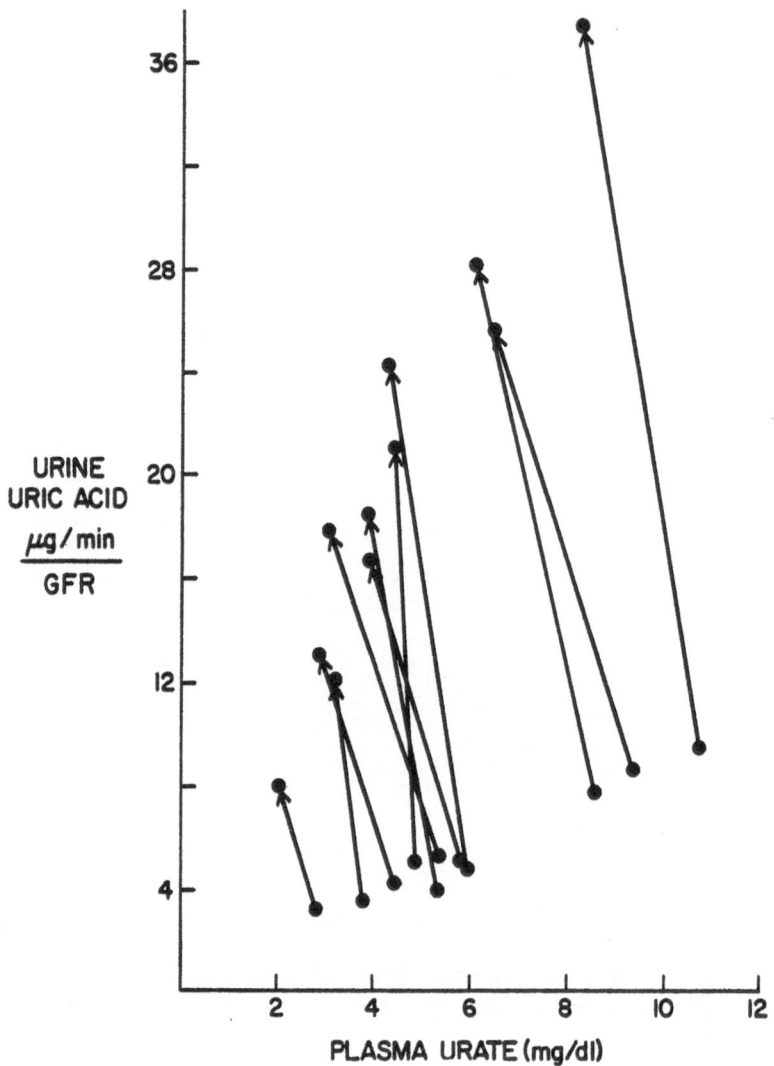

Fig. 2. Baseline excretion and maximum uricosuric response to
         80 mg benzbromarone in eight healthy men and three gouty
         patients with over-production of uric acid. Uric acid
         secretion is expressed in μg/min per ml glomerular
         filtration rate to correct for differences in renal
         functional areas.

response to benzbromarone for eight normal volunteers and three
patients with primary gout who had increased production of uric
acid as determined in turnover studies with ^{14}C-uric acid.
Baseline urate excretion values during the control period and
maximum uricosuric responses are plotted versus plasma urate values

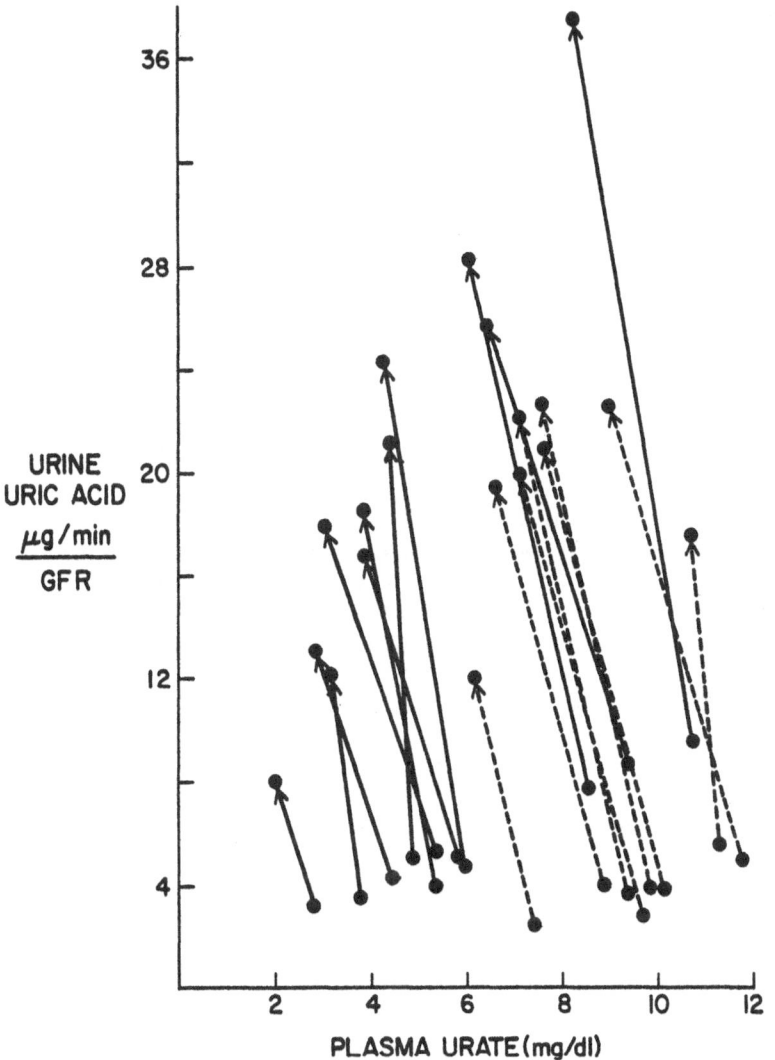

Fig. 3.   Baseline excretion and maximum uricosuric response to
          80 mg benzbromarone in eight gouty patients with normal
          production of uric acid (stippled lines) superimposed
          upon the control data depicted in Figure 2 (solid lines).
          Uric acid excretion is expressed in μg/min per ml
          glomerular filtration rate to correct for differences in
          renal functional mass.

over a range between 2.05 mg/dl to 10/72 mg/dl. These studies
demonstrate a functional relationship between urate secretion and
plasma urate concentration (r = 0.956; p < .005). Since benzbromarone selectively inhibits reabsorption of secreted urate, the
difference between secreted and excreted uric acid becomes a
measure of urate reabsorption.

We have used the "benzbromarone test" to define the renal
defect in primary gout characterized by a normal production of uric
acid. The results of studies in eight such patients are shown in
Figure 3, superimposed upon the values obtained for the control
population reported above. In conformity with previous observations, it can be seen that gouty normo-producers require a higher
plasma urate concentration to excrete a normal quantity of uric
acid in the urine. Although all of the patients showed a significant uricosuric response to benzbromarone, urate secretion is distinctly lower in gouty normo-producers than in their counterparts
with over-production of uric acid, indicating that the defect in
their renal handling is related to a decreased secretory response
for a given plasma concentration of urate.

The interpretation of our data is based on the assumptions
that pyrazinamide causes a selective and complete block in urate
secretion distal to the nephron segment where filtered uric acid
is reabsorbed and that benzbromarone in the dosage used causes
virtually a complete inhibition of reabsorption of secreted urate.
Although the validity of both assumptions may be questioned, the
use of pyrazinamide and benzbromarone represents the most sensitive
tools available at present for the pharmacological characterization
of bidirectional renal urate transport.

In summary, we have determined the magnitude of urate secretion
by measuring the maximum uricosuric response to benzbromarone. In
normal subjects and gouty patients with over-production of uric
acid, urate secretion is a function of plasma urate concentration.
Patients with primary gout who have normal production of uric acid
appear to have diminished tubular secretion as the underlying basis
for their hyperuricemia. The anatomic sites for the bidirectional
tubular urate transport in man remain to be clarified.

REFERENCES

1. J.S. Bennett, J. Bond, I. Singer, and A.J. Gottlieb, Hypouricemi in Hodgkin's disease, Ann. Intern. Med. 76:751 (1972).

2: D.M. Wilson, and N.P. Goldstein, Renal urate excretion in patients with Wilson's disease, Kidney Internat. 4:331 (1973).

3. T.H. Steele, and G. Boner, Origins of uricosuric response, J. Clin. Invest. 52:1368 (1973).

4. H.S. Diamond, and J.S. Paolino, Evidence for a post-secretory reabsorptive site for uric acid in man, J. Clin. Invest. 52:1491 (1973).

5. G. Lemieux, P. Vinay, and A. Gougoux, Nature of the uricosuric action of benziodarone, Am. J. Physiol. 224:1440 (1973).

6. L.B. Sorensen, and D.J. Levinson, Evidence for four components in the renal handling of uric acid in man, in: "Amino Acid Transport and Uric Acid Transport," S. Silbernagl, F. Lang and R. Greger, eds., George Thieme, Stuttgart (1976).

7. D.J. Levinson, and L.B. Sorensen, Renal handling of uric acid in man. Evidence for a four component system, Ann. Rheum. Dis. in press.

8. E. Praetorius, and J.E. Kirk, Hypouricemia: with evidence for tubular elimination of uric acid, J. Lab. Clin. Med. 35:865 (1950)

9. A.B. Gutman, T.F. Yü, and L. Berger, Tubular secretion of urate in man, J. Clin. Invst. 38:1778 (1959).

10. A.B. Gutman, and T.F. Yü, A three-component system for regulation of renal excretion of uric acid in man, Trans. Assoc. Amer. Physns. 74:353 (1961)

11. L.B. Sorensen, and D.J. Levinson, Isolated defect in post-secretory reabsorption of uric acid, Ann. Rheum. Dis. in press.

12. R. Podevin, F. Paillard, and C. Amiel, Action de la benz-bromarone sur l'excrétion rénale de l'acide urique, Rev. Franc. d'Etudes Clin. Biol. 12:361 (1967).

RIBOSE TOLERANCE IN GOUTY PATIENTS

M. Pizzichini[1], R. Marcolongo[2] and E. Marinello[3]

1 - Dept. Biol. Chem., Univ. Siena, Italy
2 - Rheum. Service, Univ. Siena, Italy
3 - Dept. Biol. Chem., E.U.L.O. Brescia, Italy

INTRODUCTION

Ribose metabolism is well known in mammalian. Ribose-5-phosphate (R-5-P) is formed in pentose phosphate pathway, in the reaction:

1) 6-phosphogluconic acid $\longrightarrow$ ribulose-5-phosphate (Ru-5-P)

Ru-5-P undergoes subsequent isomerization to R-5-P.

Also transketolase may lead to R-5-P formation, through the reversible reactions:

2) sedoheptulose-7-phosphate + glyceraldehyde-3-phosphate $\longrightarrow$
   xylulose-5-phosphate (Xu-5-P) + ribose-5-phosphate (R-5-P)

3) fructose-6-phosphate (F-6-P) + glyceraldeyde-3-phosphate $\longrightarrow$
   xylulose-5-phosphate (Xu-5-P) + erythrose-4-phosphate (E-4-P)

Xu-5-P is then epimerized to R-5-P.

Also glucuronic acid cycle produces D-xylulose, which can be phosphorylated to xylulose-5-phosphate (Xu-5-P), then isomerized to Ru-5-P and R-5-P.

R-5-P formed through several mechanisms is the "active" form of ribose metabolism. If, by any chance, free ribose is formed or introduced, it is phosphorylated to R-5-P by a specific kinase, demonstrated by Agranoff and Brady (1). R-5-P is further utilized in pentose phosphate pathway, or is channelled to purine and pyrimidine

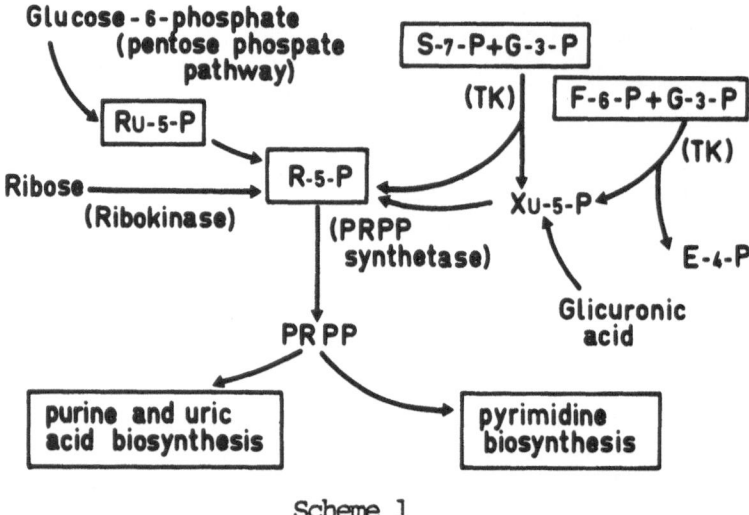

Scheme 1

biosynthesis, through well known metabolic pathways, starting from 5-phosphoribosyl-1-pyrophosphate (PRPP) (2). Formation and utilization of ribose, can so be represented(see Scheme).

Pentose phosphate pathway or ribose-5-phosphate formation from fructose-6-phosphate, PRPP formation in erythrocytes (3) are operative in man (4).

Wuest and Solmssen (5) have given an oral load of ribose to man, evaluating the urinary excretion of the sugar, which resulted 7.8-16.6%; the low recovery was ascribed to a poor intestinal absorption.

In gouty patients (6) many researches have been carried out on PRPP content in erytrocytes: although an increased incorporation of glycine in uric acid and an increased purine biosynthesis (7) have been demonstrated, no date are available at the moment on ribose metabolism in such patients.

For this reason, normal subjects and gouty patients have been by us submitted to intravenous load of ribose, evaluating the behavior of ribose in blood and urines; the behavior of blood glucose, lactic acid, pyruvic acid, uric acid have also been evaluated.

MATERIALS AND METHODS

20% D(-)Ribose sterile solutions were prepared by the Istituto Sieroterapico e Vaccinogeno Toscano Sclavo of Siena (Italy).

Table 1. Urinary excretion (g/h) of ribose after intravenous infusion of ribose in normal subjects and gouty patients. Reported values are the average of four subjects.

times	normal persons	gouty patients
0	0.01	0.01
1 h	2.17	2.58
2 h	1.16	1.13
3 h	0.69	0.33
4 h	0.16	0.20
5 h	0.01	0.01

Four normal persons and four gouty patients received intravenously 0.5 g of D(-)Ribose/Kg b.w. At various times after load, blood was taken from them, for the different determinations: urines were taken also every hour, for five hours. On serum, glucose was estimated by the enzymatic method of Trinder (8), ribose according to Roe (9), uric acid according to Liddle (10).

Table 2. Ribose recovered in urines after intravenous infusion of ribose in normal subjects and gouty patients.

normal subjects	weight (Kg)	g ribose infused	% recovery
" "	48	24.0	17.6
" "	50	25.0	16.0
" "	70	35.0	4.9
" "	38	19.0	20.3
gouty patients	65	32.5	13.6
" "	80	40.0	5.7
" "	75	37.5	16.1
" "	70	35.0	11.8

RESULTS AND DISCUSSION

The results are reported in the following figures and tables.

As shown in table 1, ribose was excreted within the first four hours after load, with a maximum at the first hour. No ribose was present in urine at the fourth hour.

The amount of ribose recovered in the urines represents 10-15% of the administered dose, in normal subjects and in gouty patients (table 2).

Ribose load induces hypoglycemia, which is more persistent in gouty patients than in normal subjects (fig.1).

The mechanism of ribose induced hypoglycemia has been still discussed by Segal (11).

There were no significant changes in the concentration of blood uric acid, as also lactic and pyruvic acid.

The only evident difference was the behavior of ribose in the blood of gouty patients, in comparison to the normal (figure 2): ribose shows a peak at 30 min after load in both groups, but its persistance was more evident, even at 60 min, in the blood of gouty patients.

It seems, from these preliminary results, that "ribose tolerance" of gouty patients is lower than in normal subjects; some troubles of ribose metabolism might be demonstrated in gouty patients, and have, as a consequence, an alteration of purine biosynthesis. Such alterations, however, are not reflected in uric acid levels after ribose load, and need further investigations.

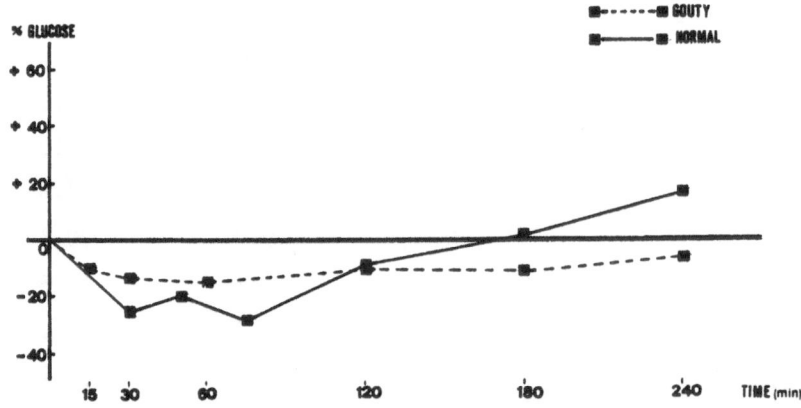

Fig. 1- Percent variation of blood glucose after ribose infusion.

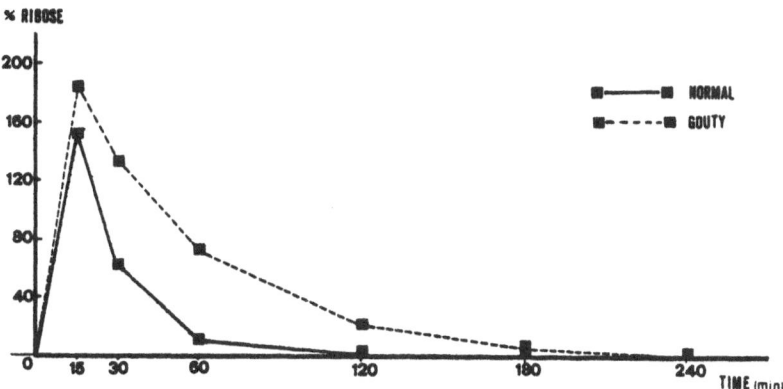

Fig. 2 - Percent variation of blood ribose after ribose infusion.

References:

1. B.W.Agranoff, R.O.Brady, Purification and properties of calf
   liver ribokinase. J.Biol.Chem. 219:221 (1956).

2. W.N.Kelley, E.W. Holmes, M.B.Vand der Weyden, Current concepts
   on the regulation of purine biosynthesis de novo in man.
   Arthritis and Rheumatism 18:673 (1975).

3. A.Hershko, A.Razin, J.Mager, Regulation of the synthesis of
   5-phosphoribosil-1-pyrophosphate in inctac red blood cells
   and in cell-free preparations. Biochem.Bioph.Acta 184:64
   (1969).

4. H.H.Hiatt, Studies of ribose metabolism. Pathways of ribose
   synthesis in man. J.Clin.Invest. 37:1461 (1958).

5. H.M. Wuest, U.V.Solmssen, The urinary excretion of d-ribose in
   man. Arch.Biochem. 11:199 (1946).

6. I.Fox, W.N. Kelley, Phosphoribosylpyrophosphate in man: biochemical and clinical significance. Ann.Intern.Med. 74:424(1971).

7. J.D.Benedict, T.F. Yu, E.J.Bien, A.B.Gutman, D.W.Jr.Stetten, Incorporation of glycine nitrogen into uric acid in normal and
   gouty man. Metabolism 1:3 (1952).

8. P.Trinder, Determination of glucose in blood using glucose oxidase with an alternative oxigen acceptor.Ann.Clin.Biochem. 6:
   24 (1969).

9. J.H.Roe, E.W.Rice, A photometric method for the determination
   of free pentoses in animal tissues. J.Biol.Chem. 173:507(1948).

10.L.L.Liddle, J.E.Seegmiller, L.Laster, The enzymatic spectropho-
    tometric method for determination of uric acid.J.Lab.Clin.Med.
    54:903 (1959).

11.S.Segal, J.Foley, J.B.Wyngaarden, Hypoglicemic effect of d-ri-
    bose in man. Proc.Soc.Exp.Biol.Med. 95:551 (1957).

URINARY URATE AND URIC ACID RELATIVE SATURATION IN NORMOURICURIC CALCIUM OXALATE STONE FORMERS WITH NORMAL URINARY CALCIUM OXALATE SATURATION

M. Labeeuw, C. Gerbaulet, N. Pozet, P. Zech, J. Traeger

Edouard Herriot Hospital
69374 Lyon Cedex 2, France

Recent theories on the physiopathology of Calcium Oxalate (Ca. Ox) nephrolithiasis proposed the supersaturation of urine for Ca. Ox as the main mechanism of stone formation. (1,2). It is obvious however that even in normals, urine might be saturated with Ca. Ox (3). An additionnal mechanism is then needed. The role of Uric Acid (U.A) or Urate has been suspected for several years. Three theories are currently under investigation :
    - Heterogeneous nucleation with occurence of epitaxial growth of Ca. Ox on a nucleus made of a crystal of Monosodium (Na.U) or Ammonium Urate (NH4.U) (4) ;
    - Excessive aggregation of Ca.Ox crystals due to a decrease in the normal inhibitory activity, this decrease being related to an increased urinary (total or colloidal) Na. U concentration (5).
    - Facilitation of Ca. Ox precipitation due to a decreased formation product for Ca. Ox, depending on an increased U.A. urinary elimination (6).

These hypothesis are supported by in vitro studies or observations made in hyperuricuric patients. However hyperuricosuria alone occurs in less than 15 % of recurrent Ca. Ox stone formers (down S.F.). Thus a number of patients, not yet appreciated in large series, should exhibit both normal uricosuria on a 24 hours basis and a Ca. Ox relative saturation (RS) identical to non forming stone subjects. It has been suggested that Allopurinol might be of some benefit in preventing the recurrence of Ca. Ox stones even in those patients. This study was undertaken to determine whether, in such patients, subtle changes in urinary U.A. or Urate were present and could support one of the above mentionned theories.

PATIENTS AND PROTOCOL

     Twelve patients were selected on the following criteria. Acti-
ve Ca. Ox stone disease (more than one stone formed every year, the
last one within the last 6 months), no urological abnormality, no
urinary tract infection, no treatment, normal U.A blood level and
U.A 24 hours elimination while under their own diet, normal renal
function, and Ca. Ox RS identical to controls. The twelve controls
were age and sex matched and studied during the same period of the
year (table I). Subjects were instructed not to change their dietary habits prior to the study. Urine was collected following a protocol which intended to reflect as far as possible conditions occurring spontaneously : fasting and water restriction (U1 : 7am
to 9am), hydratation by a oral 300 ml water load (U2 : 9am to
11am), post prandial state (meal with 300 ml of water : U3 : 11am
to 1pm, U4 : 1pm to 3pm). The purine content of the meal was
37 mg   . Urine volume and pH were immediately recorded, urine
screened for crystalluria and saved for biochemical measurements.

METHODS

     Ca. Ox, Na.U, NH4.U, and U.A relative saturation were deter-
mined according to Marshall and Roberston (8). A value of R.S. $\geq$ 1
means urinary supersaturation and indicates a risk of spontaneous
crystallisation, the obligatory first step of heterogeneous nucleation. The formation product of Ca. Ox was not measured. However,
as it is correlated with the state of saturation of the urine for
Na. U (6), the values of the Activity Product (A.P) or of RS were
used. No attempt was made to appreciate the fraction of colloidal
urate. It was shown that, in few samples, colloidal urate was higher in urine with high U.A content than in Urine with a low U.A
concentration. Therefore, total U.A was only determined, assuming
some degree of correlation with colloidal urate. U.A, Na^+, $NH4^+$,
and Ca^{++} were determined by conventional laboratory methods. Oxalate was measured according to Dubuque (9).

TABLE I

(	Age years	Males	G. F. R. ml/s	24H U.A mM	Ca. Ox RS)
(Ca. Ox SF	33.2	8	1.72	3.86	.99)
(n = 12	± 8.2		±.28	±.76	±.28)
(					)
(Controls	31.8	8	1.94	4.30	.94)
(n = 12	± 8.5	8	±.17	±.97	±.35)
(					)

RESULTS

The values of Ca. Ox RS obtained in stone formers were not different from those from controls (see table I). The comparaison of the two groups for each period of time revealed no difference. Therefore the Ca. Ox RS were assumed to be identical.

As expected, the U.A excretion rate was identical in both groups. However, the urinary flow rate being slightly lower in controls, the urinary U.A concentrations were higher in SF, although this difference became significant (Wilcoxon rank test) only in post prandial samples (table II).

As a consequence, the urinary pH being not different between the two groups, U.A RS was lower ($p < .02$) in SF than in controls, mainly in post prandial samples (table II).

The values of RS for Na. U (fig.1) and NH4.U (fig.2) were always lower in SF than in C. For Na. U, the difference was only suggested, the different time periods being considered as a group or separately. For NH4. U, the difference was significant only in U 3 ($p < .05$). This difference accounts for the value of p ($p < .025$) when all samples were analyzed together. PH values were not different between the two groups and are not expected to account for these results since the effect of pH on the U.A and Na. U RS are divergent. Na^+ and $NH4^+$ concentrations were not statistically different.

Table II

	[U.A]	(mM)	[Na]	(mM)	U.A R.S	
	SF	C	SF	C	SF	C
U 1	3.2 ± 1.3	4.1 ± 2.3	96.7 ± 36.4	101.0 ± 50.0	-1.23 +2.00	- .51 + 1.69
U 2	1.9 ± 1.0	2.9 ± 2.3	63.8 ± 33.3	57.8 ± 34.5	-2.43 +2.16	- 1.57 + 1.23
U 3	2.6 ± 1.1	4.3 + 1.9**	76.3 ± 32.7	67.4 + 31.2	-1.52 +1.53	.38** + 1.31
U 4	3.1 ± 1.5	5.2 ± 2.1*	77.5 ± 18.4	96.8 ± 52.2	-1.28 +1.35	- .17* + 1.59

* p < .05; ** p < .02

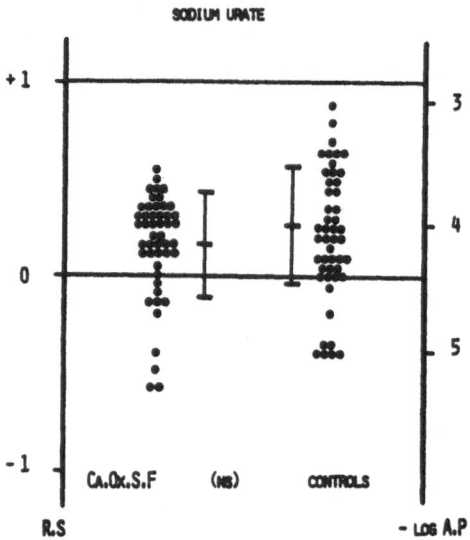

Figure 1.   Urinary monosodium urate relative saturation in recurrent calcium oxalate stone formers and controls.

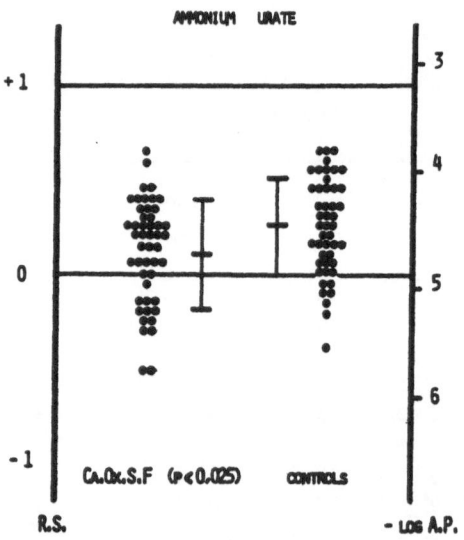

Figure 2.   Urinary ammonium urate relative saturation in recurrent calcium oxalate stone formers and controls.

COMMENTS

Since under the same conditions, Ca. Ox RS was identical in
SF and controls, supersaturation of urine with Ca Ox should not by
itself be responsable for the active stone disease present in all
patients. Among the necessary additionnal factors, an excessive
24 h U.A elimination was excluded from the beginning. However, the
risk of crystal formation does not depend only on the urinary elimination but also on the concentrations of U.A, Na, NH4 and pH.
The state of saturation (appreciated by the RS) better reflects
the risk of spontaneous crystallisation which is the first step of
subsequent epitaxial growth of Ca. Ox. In our patients, this risk
was not higher, both for Na. U and NH4,U, in SF than in controls.
Thus epitaxial growth on a Na. or NH4 urate crystal is not likely
to occur more easily in SF than in controls. These results do not
exclude the possibility of heterogeneous nucleation : if it occurs, it should be mediated by other abnormalities.

The inhibitory activity of Ca. Ox was proposed as an important
additionnal factor. It is suggested that its decrease in SF is related to an increased U.A or Na. U urinary content. Table II shows
that such an mechanism is unlikely in our patients. However, if
colloidal rather than crystalline urate is to be considered, we
cannot exclude the possibility of a colloidal/total urate ratio
different in SF and in controls.

In hyperuricuric Ca. Ox SF, the formation product (FP)for Ca.
Ox was found to be decreased, making its precipitation easier.(6)
FP was negatively correlated with the state of saturation for Na.
U. In our patients, the Na. U RS was, if something, lower than in
controls. Should a decreased Ca. Ox FP exist, it would not depend
on the Na. U RS.

In several cases, the values of RS were higher in controls,
mainly because of a lower urinary flow rate. The protocol was designed in order to study patients in conditions as close as possible to their usual conditions of life. The fact that they were ac-
tively forming stones under these conditions suggests that the U.
A or Na. U RS is not the most critical point to consider.

Our results do not completely rule out U.A or urate as an important factor in Ca. Ox nephrolithiasis. Only normouricuric SF,
with a Ca. Ox RS indentical to their controls were studied. In
these patients, the studied parameters were identical to normals
and do not support the theories mentionned at the beginning. If
Allopurinol acts in such patients, it does not act only by decreasing the U.A or urate urinary excretion. The influence of the colloidal fraction of Na.U might be important and cannot by ruled out
by our study.

REFERENCES

1. B. Finlayson, 1978, Physicochemical aspects of urolithiasis,
    Kidney Int., 13 : 344.
2. C. Pak, 1978, in : Calcium Urolithiasis - Plenum medical book
    Ed. p. 5.
3. W.G. Roberston, M. Peacock, B.E.C. Nordin, 1971,Calcium oxala-
    te crystalluria and urine saturation in recurrent renal stone
    formers, Clin. Sci., 40 : 365.
4. C.Y. Pak, O. Waters, L. Arnold, K. Holt, C. Cox, D. Barilla,
    1977, Mechanism for calcium urolithiasis among patients with
    hyperuricosuria, J. Clin. Invest. 59 : 426.
5. W.G. Roberston, F. Knowles, M. Peacock,Urinary acid mucopoly-
    saccharide inhibitors of calcium oxalate crystallization, In
    "Urolithiasis research" Plenum Press Ed. New York, 1976,
    p. 331.
6. C.Y. Pak, D.E. Barilla, K. Holt, L. Brinkley, R. Tolentino,
    J.E. Zerwekh, 1978, Effect of oral purine load and allopuri-
    nol on the crystallization of calcium salts in urine of pa-
    tients with hyperuricosuria calcium urolithiasis,Amer. J.
    Med. 65 : 593.
7. F.L. Col, 1977, Treated and untreated recurrent calcium ne-
    phrolithiasis in patients with idiopathic hypercalciuria,
    hyperuricosuria, or no metabolic disorder, Am. Intern. Med.
    87 : 404.
8. R.W. Marshall, W.G. Roberston, 1976, Nomograms for the esti-
    mation of the saturation of urine with calcium oxalate, cal-
    cium phosphate, magnesium, ammonium phosphate, uric acid,
    sodium acid urate, ammonium acid urate and cystine, Clin.
    Chem. Acta 72 : 253.

9. M. TH. Dubuque, J.M. Melon, J. Thomas, E. Thomas, R. Pierre,
    C. Charransol, P. Desgrez, 1970, Am. Biol. Clin. 28 : 95.

CORRELATION BETWEEN THE URIC ACID AND CALCIUM CONCENTRATION IN

URINE. RESULTS OF A LONG TERM STUDY ON RECURRENT STONE-FORMERS

AND HEALTHY CONTROLS

P. Leskovar, R. Hartung and M. Kratzer

Urologische Klinik und Poliklinik r.d.Isar der Techn.
Universität München (Direktor: Prof.Dr.W.Mauermayer)
D-8000 München 80, Ismaningerstraße 22

1. INTRODUCTION

Clinical observations of the last 10 - 15 years called attention
to the interrelation between the raised uric acid levels in serum
respectively in urine and the oxalate lithiasis. In 1968, Prien
et al.(20) observed an unexpectedly high incidence of calcium
nephrolithiasis with gout patients. Gutman and Yü (14) as well
found in 12 % of the examined gout patients calcium oxalate stones.
On the other hand, Dent and Sutor (11) observed that the calcium
oxalate stone-formers showed surprisingly often a raised uric
acid level in serum. Later studies confirmed this observation. So,
Braun, May and Birtel (3), further Coe (6, 7), Coe and Raisen (8),
Coe and Kavalach (9), Hartung (15) as well as Eisen et al. (12)
found in a considerable percentage of their oxalate-patients
significantly raised uric acid concentration in serum, respectively in urine, the hyperuricosuria being partly accompanied by
a distinct hypercalciuria. Mediately, interrelations between
oxalate lithiasis and hyperuricaemia respectively hyperuricosuria
can be deduced from the remarkable therapeutical effect of the
uricostaticum Zyloric not only in the uric acid but also in the
oxalate nephrolithiasis. Here, only the basic studies, concerning
the allopurinol-effect on oxalate lithiasis, by Smith (23), Coe
and Raisen (8), Zöllner and Schattenkirchner (25), Schwille (21)
as well as Brien and Bick (4) should be cited. In our study, we
were especially interested in the eventual correlation between the
uric acid and calcium in the urine of recurrent oxalate stone-
formers and healthy controls. We suppose that the uric acid overproduction, especially in the tophaceous gout, leads to the
mobilisation of the bone base reserve and indirectly to the

93

release of the bone calcium. In our opinion, the uric acid,although representing a weak organic acid, is strong enough to react with the bicarbonate ion in blood, reducing in this way the alcaline reserve. For the pK_1 of uric acid values from 3,9 (F. Klages: Organic chemistry) to 5,4 respectively 5,7 have been quoted. As the pK_1 of the carbonic acid is 6,1, so a direct reaction between the de novo synthetized uric acid and the bicarbonate ion in blood must be expected (In vitro, such a reaction could be observed gasometrically).

2. MATERIALS AND METHODS

In our long-term study, ten recurrent oxalate stone-formers and eleven control persons took part. On three days a week, during a period of 4 - 6 weeks, the morning, midday and evening urinary samples were collected in plastic vessels, containing toluol respectively 15 % thymol/propanol as preserver.
The calcium concentration in urine was determined by the atomic-absorption-spectrophotometry (modell 400, Perkin & Elmer/Überlin-gen). The determination of the uric acid concentration in urine was carried out by the Boehringer test-combination (Art.15 865, Boehringer/Mannheim). The urinary osmolarity was measured by the Wescor vapour-pressure osmometer (modell 5130, Wescor Inc.,Utah, USA).

3. RESULTS

The mean values of the uric acid concentration in the morning, midday and evening urinary samples as well as the uncorrected and the osmolarity - corrected daily means are summarized in Table 1. The corresponding values for the urinary calcium are presented in Table 2. In accordance with Baltzer, and Rottmann (1) as well as Crassweller and Oreopoulos (10), but in discordance with Hodgkin-son (16), Braun et al. (3),Eisen et al. (12), Coe and Raisen (8) and some others, we found no higher excretion of uric acid in oxalate stone-formers compared with their healthy controls; after the exclusion of the dilution factor in the group of stone-formers by the osmolarity-correcture, both groups showed the same mean uric acid levels in the urine. It must however be underlined that we didn't measure the 24-hour-excretion but the actual concentra-tion of uric acid in urine. Differences between both groups became evident when the percentage of single urinary samples exceeding an arbitrarily chosen upper limit of the normouricosuria was calculated. So, 10 % of the patients' urines but only 2 % of the controls' urinary samples exceeded the arbitrary treshold of 800 mg/l (in the case of patients) respectively of 1120 mg/l (at control persons), the higher limit's value of 1120 mg/l taking into account the dilution factor (1,4) of patients' urine. Analogously, 8 % of the patients' but none of the controls'

Table 1: The average uric acid concentrations (mg/1) in the morning, midday and evening urinary samples as well as the average osmolarity and the osmolarity-corrected uric acid concentration in the 12-hour-urine of ten oxalate stone-formers and eleven controls.

Examined group	Number n	Morning x̄	s	Midday x̄	s	Evening x̄	s	Day x̄	s	Osmolarity x̄	s	Uric acid/Osmolarity
Oxalate stone-formers	10	426,06	167,34	484,46	229,20	414,10	187,16	431,66	190,95	622,2	223,5	0,70
Controls	11	607,29	77,56	626,23	114,31	632,01	128,52	616,23	88,94	878,0	106,2	0,70

Table 2: The average calcium concentrations (mg/1) in the morning, midday and evening urinary samples as well as the average osmolarity and the osmolarity - corrected calcium concentration in the 12-hour-urine of ten oxalate stone-formers and eleven control persons.

Examined group	Number n	Morning x̄	s	Midday x̄	s	Evening x̄	s	Day x̄	s	Osmolarity x̄	s	Calcium/Osmolarity
Oxalate stone-formers	10	4,0	1,9	3,9	2,08	3,85	2,37	3,94	2,09	622,2	223,5	0,0060
Controls	11	4,6	1,1	3,8	0,98	4,25	1,08	4,23	1,0	878,0	106,2	0,0048

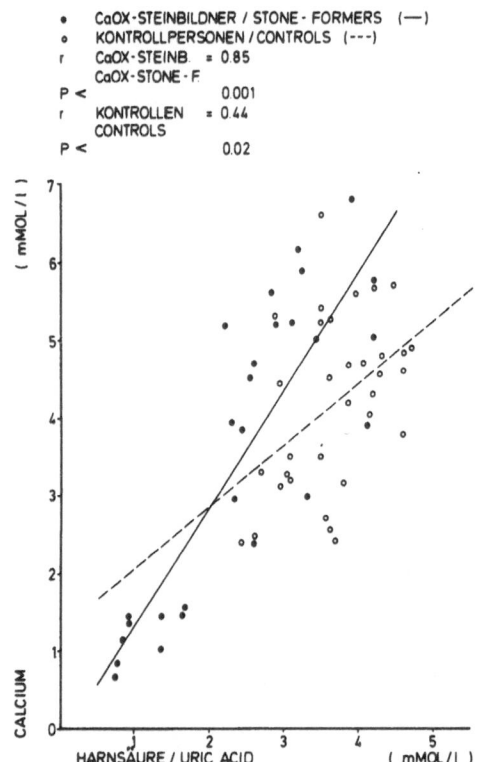

Figure 1.   Correlation between the actual uric acid and calcium
            concentrations in the morning, midday and evening urine
            of recurrent oxalate stone-formers and control persons.
            Linear regression: (a) stone-formers: Ca = 1.54 x uric
            acid -0.14;   (b) controls: Ca = 0.79 x uric acid +1.31.

urinary samples exceeded the arbitrarily chosen upper treshold of
8,25 mmoles/1 calcium. Further it could be shown that 14,8 % of
these 8 % single urinary samples contained more than 800 mg/1 uric
acid. On the other hand, 14,3 % of the samples, being characterized
by a raised uric acid concentration, showed a calcium concentration
exceeding 8,25 mmoles/1. To find out if there exist parallelisms
between the urinary calcium and uric acid excretion, we calculated
on the basis of the uric acid and calcium concentrations in actual
urinary samples the linear regression as well as the coefficient
of correlation. The linear regression in the group of oxalate
stone-formers was: Ca = 1,54 x uric acid - 0,14 and in the group
of the healthy controls: Ca = 0,79 x uric acid + 1,31 (Fig. 1).
The stone-formers showed a highly significant correlation with
r = 0,85 (P<0,001). In the group of control persons, the calculated
correlation was essentialy lower, r being 0,44, although the
correlation was significant as well (P<0,02). Both groups, the

stone-formers and the controls, showed a considerably different
slope of the regression curve; in the group of stone-formers, the
calcium concentration increased, owing to a higher m - value
(1,54 instead of 0,79), nearly twice as fast with increasing
uric acid concentration as in the group of control persons.

4. DISCUSSION

Terhorst (24) as well as Bastian (2) reported about a clear
relationship between the raised uric acid level in plasma respectively urine and the urinary Ca-excretion in 30 - 40 % of all
recurrent oxalate stone-formers examined. Frank et al (13) called
attention to the frequent hypercalcaemia in patients with uric
acid lithiasis. May et al. (17) found in 1/2 of their patients
the hypercalciuria to be accompanied by the hyperuricosuria.
Among the patients with hyperuricosuria, about 30 % showed an
increased Ca-excretion. Coe (5) also observed a relationship
between the hyperuricosuria and the idiopathic hypercalciuria.
Mintz et al. (18), further Ory et al. (19) as well as Scott et
al (22) called attention to the frequent coexistence of hyperuricaemia and primary hyperparathyreoidism. Schwille (21) on
the contrary couldn't find a direct correlation between the uric
acid and calcium excretion and was unable to reduce the Ca-
excretion by allopurinol.
The high correlation between the urinary calcium and uric acid
we found during our long-term observation of recurrent oxalate
stone-formers (r = 0,85), could be of a special interest in
connection with our assumption that the overproduced uric acid
mobilizes the buffer system of bone and mediately the bone
calcium.

5. REFERENCES

(1) Baltzer, G., and Rottmann, G.: Serum-Harnsäure und Harnsäureausscheidung bei Calcium-Oxalat-Urolithiasis, in: Pathogenese und
Klinik der Harnsteine V, Bd. 9 (W. Vahlensieck und G. Gasser,Hrsg)
(1977), Dr.D.Steinkopff, Darmstadt, p. 150   (2) Bastian, H.P.,
Dtsch. med. Wschr. 98: 1306 (1973)   (3) Braun, S.J., May, P.,and
Birtel, R.; Nieren- und Hochdruckkrankheiten 3: 116 (1974)
(4) Brien, G., Bick, C., Eur. Urol. 3: 35 (1977)   (5) Coe, F.L.,
Clin. Res. 20: 589 (1972)   (6) Coe, F.L., Ann.Int. Med.87:404
(1977)   (7) Coe, F.L., J. Chron. Dis. 29: 793 (1976)   (8) Coe,
F.L., Raisen, L., Lancet 1: 129 (1973)   (9) Coe, F.L., Kavalach,
A.G., New Engl. J. Med. 251: 1344 (1971)   (10) Crassweller, P.O.,
Oreopoulos, D.G., Toguri, A., Husdan, H., Wilson,D.R., and Rapo-
port, A., J. Urol. 120: 6 (1978)   (11) Dent, C.E., and Sutor,
D.J., Lancet 2: 775 (1971)   (12) Eisen, M., Dosch, W., Altwein,
J.E., Hohenfellner, R., Nieren- und Hochdruckkrankheiten 4: 174
(1975)   (13) Frank, M., Lazebnik, J., de Vries, A., Urol. int.
25: 32 (1970)   (14) Gutman, A.B., and Yü, T.F., Amer.J.Med.
45: 756 (1968)   (15) Hartung, R., Münch. med. Wschr. 117: 387
(1975)   (16) Hodgkinson, A., Marshall, R.W., Cochran, M., Isr.J.
Med. Sci. 73: 1230 (1971)   (17) May, P., and Braun, J.S.: Serum-
- und Urinanalysen bei Patienten mit calciumhaltigen Harnsteinen,
in: Pathogenese und Klinik der Harnsteine III, Bd. 4 (W. Vahlensieck und G. Gasser, Hrsg.) (1975), Dr. D. Steinkopff, Darmstadt,
p. 15   (18) Mintz, D.H., Canary, J.J., Garreon, G., and Kyle,
L.H., New Engl. J.Med. 265: 112 (1961)   (19) Ory, E.M., Hisey,
D.P., and Redmond, E.D., Southern Med. J. 63: 194 (1970)
(20) Prien, E.L., and Prien, E.L.,Jr., Amer.J.Med. 45: 654 (1968)
(21) Schwille, P.O., Med. Welt 23: 1478 (1972)   (22) Scott,J.T.,
Dixon, A.S.J., and Bywaters, E.G.L., Brit. Med. J. 1: 1070 (1969)
(23) Smith, M.J.V., and Boyce, W.H., J.Urol., 102: 750 (1969)
(24) Terhorst, B., Dtsch.med.Wschr. 98: 1306 (1973)   (25) Zöllner
N., and Schattenkirchner, M;Dtsch. med. Wschr. 92: 654 (1967)

INTERACTION OF HYPERURICURIA AND HYPEROXALURIA

ON RENAL CALCIUM OXALATE STONE FORMATION

Franz Hering, Karl-Heinz Bigalke and
Wolfgang Lutzeyer
Dep. of Urology and Dep. of Pathology
Med. Faculty, RWTH Aachen
Goethestraße 27/29 D-5100 Aachen

There is increasing evidence that excessive urinary
excretion of uric acid plays a role in the genesis of
renal calcium oxalate stones. 17 percent of our stone
formers show a hyperuricuria, a much higher frequency
than in normal people. Robertson (1) describes an
inhibitory effect on acid mucopolysaccarides, which
serve as inhibitors of crystal aggregation.

The report presented details the results of our investigation on the influence of hyperuricuria and
hyperoxaluria in renal stone formation and crystallisation in renal tissue.

Materials and Methods

Experiments were carried out with male Wistar rats
with an approximate weight of 300-350 g. (Table 1)

A Hyperuricosuria was induced by oxonic acid, a
specific blocker of uricase.
Contrary to human beings uric acid in rats is metabolized to allantoin by the enzyme uricase (2). By
blocking this enzyme in rats a mild hyperuricemia and
hypercuricuria will be caused.

Oxonic acid, 2 percent, was given by a duodenal tube
twice a day. Ethylene glycol-a metabolic precursor of

Table 1.   Experimental procedure and test specimen

PROCEDURE

male Wistar rats , approximate   weight 350g

Controls sham treated	2%Oxonic acid by duodenal tube	0.8%Ethylene glycol in food (water)	0.8%Ethyl. glycol + 2%Oxonic acid

Test specimen:

1) Oxalate excretion in urine
2) Uric acid excretion in urine
3) Histochemistry and planimetry of
   oxalate-and urate deposits in
   renal tissue

4) T E M
5) Stone - frequency
   - analysis
   - weight

oxalate-was added to water in a daily dose of 0.8 per cent.
For a periode of 30 days these two drugs were given separetely or in combination. A control group of ten animals were shamtreated.
Animals were kept in metabolic wards.

In a ten day interval five animals of each group were killed by an overdosage of barbital and the following parameters were examined:

1) Urinary excretion of oxalate and uric acid.

2) Histochemistry and planimetry of oxalate and urate crystal deposits in renal tissue.

3) Transmission electron microscopy of renal tissue.

4) Stone frequency, -analysis and- weight.

Results

During the combination of both drugs there was obtained a significant hyperoxaluria 20 days after onset of treatment, and only a mild hyperuricuria as well as with the combination of oxonic acid and ethylene glycol as with oxonic acid treatment alone. (Fig. 1)

According to Pearses (3) method of histochemical staining of oxalate crystal deposits were investigated in renal tissue.

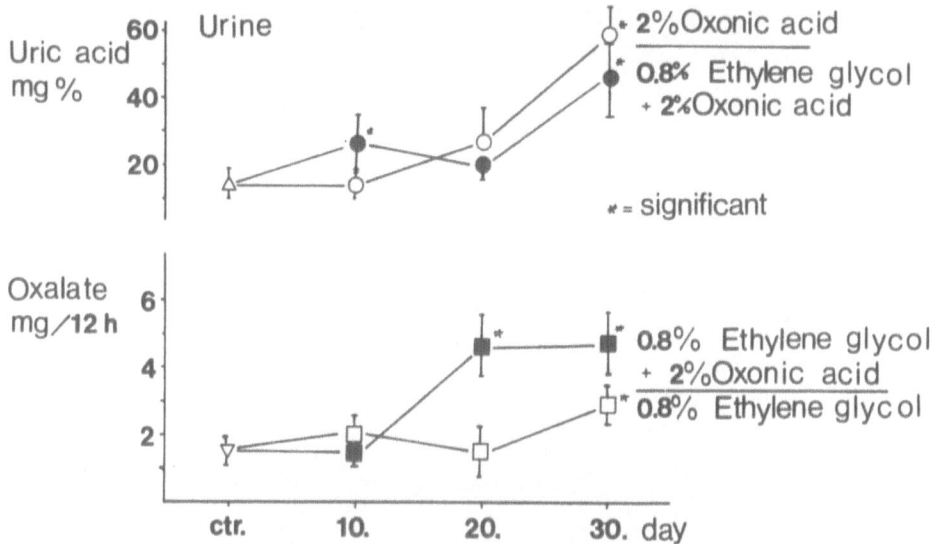

Fig. 1   Urinary uric acid- and oxalate excretion
          under treatmeat with ethylene glycol, oxonic
          acid and their combination.

The next figure shows a typical pattern of oxalate
crystalls in a rat kidney treated with ethylene glycol
alone for 30 days. (Fig. 2)

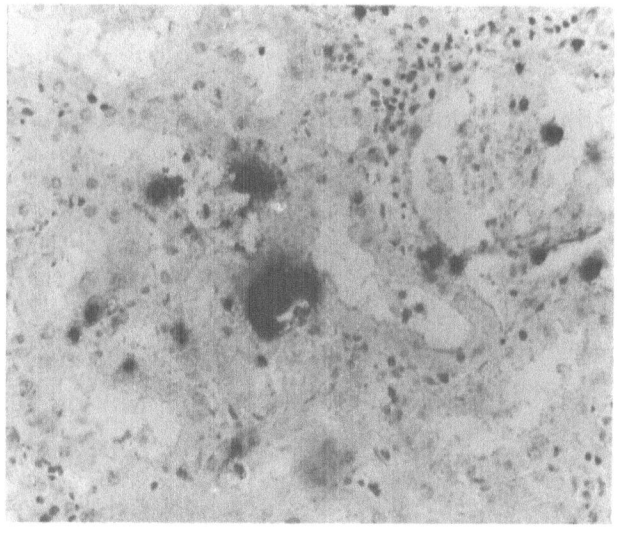

Fig. 2   Oxalate crystalls in a rat kidney 30 days
           treated with ethylene glycol alone - Histo-
           chemical staining of oxalate

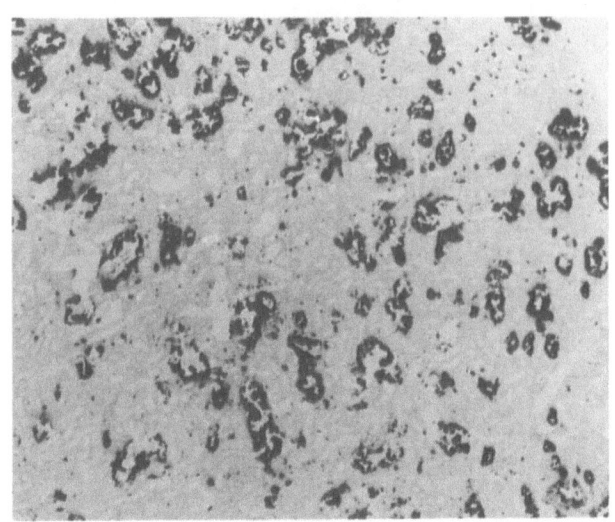

Fig. 3   Oxalate crystalls in a rat kidney 30 days
          treated with the combination of oxonic
          acid and ethylene glycol -Histochemical
          staining of oxalate

In contrast much more crystalls were seen in rats
treated 30 days with the combination of ethylene
glycol and oxonic acid. (Fig. 3)

Remarkable that there are mainly peritubular deposits
of oxalate and fewer intratubular casts. No oxalate
crystalls were seen during treatment by oxonic acid
only. (Fig. 4)

By transmission electronmicroscopy of renal tissue
calcium oxalate could be identified as well as intratubular as in the peri- or intertubular space. (Fig. 5)

The monolayer epithelium of the tubule is destroyed.
At present, it is impossible to identifie the primary
site of crystallisation, however there is a possible
transtubular transport of crystalls.

Only a few urate crystall deposits were observed in
rats treated with oxonic acid alone or in combination
with ethylene glycol. (Fig. 6)

Urate crystalls were stained according a histochemical
method described by Berg (4).

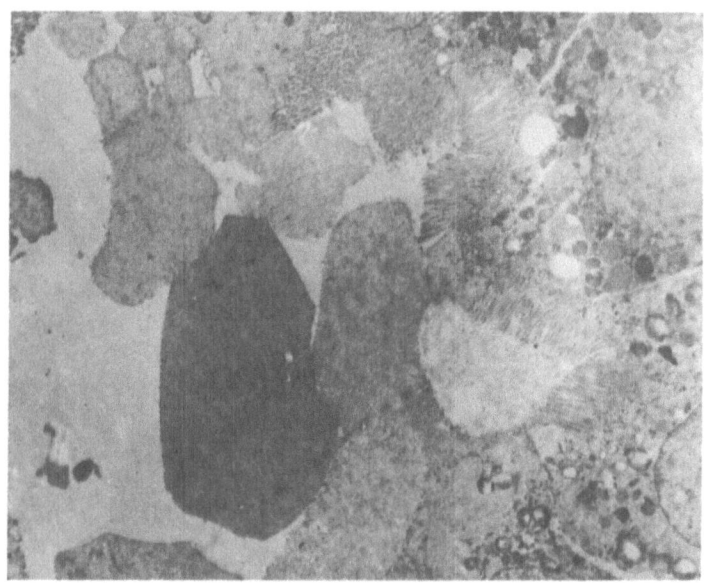

Fig. 4   Renal tissue of a rat 30 day treated with
           both drugs- Ultrathin section - photomicro-
           graph.

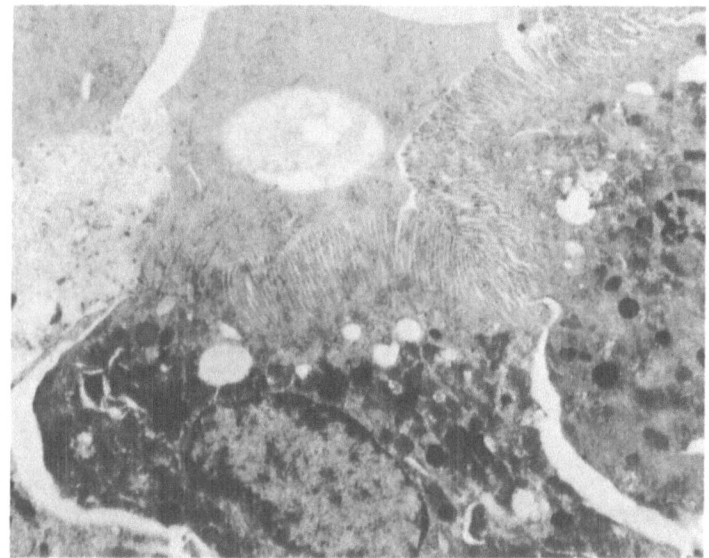

Fig. 5   Transmission electron microscopy of renal
           tissue of a rat 30 days treated with the
           combination of oxonic acid and ethylene
           glycol.

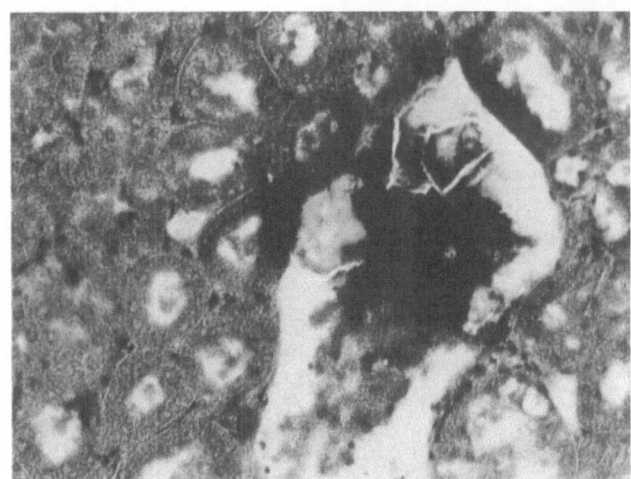

Fig. 6   Renal tissue of a rat 30 days treated with
           oxonic acid alone ⌐Histochemical staining of
           urate.

A planimetric study of oxalate crystall deposits in
renal tissue was related to treatment, time and site
of crystallisation with respect to renal cortex or
medulla. (Fig. 7)

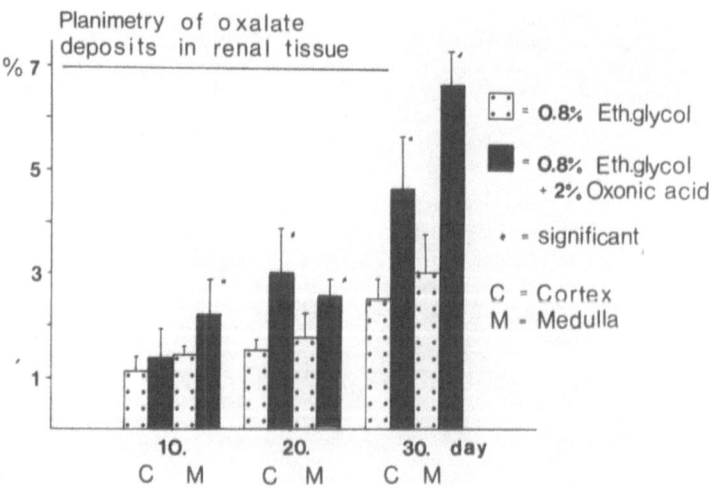

Fig. 7   Planimetric study of oxalate deposits in
           renal tissue in relation to treatmeat,
           duration of treatment and site of crystalli-
           sation.

In relation to duration of treatment significantly more oxalate crystalls were seen in renal tissue, preferably in the region of the medulla.

In relation to treatment the combination of both drugs produced significantly more oxalat crystalls.

Regarding the stone formation rats treated with both drugs showed as well as a higher incidence of stone frequency per animal as a higher stone weight. (Table 2)

No stone formation was observed under treatment with oxonic acid alone. Stones were analyzed by infrared spectroscopy: they consist of pure calcium-oxalate-dihydrate.

A typical kidney specimen with a stone is shown in the last figure.(Fig. 8)

Table 2. Stone frequency, -weight and -analysis in relation to treatment.

Stone-	0.8% Ethylene glyc. 2% Oxonic acid	0.8% Ethylene glyc.	2%Oxonic acid
frequency/rat	1.36	0.56	—
weight	53.1 mg	34.6 mg	—
analysis	Calciumoxalatedihydrate		—

Fig. 8 Rat kidney with a stone - rat treated with the combination of oxonic acid and ethylene glycol.

Remarkable are the yellowish-brownish crystall deposits seen in this specimen predominantly in the
renal cortex.

Discussion and Conclusions

The results presented here indicate that the presence
of uric acid in a metastable supersaturated solution
of calcium oxalate facilitates the crystallisation of
calcium oxalate presumably by the expitaxially induced
nucleation of calcium oxalate by the foreign crystalline phase.
Our data of x-ray diffraction and scanning electron
microscopy studies made in urine of stone formers
showed that the calculated misfit between the lattice
dimensions is less than 2 percent. This is one of the
closest lattice matches of common urinary stone components. This data confirms the results Meyer published
in Investigative Urology 1976 (5).

Other investigations by Robertson (1) demonstrated
that uric acid acts as inhibitor of acid mucopolysaccharides, which act as inhibitors of growth and
aggregation of calcium oxalate crystalls rather than
of their nucleation.
These data confirm the results of Finlayson (6) who
demonstrated the importance of uric acid by computer-
ized calculation of the solubility product.

In tnis study presented here, in renal tissue only a
mild hyperuricuria in combination with a hyperoxaluria produced a higher incidence of oxalate crystallisation in combination with a higher incidence of
stone formation.

Literature

1. W.G. ROBERTSON, M. PEACOCK and B.E.C. NORDIN
   Clin. Chim. Acta 43, 31, 1973

2. D.P. MERTZ
   Hippokrates 39, 5, 1968

3. A.G.E. PEARSE
   Histochemistry
   Churchill Livingstone, Edinburgh and London 1972

4. G. BERG
   Histologische Labortechnik
   J.F. Lehmanns Verlag, München 1972

5. J.L. MEYER, J.H. BERGERT and L.H. SMITH
   Invest. Urol. 14, 115, 1976

6. B. FINLAYSON and F. REID
   Invest. Urol 15, 489, 1978

URIC ACID/CALCIUM OXALATE NEPHROLITHIASIS. CLINICAL AND BIOCHEMICAL FINDINGS IN 86 PATIENTS

A. Rapado, J.M. Castrillo, M. Diaz-Curiel, M.L. Traba,
M. Santos, L. Cifuentes-Delatte

Metabolic Unit. Fundación Jimenez Díaz
Madrid (Spain)

INTRODUCTION

Uric acid alterations produced in hypercalciuric renal lithiasis is a frequent fact described in literature,[1-4], and it confers a special pathogenic, as well as therapeutic approach to this group of calcium-stone formers.

Nevertheless, little mention is made in literature to the real incidence of patients who form mixed uric acid and calcium oxalate calculi. Even though these patients are frequently considered carriers of a variety of uric lithiasis, a high proportion of them has no calcium or uric acid metabolic alterations. Aside from gout, an infrequent disease in this group and in which the existence of mixed stones is described,[5,6], little is known about the physiopathologic abnormalities that are associated with this group of lithiasic patients.

We present the biochemical and clinical characteristics of 86 patients which passed or removed at least one mixed uric acid and calcium oxalate stone.

MATERIAL AND METHODS

In the Urolithiasis Laboratory of this Metabolic Unit, 158 mixed calculi were selected, corresponding to 86 patients; 48 men and 38 women. Their clinical protocol which included their personal history and associated diseases, family history of lithiasis and gout, age of the first colic, number of calculi per year, the bilaterality of the stones, the existence of urinary infections, the

number of surgical interventions and the existence of previous treatment was analyzed. Different parameters in blood and urine were
studied, according to a protocol previously exposed,[7]. The lowest
urinary pH was chosen, in the case of having determined it more than
once. The composition of the calculi was analyzed using a combination
of polarized light crystallography and infrared spectrophotometry.

As control, the clinical and biochemical protocols of a group of
100 consecutive calcium oxalate stone-former patients from our laboratory were analyzed. The results were compared with the mixed-
stones group in a statistical study using the Student t.

RESULTS

Table 1 shows the clinical characteristics of our 86 patients.
Our attention is called by the advanced age of appearance and the
seriousness of the clinical picture, as indicated by the number of
colics per year, bilaterality and recurrence of the stones, as well
as the need for surgical interventions.

On comparing abnormal biochemical facts (table 2) of blood and
urine with the control group of calcium oxalate stone patients, we
observe that there does not exist any statistical difference between
both groups, excepting the lowest urinary pH which is statistically
lower in the group of mixed stone lithiasics, and the higher excretion of uric acid in the control group.

DISCUSSION

Lithiasic patients who form mixed calculi of uric acid and
calcium oxalate because of its distinctive clinical character
(number of infections, surgical operations, etc.) appear to constitute a specific subgroup of renal lithiasis. There are facts in which
they clinically resemble the calcium oxalate stone-former group
associated with hyperuricosuria,[8], such as the number of colics per
year, bilaterality, etc. even though, as a group they have more hypercalciuria and hyperuricosuria than the mixed stones group. The
age of onset resembles that of uric acid lithiasis, of which some
people believe it is a subgroup. That would be supported by the
number of gouty patients in our cases (5.8% with respect to 0% of
the control group).

In literature, there exist few references to the real incidence
of this type of nephrolithiasis. Thus, Coe,[9], finds it in 23 cases
out of 539 (4.3%). We have found 158 calculi out of a total of 3,158,
which means 5%.

Among the biochemical characteristics of these patients, our
attention is called by the urinary pH which is lower than in the

Table 1. Calcium Oxalate / Uric Acid Nephrolithiasis

(Number of Patients)

Age of consultation:	51.5 ± 12.7	years	(20 - 70)
Age of first colic:	41.6 ± 13.8	Years	(7 - 70)
Stones / year:	1.3		

Associated Diseases

Arterial Hypertension	18	Diabetes	5
Gout	5	Hyperparathyroidism	1

Familial History

Renal Lithiasis	29
Gout	2

Urinary tract infections	29	Surgical Operation	53
Bilateral lithiasis	41	Nephrectomy	14

Renal Stones Composition

Mixed	158
Calcium oxalate	64
Uric Acid	46
	268

Table 2. Biochemical Data in Mixed Calcium / Uric Acid
Nephrolithiasis

(Percentage of Cases)

	Nephrolithiasis:		
	Calcium/Uric Acid		Calcium Oxalate
	86 Cases		100 Cases
< 70 ml/min. Ccr	36		23
24-Hour Urine pH 5.5	48	*	6
Hyperuricemia	40		38
Hyperuricosuria	33	*	50
Cur/Ccr Ratio	62		57
Hypercalcemia	2		4
Hypercalciuria	9		14
Hyperuricosuria + Hypercalciuria	5		12

* p < 0.01

lithiasic control group, and by the existence of metabolic alterations in 64%, which coincides with Coe (65%). However, Coe finds a
wide group of hypercalciuric-hyperuricosuric (39%) patients, only
presented in 6.2% of our patients. On the other hand, our cases
showed 33% of hyperuricosuria with respect to 0% in Coe's study.
This can partially be explained by the composition of Spanish diet,
where a lower incidence of hypercalciuric patients in our stone-
former population than in Anglosaxon literature has been found.

The metabolic alterations of these patients which include hyperuricosuria, hypercalciuria as well as pH acid are the biochemical
bases for the formation of mixed calculi. There exist crystallographic evidences which show similarity among the crystals of nonhydrate
uric acid and calcium oxalate, which due to epitaxial growth or to
heterogeneous nucleation would promote the formation of mixed calculi,[3]. On the other hand, there exists some evidence that uric acid
can interfere with some of the inhibitors of the formation of calcium oxalate crystals, increasing the risk of lithiasis,[2]. The good
clinical response attained in these patients on lowering therapeuti-
cally the amount of uric acid in urine,[10], supports the above mentioned postulates.

Supported by this clinical response and by the existence of
hyperuricosuria alone or mixed in at least 40% of the patients,
treatment with allopurinol has been recommended, and Coe has shown
a lower incidence of calculi per year, statistically significant,
in patients treated with this drug. In view of these results, we
recommend as well to increase the urinary pH in these patients.

REFERENCES

1. F.L. Coe, Hyperuricemia calcium oxalate nephrolithiasis. Kidney
   Int. 13, 418, 1978.
2. F.L. Coe, Nephrolithiasis. Pg. 95. Year Book Pub. Chicago, 1978.
3. C.Y.C. Pak, Calcium nephrolithiasis. Pg. 120. Plenum Medical
   Book Comp., New York, 1978.
4. M. Diaz-Curiel, J.M. Castrillo, and A. Rapado. Alteraciones del
   fósforo y acido urico en la litiasis hipercalciurica. Nefrolo-
   gia. 2, 9, 1978
5. J.H. Talbott and T.F. Yu, Gout and uric acid metabolism. Pg. 85,
   Georg Thieme Pub. Stuttgart, 1976
6. L. Cifuentes, J. Bellanato, M. Santos and J.L. Rodriguez-Miñon.
   Monosodium urate in urinary calculi. Eurp. Urol. 4, 441, 1978.
7. A. Rapado, M.L. Traba, J.M. Castrillo et al. El laboratorio en
   el estudio de la litiasis renal. Pg. 215. En: Problemas actuales
   de Urología. Edit. by L. Cifuentes et al. Salvat Barcelona,
   1977.
8. C.Y.C. Pak, C. Fetner, J. Townsend et al: Evaluation of calcium
   urolithiasis on ambulatory patients. Am. J. Med. 64, 979, 1978.

9. F.L. Coe: Calcium-uric acid nephrolithiasis. Arch. Int. Med. 138, 1090, 1978.

10. C.Y.C. Pak, D.E. Barilla, K. Holt et al: Effect of oral purine and allopurinol on the crystallization of calcium salts in urine of patients with hyperuricosuric calcium urolithiasis. Am. J. Med. 65, 593, 1978.

11. M.J.V. Smith, L.D. Hunt, J.S. King and W.M. Boyce: Uricemia and urolithiasis. J. Urol. 101, 637, 1969.

THE URIC ACID: CYSTINE CORRELATION IN THE URINE OF RECURRENT

CALCIUM OXALATE STONE-FORMERS AND HEALTHY CONTROLS

P. Leskovar, R. Hartung and M. Kratzer

Urologische Klinik und Poliklinik r.d.Isar der Techn.
Universität München (Direktor: Prof.Dr.W. Mauermayer)
D-8000 München 80, Ismaningerstraße 22

1. INTRODUCTION

The relation between cystinuria and uric acid lithiasis, though
already known for a long time (8,1), recently finds new
attention in the concerning literature (5) caused by the increasing
observations of primary hyperuricaemia and simultaneous cystinuria
This relation was pointed out by Meloni and Canary in 1967 (6).
They supposed the presence of two inheritable metabolic deficiencies, but they could not precise a possible relation between both
of .them. The hyperuricaemia was however not the consequence of a
renal damage by cystinuria. Three years later Vergis and Walker
(9) described a case of a 19-year-old woman with cystinuria coincident with hyperuricaemia. But in fail of renal function analysis a renal caused increaese of uric acid cannot be excluded.
More patients with cystinuria and hyperuricaemia have been observed by King and Wainer (3) and by Krizek (4). Details of these
cases are described in a clear way by Newcombe (7), who concludes,
that the reported cases suggest a possible relationship between
cystinuria and hyperuricaemia. Marketos, S. et al. (5) examined 5
patients with primary hyperuricaemia, cystinuria and urolithiasis.
They found it proved, that people with cystine calculi show an
increased excretion of uric acid and also tend to disturbed urate
homeostasis. Hyperuricaemia caused by renal disfunction could be
excluded in this case by normal glomerular filtration rates. The
authors therefore suggest, that these disturbances are caused by
one or two inborn errors of metabolism. Our own examinations
treated the comparative measurement of actual uric acid and
cystine concentrations in the urine of idiopathic calcium oxalate
stone-formers and healthy controls. They do therefore not follow

exactly the above reported cases of people suffering from cystinuria, but a rather important correlation between our cystine and
uric acid values urged us to present our results in association
with the above mentioned facts.

2. PATIENTS AND CONTROLS

Ten recurrent oxalate stone-formers we compared with eleven
healthy controls. Three of the patients and four of the controls
were women. Each of the parameters in question were measured over
a period of 4 - 5 weeks. Taking into consideration that stone-
patients get their repeated stone-recurrences not under a special
diet, but under their individual habits, we didn't prescribe them
any uniform diet. The patients and the controls collected their
urine thrice a day (morning, midday, evening) in special plastic
containers, with 15 % thymol/propanol respectively toluol as
preserver.

3. METHODS

The determination of cystine was carried out following a modified
method of Shinohara and Padis (2). Cystine is reduced by sodium
bisulfite to cysteine; in addition, the sulfhydryl-group of
cysteine changes the added phosphotungstic acid to the tungstic
blue which can be evaluated (spectro)photometrically. The normal
cystin excretion is 10 - 100 mg/die. Our results were related to
a cystine standard of 400 mg/l. The uric acid was determined by
the Boehringer test-combination (Boehringer/Mannheim, Art.15 865/
15 866). The actual urinary concentrations of cystine and uric
acid were the basis for the calculation of the linear regression
curves and the coefficients of correlation. The daily mean value
was calculated from the average daily concentration of all
patients and controls and not from the mean value of the morning,
midday and evening urine, which may explain the small numeric
difference.

4. RESULTS AND DISCUSSION

As it can be seen in Table 1 and Table 2, the (uncorrected) concentrations (not 24-h-excretions) of cystine and uric acid were
lower in the group of stone-formers than in that of control persons. The consideration of the equally lower urine osmolarity
of patients however compensated this difference so that no significant differences between patients and controls could be
ascertained. The calculation of the linear regression curve
showed for the patients the equation: cystine = 0,067 x uric acid
+ 14,7, for the controls: cystine = 0,065 x uric acid + 18,7.
The curves are - in virtue of the nearly identic slope - very
similar in their course (Fig. 1). The patients' values showed a
highly significant positive correlation with r = 0,68 (P<0,001),

Table 1: The average uric acid concentrations (mg/l) in the morning, midday and evening urinary samples as well as the average osmolarity and the osmolarity-corrected uric acid concentration in the 12-hour-urine of ten oxalate stone-formers and eleven controls.

Examined group	Number n	Morning		Midday		Evening		Day		Osmolarity		Uric acid Osmolarity
		X̄	s	X̄	s	X̄	s	X̄	s	X̄	s	
Oxalate stone-formers	10	426,06	167,34	484,46	229,20	414,10	187,16	431,66	190,95	622,2	223,5	0,70
Controls	11	607,29	77,56	626,23	114,31	632,01	128,52	616,23	88,94	878,0	106,2	0,70

Table 2: The mean and the standard deviation of the cystine concentration (mg/l) in actual morning, midday and evening urine samples as well as the mean of the osmolarity and the mean of the osmolarity-corrected cystine concentration in the 12-hour-urine of 10 oxalate stone-formers and 11 controls.

Examined group	Number n	Morning		Midday		Evening		Day		Osmolarity		Cystine Osmolarity
		X̄	s	X̄	s	X̄	s	X̄	s	X̄	s	
Oxalate stone-formers	10	43,07	16,27	43,92	19,00	44,46	18,60	42,60	15,5	622 ?	223,5	0,068
Controls	11	57,05	12,60	59,86	10,8	61,26	12,73	59,65	8,87	878,0	106,2	0,068

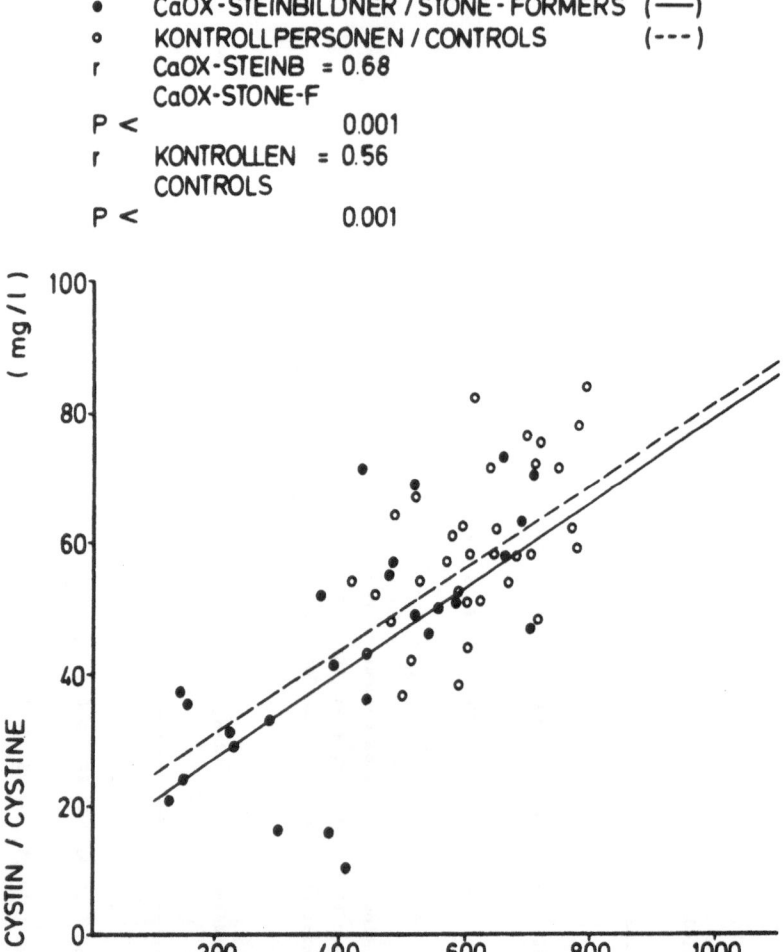

Figure 1. Correlation between the actual uric acid and cystine concentrations in the morning, midday and evening urine of recurrent oxalate stone-formers and control persons. Linear regression: (a) stone-formers: cystine = 0.067 x uric acid + 14.7; (b) controls: cystine = 0.065 x uric acid + 18.7.

the controls' values a some what lower positive correlation with r=0,56 and P<0,001. In both groups, the significance of the ascertained correlation was high. Comparable examinations are not known to the authors. We couldn't find any literature references to the relation between cystine and uric acid in the urine of calcium stone-formers. It would be desirable to get more inquiries concerning this problem.

5. REFERENCES

(1) Bostrom, H., and Hambraeus, L.: Cystinuria in Sweden. VII.
Clinical, histopathological, and medico-social aspects of the
disease. Acta Med. Scand. (Suppl.) 411: 1-128 (1964).   (2)Henry,
R. J., Cannon, D.C. Winkelman, J.W. eds.: Clinical Chemistry:
Principles and Technics, p. 589 (1974).   (3) King, J.S.,Jr.,and
Wainer, A.: Cystinuria with hyperuricemia and methioninuria. Bio-
chemical study of a case. Am.J.Med. 43: 125-130 (1967). (4)Krizek,
V.: Uricemia in cystinuria. Horm. Metab. Res. 4: 51-53 (1972).
(5) Marketos, S., Mountokalakis, Th., Halazonitis, N., and
Merikas, G.: Primary hyperuricaemia in cystinuria. p. 469-471 in
Urolithiasis Research, Ed. by H.Fleisch, W.G. Robertson, L.H.
Smith, and W. Vahlensieck, Plenum Press, New York and London(1976)
(6) Meloni, C.R. and Canary, J.J.: Cystinuria with hyperuricemia.
J. Amer. Med. Ass., 200, 257-259 (1967).   (7) Newcombe, D.S.,
M.D.: Inherited biochemical disorders and uric acid metabolism.
HM + M Medical and Scientific Publishers, England, (1975).
(8) Renander, A.: The roentgen density of the cystine calculus.A
roentgenographic and experimental study including comparison
with more common uroliths. Acta Radiol. (Suppl.) 41: 1-148(1941).
(9) Vergis, J.G., and Walker, B.R.: Cystinuria,hyperuricaemia
and uric acid nephrolithiasis: case report. Nephron 7:577-579
(1970).

THE ROLE OF URATE IN IDIOPATHIC CALCIUM UROLITHIASIS

S.R. Silcock

The Wellcome Research Laboratories

Beckenham, Kent, BR3 3BS, UK

Upper urinary tract stone incidence has increased markedly in the 20th century, coincident with a decline in bladder stones. The pattern appears to be related to rising living standards (Blacklock, 1976).

All kidney stones form from relatively insoluble substances which crystallize from urine. In section, they usually show ordered growth rings consisting of compact crystalline masses that reflect phases of crystal deposition. Histochemical techniques reveal the presence of an organic matrix, interspersed with the crystals. The crystals may be of one type only (apparently "pure" stones), or they may be mixed. The predominant crystalline component is calcium oxalate, and over two-thirds of kidney stones in developed regions consist of calcium oxalate, with or without a calcium phosphate admixture (Williams and Prien, 1977). Such stones are often loosely termed "calcium stones" (Prien, 1974).

The genesis of most other stones, e.g. magnesium ammonium phosphate ("struvite" or "infection" stones) and uric acid stones, is well understood. This is not true for most calcium stones. Only some 10 to 20% of them are attributable to well-recognised causes, such as hyperparathyroidism, sarcoidosis, etc. These conditions give rise to excessive urinary calcium excretion, an important factor in calcium stone formation. Calcium stones without recognisable associated pathology have been termed "idiopathic", which includes a high proportion of unexplained hypercalciurias.

In addition to hypercalciuria, which, in the absence of other abnormal findings, may be classed as a "metabolic disorder", attention has focussed on the association of idiopathic calcium

stones with hyperuricaemia and hyperuricosuria. Occasional unexplained hyperoxaluria has also been recognised (Williams, 1978). Both
hypercalciuria and hyperoxaluria have an obvious link with calcium
oxalate stones. However, the connection with urate disorder is less
immediately apparent. Very few calcium stones contain readily detectable amounts of urate (Prien, 1974), though a small percentage of
patients form mixed stones containing calcium and uric acid (Coe,
1978a) and monosodium urate crystals have recently been identified
in 1.66% of kidney stones examined in thin sections (Cifuentes-
Delatte et al, 1978).

In stone-forming populations whose biochemistry has been
studied in detail, hyperuricaemia and/or hyperuricosuria are present
in about 30% of idiopathic calcium stone formers. No satisfactory
explanation has been advanced for an influence of hyperuricaemia,
but there is good evidence for a physicochemical explanation of the
link between hyperuricosuria and calcium stone formation (Coe,
1978b).

High concentrations of urinary urate encourage the formation,
growth and aggregation of calcium oxalate crystals. These are
important steps in calcium stone formation.

Calcium oxalate crystals can only form in urine which is
already supersaturated with calcium oxalate. The same applies to
any substance capable of forming crystals (i.e. a crystalloid), a
number of which are present in urine. "Supersaturated" describes
a solution whose crystalloid concentration is above the maximum
concentration reached by dissolving the solid crystalloid in water
(the "saturated" state). Supersaturation with various crystalloids
occurs readily in urine. It is a relatively unstable state, in
that crystal formation becomes more likely as supersaturation increases. The scale of undersaturation, saturation and supersaturation in complex solutions cannot be related directly to
crystalloid concentration, but is expressed in terms of its
"activity product". The activity product of calcium oxalate, for
example, is "active" calcium ion concentration X "active" oxalate
ion concentration. In urine, "active" ion concentrations are influenced by (i) the presence of other ions, (ii) the ionic strength,
(iii) pH, and must be calculated taking these factors into account.
The activity product at which crystals form (the "formation product")
is variable in urine (Pak, 1978a), though it is constant for a
single crystalloid in water under defined physicochemical conditions. It is raised by crystal inhibitors and depressed by
crystal promoters, both of which may be present in urine (see
Fig. 1).

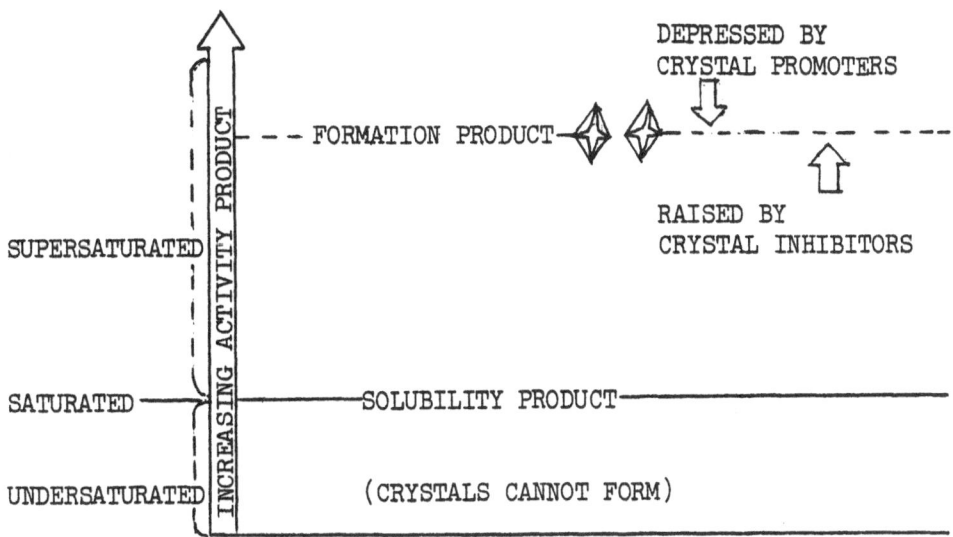

Fig. 1.   Supersaturation and Crystal Formation

The presence of crystal inhibitors in urine is well established (Robertson et al, 1976a) and a number of them (e.g. citrate, pyrophosphate, acid mucopolysaccharides) have been characterised. They inhibit crystal formation, crystal growth, and crystal aggregation, though not all inhibitors may inhibit all these stages. The concept of crystal promoters is less well established (Pak, 1978a) and their nature remains speculative. Precursors of stone matrix, such as uromucoid (Hallson and Rose, 1979) may act in this way. The absence of promoters will have a net crystal "inhibitory" effect, and the absence or blockage of inhibitors will have a net crystal "promoting" effect (see Fig. 2).

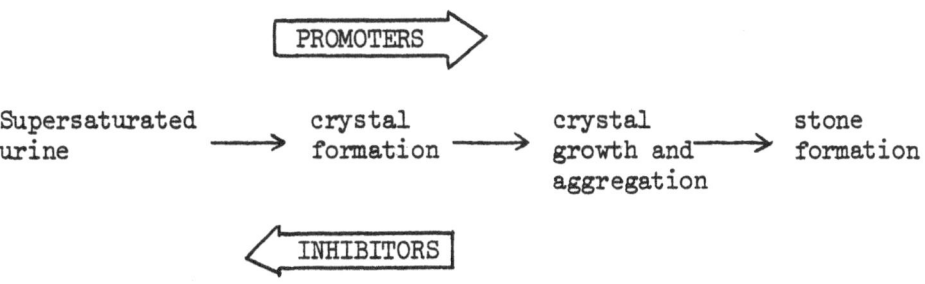

Fig. 2.   Promoter and Inhibitor Effects on Stone Formation

On the basis of these concepts, it is possible to explain how urate is thought to influence calcium oxalate crystallization in urine.

Above a pH of 5.47 (Finlayson and Smith, 1974), uric acid begins to dissociate into urate ions and hydrogen ions. This dissociation increases as the pH increases. In the presence of sodium ions and at the appropriate pH range, urate behaves as sodium urate rather than uric acid. Pak et al (1977) found that at a urine pH of around 6.0 - the average pH in idiopathic calcium stone formers - monosodium urate predominates over uric acid.

Calcium oxalate has been found to crystallize more readily in urine with a high urate concentration than in urine with a low urate concentration (Pak et al, 1978). More specifically, when urine was undersaturated for sodium urate, the urine had to be made more than 11-fold supersaturated for calcium oxalate before calcium oxalate crystals formed. Conversely, when the urine was made 2-fold supersaturated for sodium urate, urine had only to be made 7-fold supersaturated with calcium oxalate in order for calcium oxalate crystals to form. Increasing supersaturation of urine with sodium urate thus lowered the formation product for calcium oxalate in urine, though it did not influence the actual activity product, which determines the levels of supersaturation with calcium oxalate. The promoting effect of urate on calcium oxalate crystallization in urine could be direct (heterogeneous nucleation) or indirect (blocking of one or more crystal inhibitors).

The evidence suggests that either or both mechanisms may operate. In urine supersaturated with monosodium urate, colloidal or crystalline monosodium urate may form. Either phase could provide a surface on which calcium oxalate crystallization would readily occur (heterogeneous nucleation). Earlier research had shown that sodium urate crystals accelerated calcium oxalate crystallization in vitro. Colloidal urate appears to bind an important group of crystal inhibitors, the acid mucopolysaccharides (Robertson et al, 1976b). The physicochemical feasibility of this system has been substantiated using finely divided monosodium urate (Finlayson and Du Bois, 1978). The precise physical form of urate in the urine of calcium stone formers has not been identified. However, the overall scheme of action can be summarised as follows (Pak et al, 1978):
High urinary concentrations $\longrightarrow$ supersaturation with monosodium urate (MSU) $\longrightarrow$ formation of colloidal or crystalline MSU $\longrightarrow$ promotion of calcium oxalate crystal formation $\longrightarrow$ calcium stone formation favoured (see Fig. 3).

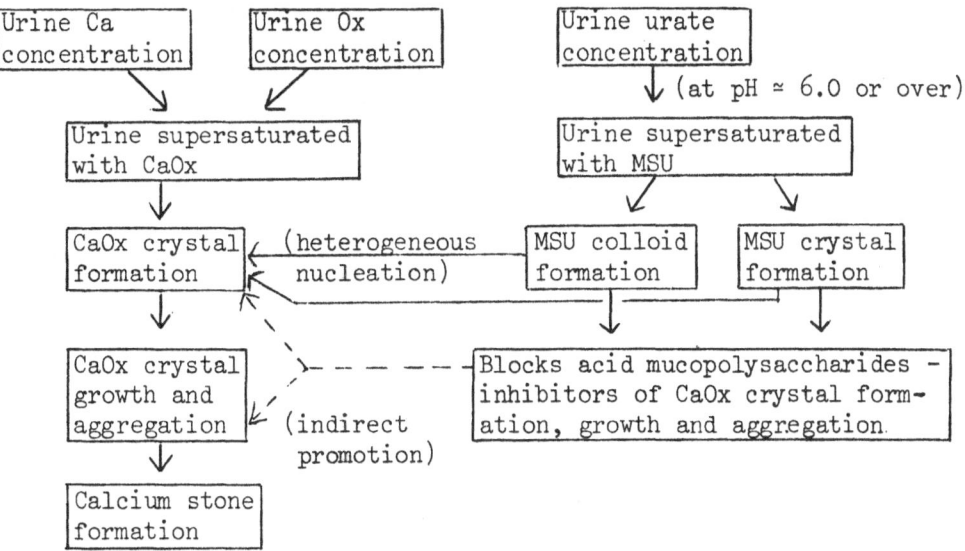

Fig. 3.   The Influence of Urinary Urate on Calcium Oxalate
          Crystallization

Robertson et al (1978) have characterised the main risk factors in
calcium stone formation as calcium, oxalate, pH, acid mucopolysaccharides and urate.

     Clearly, supersaturation of urine with crystalloids is a
critical determinant of stone formation, though the final outcome
may be influenced by both crystal promoters and crystal inhibitors
in urine. However, the concentrations of calcium, oxalate and
urate in urine are only partial determinants of supersaturation
with calcium oxalate and sodium urate, since the activity products
of these crystalloids are profoundly modified by urine composition
and pH. While the clinical criteria of hypercalciuria, hyperoxaluria and hyperuricosuria as commonly defined (urinary output
in milligrams or millimoles per 24 hr) may indicate "metabolic
abnormality", they do not express concentration. Still less can
they predict or reflect supersaturation, which has been identified
in a significant number of patients with "normocalciuria" (Weber
et al, 1979). Likewise, urine was invariably found to be supersaturated with monosodium urate at total urate concentrations of
more than 300 mg/litre (Pak et al, 1977). Such concentrations
could easily occur at total urate excretion values well below the
usual threshold for "hyperuricosuria" (750 to 800 mg per 24 hr)
if urine output were less than 2 litres.

     Despite these discrepancies, it is appropriate to summarise
the factors leading to excessive excretion of calcium, oxalate and
urate within the context of idiopathic calcium stone formation.
Hypercalciuria may be of dietary, absorptive, and/or renal origin

(Pak, 1978b; Coe, 1978c). Hyperoxaluria may be of dietary, absorptive, and/or metabolic origin (Pak, 1978c; Coe, 1978d), or possibly
of renal origin (Pinto et al, 1974). Hyperuricosuria may be of
dietary, metabolic and/or renal origin. The urate load reaching
the kidney depends on dietary purine intake, de novo purine biosynthesis, the rate of tissue nucleotide breakdown, the efficiency
of the purine salvage pathway, and the rate of purine conversion to
urate. A common cause of hyperuricosuria in idiopathic calcium stone
formers is high dietary purine intake, but some show evidence of
urate overproduction (Coe, 1978b). The final urate load excreted by
the kidney depends on urate filtered at the glomerulus, which then
undergoes complex reabsorption and secretion within the tubule.

Clinically, hyperuricosuria seems to lead to a particularly
severe form of stone disease (Coe et al, 1977). Its correction with
allopurinol dramatically reduces or arrests new stone formation
(Coe, 1977).

REFERENCES

Blacklock, N.J., 1976, Epidemiology of urolithiasis, in: "Scientific
    Foundations of Urology", Vol. I Renal Disorders, Infections and
    Calculi, Williams, D.I., Chisholm, G.D. eds. W. Heinemann, London

Cifuentes-Delatte, L., Bellanato, J., Santos, M. and Rodriguez-Miñon,
    L., 1978, Monosodium urate in urinary calculi, Eur. Urol., 4:441-
    447.
Coe, F.L., 1977, Treated and untreated recurrent calcium nephro-
    lithiasis in patients with idiopathic hypercalciuria, hyper-
    uricosuria or no metabolic disorder, Ann. Int. Med., 87:404-410.
Coe, F.L., 1978a, Calcium-uric acid nephrolithiasis, Arch. Int. Med.,
    138:1090-1093.
Coe, F.L., 1978b, Hyperuricosuric calcium oxalate nephrolithiasis,
    Kidney Int., 13:418-426.
Coe, F.L., 1978c, Idiopathic hypercalciuria, in: "Nephrolithiasis,
    pathogenesis and treatment", Year Book Medical Publishers, Inc.,
    Chicago, London.
Coe, F.L., 1978d, Hyperoxaluric states, in: ibid.
Coe, F.L., Keck, J., Norton, E.R., 1977, The natural history of
    calcium urolithiasis, J. Am. Med. Ass., 238:1519-1523.
Finlayson, B. and Smith, A., 1974, Stability of first dissociable
    proton of uric acid, J. Chem. Eng. Data, 19:94-97.
Finlayson, B. and Du Bois, L., 1978, Adsorption of heparin on sodium
    acid urate, Clin. Chim. Acta, 84:203-206.
Hallson, P.C. and Rose, G.A., 1979, Uromucoids and urinary stone
    formation, Lancet, 1:8124.
Pak, C.Y.C., 1978a, Physical Chemistry of Stone Formation, in:
    "Calcium Urolithiasis, Pathogenesis, Diagnosis and Management",
    Alvioli, L.V., ed., Plenum Medical Book Company, New York, London.

Pak, C.Y.C., 1978b, Hypercalciurias, in:ibid.

Pak, C.Y.C., 1978c, Hyperoxaluric calcium urolithiasis, in:ibid.

Pak, C.Y.C., Barilla, D.E., Holt, K., Brinkley, L., Tolentino, R., and Zerwekh, J., 1978, Effect of oral purine load and allopurinol on the crystallization of calcium salts in urine of patients with hyperuricosuric calcium urolithiasis, Am. J. Med., 65:593-599.

Pak, C.Y.C., Waters, O., Arnold, L., Holt, K., Cox, C. and Barilla, D., 1977, Mechanism for calcium urolithiasis among patients with hyperuricosuria: supersaturation of urine with respect to mono-sodium urate, J. Clin. Invest., 59:426-431.

Pinto, B., Crespi, G. Solé-Balcells, F. and Barceló, P., 1974, Patterns of oxalate metabolism in recurrent oxalate stone formers, Kidney Int., 5:285-291.

Prien, E.L., 1974, The analysis of urinary calculi, Urol. Clin. N. Amer., 1:229-240.

Robertson, W.G., Knowles, F. and Peacock, M., 1976, Urinary acid mucopolysaccharide inhibitors of calcium oxalate crystallization, in "Urolithiasis Research", Fleisch, H., Robertson, W.G. et al, eds., Plenum Press, New York, London.

Robertson, W.G., Peacock, M. and Marshall R.W., 1976b, Saturation-inhibition index as a measure of the risk of calcium oxalate stone formation in the urinary tract, New Engl.J. Med., 294:249-252.

Robertson, W.G., Peacock, M., Heyburn, P.J., Marshall, D.H. and Clark, P.B., 1978, Risk factors in calcium stone disease of the urinary tract, Br. J. Urol., 50:449-454.

Weber, D.V., Coe, F.L., Parks, J.H., Dunn, M.S.L. and Tembe, V., 1979, Urinary saturation measurements in calcium nephrolithiasis, Ann. Int. Med., 90:180-184.

Williams, H.E., 1978, Oxalic acid and the hyperoxaluric syndromes, Kidney Int., 13:410-417.

Williams, H.E. and Prien, E.L., 1977, Nephrolithiasis, in "Metabolic bone disease", Alvioli, L.V., Krane, S.M. eds., Academic Press, New York, San Francisco, London.

MINERALOGIC COMPOSITION OF 66 MIXED URINARY CALCULI OF CALCIUM OXALATE AND URIC ACID

J.R. Miñón-Cifuentes, M. Santos and L. Cifuentes-Delatte

Laboratorio de Urolitiasis
Fundación Jimenez-Díaz
Madrid

The study and knowledge of the composition of urinary calculi is of great importance in the prophylactic medication of patients suffering from nephrolithiasis.

Generally, the various types of analysis yield qualitative and quantitative data of the composition of the whole calculi. The composition of the calculi, however, changes in the course of its development due probably to differences in conditions of formation.

We know, for example, that patients with disorders of uric acid metabolism can develop mixed calculi of calcium oxalate and uric acid, pure calculi of uric acid or pure calculi of calcium oxalate. The composition of calculi determined by polarizing microscopy yields important data of the conditions of calculi formation. These examinations and observations can be complemented by the history of the patient's disease.

MATERIAL AND METHODS

Sixty-six mixed calculi of calcium oxalate and uric acid were studied in sections of approximately $15 \mkern2mu \varkappa$ thick, under the conventional phase and polarizing microscope. Each calculus corresponded to a different patient. The calculi were carefully divided into two halves, one of which was prepared in a thin section in the Mineralogy Department of the Nuclear Energy Board Laboratories (Madrid). The other half was used for a complementary analysis with infrared spectroscopy in a Perkin-Elmer, model 457, spectrophotometer. The material was compiled from diverse layers of the calculi and its microscopic morphology compared with their respective infrared spectra.

In such sections, similar to those used for histological examination, we can study the relationship between adjacent constituents
of the calculi, the various stages of calculus formation and the
changes occuring during the development and growth of calculi. Knowledge of the structure is of special importance from the therapeutic
and prophylactic viewpoint. The evolution of a calculus as well as
its general structural pattern, can be perfectly recognized in these
thin sections.

RESULTS

As a result of our comparative microscopical and spectroscopi-
cal studies, we describe the morphological aspect of the mixed
calculi of calcium oxalate and uric acid.

Calcium oxalate monohydrate or whewellite presents intense
birefringence by polarization when it appears in its most typical
form of concentric rings, very compact bands in which radial stria-
tions appear and which correspond to their fibrous structure. In
most cases they are crossed by arched, concentric transversal lines
which, when observed under the phase microscope, appear to be denser
and to correspond to the calcium phosphate that, by infrared spectroscopy we have found always accompanying,[1].

Calcium oxalate dihydrate has a characteristic form of a
rhomboidal prism lanceolate, with several inner angles which are
remarkably well visualized under the phase microscope. Under the
polarizing microscope it shows intense birefringence with magnifi-
cent chromatic effects,[1].

Uric acid is easily recognized under the microscope and on
polarization it does not have birefringence. Sometimes to make a
distinction between uric acid, sodium acid urate, it is necessary
to perform analysis with infrared spectroscopy.

In this study, we have found four patterns of calculi according to the mineral composition of their central and peripheral
area.

Pattern A: 32 calculi (48.4%). The central area was formed by calcium oxalate monohydrate (Whewellite) and it was surrrounded by
uric acid in all cases (Fig.1), Whewellite was the main component
in 20 calculi and uric acid in 12.

Pattern B: 27 calculi (40.9%). The central area was formed by uric
acid and was surrounded by calcium oxalate monohydrate (Fig. 2).
Uric acid was the main component in 25 cases and whewellite in 2.

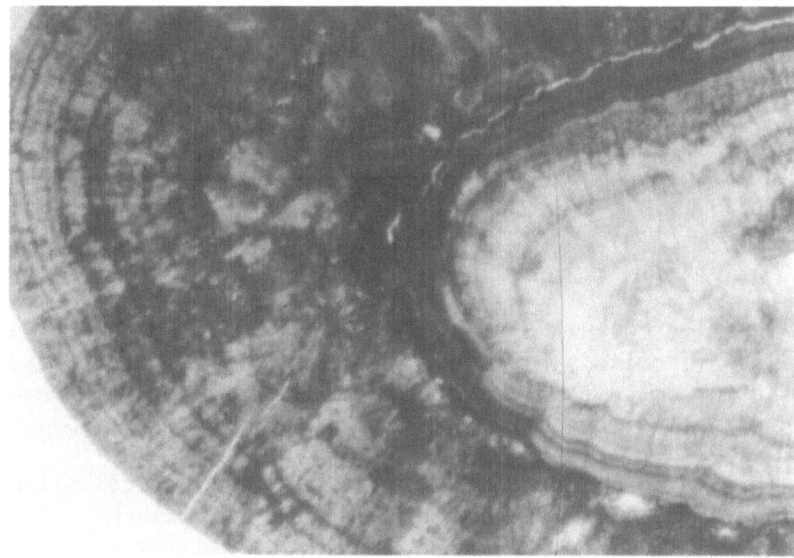

Fig. 1. Central area formed by whewellite, surrounded by uric acid.

Fig. 2. Central area formed by uric acid, surrounded by whewellite.

Pattern C: The four mixed calculi (6%) of this pattern consisted of calcium oxalate monohydrate in the central area and calcium oxalate dihydrate (weddellite) and uric acid in the periphery. Calcium oxalate monohydrate was the main component in the four cases.

Pattern D: The three mixed calculi (4.5%) of this pattern consisted of uric acid in the central area and calcium oxalate dihydrate and uric acid in the periphery. Uric acid was the main component in all 3 cases.

DISCUSSION

In mixed calculi of calcium oxalate and uric acid, the whewellite is predominant in most cases (89.3%). The most frequent pattern was that of calculi with a central area formed by calcium oxalate monohydrate and uric acid in the periphery.

Thirty-three per cent of patients with mixed calculi of uric acid and calcium oxalate had hyperuricosuria, 9% hypercalciuria and 6% both disorders: hyperuricosuria and hypercalciuria.

Lonsdale,[5], has suggested that epitaxy could play an important role in the growth of the mixed calculi. The concept of epitaxy is an attractive explanation for the growth of mixed urinary calculi. Crystallographic evidence and heterogeneous seeding experiments both suggest that uric acid and calcium oxalate crystals share sufficient structural similarities to allow epitaxis.

Meyer and cols.,[7], indicate that the presence of uric acid crystals in a metastable supersaturated solution of calcium oxalate facilitates the crystallization of calcium oxalate. An epitaxial relationship between anhydrous uric acid and calcium oxalate monohydrate would have been predicted, based upon the close dimensional similarity of the (100) face of uric acid and the (001) face of calcium oxalate phase,[6].

A number of calcium oxalate monohydrate calculi containing a nidus of uric acid have been described,[4,8], and it is possible that epitaxy was responsible for the growth of the calculi in those instances.

There have been reports that have linked hyperuricosuria to an increased incidence of calcium oxalate lithiasis,[2,3], and in fact therapy regimes designed to lower uric acid concentrations in urine have been used with reported success in preventing the recurrence of lithiasis in these patients.

There is a structural correspondence between crystals of uric acid and calcium oxalate which is sufficient to permit one to grow upon the surface of another, or to act for one another as heterogeneous seed nuclei.

These observations suggested that mixed urinary calculi of calcium oxalate and uric acid represent a distinct subgroup of calcium calculi disease with an abnormality of urine chemistry, that could link uric acid disorders to the formation of calcium calculi in the kidney.

REFERENCES

1. L. Cifuentes Delatte, A. Hidalgo, J. Bellanato and M. Santos, Polarization microscopy and infrared spectroscopy of thin sections of calculi. Urinary calculi. Proc. Int. Symp. of Renal Stone Research. Madrid, 1972, p.220.
2. F. Coe and L. Rainen. Allopurinol treatment of uric acid disorders in calcium stone formers. Lancet 1. 129. 1973.
3. F. Coe and A. Kavalach. Hypercalciuria and Hyperuricosuria in patients with calcium nephrolithiasis. N. Engl. J. Med. 291. 1344, 1974.
4. L. Herring. Observations on the analysis of ten thousand urinary calculi. J. Urol. 88. 545, 1962.
5. K. Lonsdale. Human stones. Science. 159. 1199. 1968
6. K. Lonsdale. Epitaxy as a growth factor in urinary calculi and gallstones. Nature. 217. 56. 1968.
7. J. Meyer, J. Bergert and L. Smith. The epitaxially induced crystal growth of calcium oxalate by crystaline uric acid. Invest. Urol. 14. 115. 1976.
8. E. Prien and C. Frondel. Studies in urolithiasis. I. The composition of urinary calculi. J. Urol. 57. 949. 1947.

HYPERURICEMIA AND CYSTINURIA

F. LINARI, M. MARANGELLA, B. MALFI, G. VACHA,
M. BRUNO, G. GIORCELLI, B. FRUTTERO

OSPEDALE MAURIZIANO UMBERTO I-TORINO
TORINO (ITALY)

Cystinuria is an inherited disorder of the tubular and jejunal mucosa transport of four amino acids:cys,lys, arg and ornithine (1-2). The association of cystinuria with other chronic diseases or metabolic disorders is not common. Therefore it seemed us usefull to present the association hyperuricemia and cystinuria in 7 out of 55 cystinic lithiasis patients.

MATERIAL AND METHODS

7 patients with hyperuricemia and cystinuria (aged 21-68 years) have been compared with 10 idiopathic hyperuricemic uric acid stone formers and with 10 healthy subjects of the clinical staff for creatinine clearance (autoanalyser),serum and urinary uric acid (uricase), plasma and urinary aminoacids. The aminoacid determinations were carried out with an Optica Amino Analyser (ion exchange column lithium buffer). The fasting blood and 24 h urine samples were examined within few day from the collection.(3)

RESULTS

In tab. I are shown age, creatinine clearance, serum uric acid and uric acid clearance (mean $\pm$ s.d.).

Tab. I -   10 idiopathic hyperuricemics(group I) and
             7 hyperuricemic cystinurics (group II) com-
             pared with 10 healthy controls.

	Controls	Group I	Group II
Age	28,6 ± 7.6	40.9 ± 6.9	42.0 ± 18.4
Creat.Clear.	115.9 ± 16.9	112.2 ± 15.2	116.3 ± 17.4
Serum U.A.	4.7 ± 1.4	8.2 ± 0.7	7.9 ± 1.2
U.A. Clear.	9.1 ± 1.3	10.1 ± 2.3	9.6 ± 1.3

In tab. II and tab. III are shown respectively plasma
and urinary aminoacid values for idiopathic hyperurice-
mics (group I) and hyperuricemic cystinuric patients
(group II) compared with healthy controls.
In both group there was a significant increase of total
plasma aminoacids. Common to both groups was an increase of glycine and glutamine acid whyle only in group I
there was an increase in aspartic acid and phenylalanine;
in group II there was an increase in serine,asparagine
and glutamine and a decrease in cystine. Total aminoaciduria was significantly decreased in group I. Also
in group II there was a similar behaviour of aminoaciduria if cystine and A.L.O. were excluded. Threonine,se-
rine,glutamine and leucine were decreased in both groups.
Only in group I there was an increase in triptophane and
phenylalanine excretion; only in group II there was an
increase of methionine and a decrease of isoleucine and
tyrosine.

DISCUSSION

The association of hyperuricemia with cystinuria has
been reported up to now in 4 patients by Meloni and Canary (4) and in 5 by Marketos (5). In the previous
observations and in ours hyperuricemia was not secondary
to a decrease in renal uric acid excretion. In the family studies of our patients we did not find inherited
metabolic diseases involving uric acid. 4 hyperuricemic cystinuric patients had an uric acid stone episode.

able II — Plasma aminoacids (µM/lt) in 10 hyperurice-
mics (group I) and in 7 hyperuricemic cysti-
nurics compared with 10 healthy controls.

aminoacids	group I			controls			group II	
TAU	74.1	42.6		44.1	12.6		124.6	110.4
ASP	25.6	10.6	*	4.1	3.6		15.3	17.5
THR	132.8	44.2		125.9	19.2		167.4	74.5
SER	115.2	38.7		87.5	25.1	**	114.6	18.9
ASP-NH$_2$	42.1	16.2		32.9	8.2	**	58.9	27.8
GLU	89.9	43.3	*	28.8	11.7	*	91.4	59.0
GLU-NH$_2$	462.3	79.7		432.7	65.2	***	542.2	115.1
PRO	180.6	56.7		137.2	14.9		223.3	86.2
GLY	266.1	77.3	***	188.0	43.0	*	305.7	28.4
ALA	362.5	147.7		231.2	60.9		355.9	139.7
CIT	37.2	13.8		26.4	13.0		36.7	6.9
NH$_2$-BUT	19.3	8.3		19.5	10.4		18.6	7.2
VAL	240.7	112.1		187.1	27.5		186.9	30.4
CYS	31.5	9.1		29.3	2.3	*	13.5	5.8
MET	19.5	7.5		15.2	2.8		19.6	3.4
iLEU	60.9	28.6		30.0	6.4		51.9	16.0
LEU	128.8	55.8		83.2	13.8		105.6	33.9
TYR	52.6	16.4		38.8	5.2		53.8	12.0
PHE	53.5	13.8	***	39.2	5.9		51.0	16.7
TRIPT	37.9	5.9		36.7	9.6		47.1	17.7
ORN	69.9	16.8		55.6	15.0		61.4	22.9
LYS	137.6	33.3		132.1	14.6		126.1	25.6
HIST	84.9	24.5		85.1	30.7		98.4	19.9
ARG	52.3	32.8		42.9	8.4		56.8	22.7
Total	2726.7	571.2		2106.7	260.2		2982.7	392.8
	mean ± s.d.			mean ± s.d.			mean ± s.d.	

* p 0.001
** p 0.005
*** p 0.01

Table III   – Urinary aminoacids (µM/24h.) in 10 hyper-
             uricemics (group I) and in 7 hyperuricemic
             cystinurics compared with 10 healthy con-
             trols.

aminoacids	group I			controls			group II	
TAU	562.1	606.9		508.9	101.2		326.2	394.0
ASP	43.9	29.0		20.8	3.7		25.5	7.1
THR	137.9	33.5	*	270.2	43.3	**	151.23	107.7
SER	214.0	57.5	*	379.5	34.7	*	238.7	44.3
ASP-NH$_2$	176.5	115.5		66.1	11.3		85.7	61.5
GLU	62.6	50.9		35.8	12.7		41.9	21.7
GLU-NH$_2$	364.0	34.2	*	684.0	94.2	*	227.2	192.1
GLY	1248.8	601.0		1719.5	509.6		1398.1	558.4
ALA	495.5	339.7		204.4	23.6		210.8	81.2
NH$_2$-BUT	19.3	8.3		19.4	10.4		18.6	7.2
VAL	52.4	10.6		49.8	5.2		38.6	14.3
CYS	80.9	55.9		35.6	6.6	*	1621.1	811.6
MET	63.7	43.8		28.7	7.4	*	39.5	7.4
iLEU	50.5	17.3		55.3	9.2	*	36.9	10.7
LEU	110.7	28.5		147.4	22.2	*	78.1	48.1
TYR	166.7	69.3		168.4	12.9	*	94.9	26.6
PHE	116.5	52.9	*	86.6	14.9		63.6	36.6
TRIPT	85.7	25.9	**	51.8	18.9		52.3	11.5
ORN	30.7	26.4		17.0	5.2	*	1582.7	922.8
LYS	235.8	181.2		555.8	500.0	*	2907.7	851.7
HIST	704.0	695.1		833.8	436.8		714.9	116.6
ARG	91.4	59.5		59.9	4.8	*	3068.3	1615.6
Total	5162.3	242.8		5709.7	396.1		13055.8	3137.4

          mean ± s.d.   mean ± s.d.   mean ± s.d.

*   p 0.001
**   p 0.005

According to Meloni and Canary the association could be
an inherited predisposition to two distinct metabolic disorders.   However the absence of uric acid disorders in
our patient's families and similar plasma and urinary
aminoacid patterns let us suppose that this association
could be caused by : a) common inborn tubular defect(5)
b) common aminoacid pool change responsible for hyperuricemia.

REFERENCES

1- DENT C.E.,ROSE G.A.: Quart.J.Med. 20.205.1951
2- MILNE M.D.,ASAATOR A.M.,EDWARDS K.D.G.,LOUGHRIDGE L.:
   Gut 2.323.1961
3- MONDINO A.,BONGIOVANNI G.,FUMERO S.,ROSSI L.:J.Chro-
   matogr. 74.255.1972
4- MELONI R.C.,CANARY J.J.:J.A.M.A. 200.257.1967
5- MARKETOS S.,MOUNTAKALAKIS T.,HALAZONITIS N.,MERIKAS G.
   "Urolithiasis Research" Ed. H.FLEISCH,W.G.ROBERTSON,
   L.H.SMITH and W.VAHLESIECK. Plenum Press New York

MONOSODIUM URATE MONOHYDRATE AS SPHERULITES

Justus J. Fiechtner and Peter A. Simkin

Division of Rheumatology, Department of Medicine

University of Washington, Seattle, Washington

The demonstration of urate crystals in synovial fluid and within synovial leukocytes is recognized as the most definitive criterion for the diagnosis of gout (1). The characteristic crystals have been described by scientists from Van Leeuwenhoek in the 1600's to the present as being "needle-like" in shape. These crystals have been further described by their anisotropic quality of birefringence detectable by using the polarizing microscope. Addition of a first order red quartz conpensator to the optic system helps identify monosodium urate monohydrate (MSUM) by its ability to alter the light's wavelength and change the transmitted color from red to yellow in the gamma plane of the compensator (2). MSUM is thus, by definition, negatively birefringent. Crystals which are blue when parallel to the gamma plane are defined as positively birefringent.

Gouty tophi are found in many tissues of patients with the chronic disease and consist of multicentric deposits of urate crystals and intercrystalline matrix together with the foreign body granuloma. The crystals have been identified as MSUM by X-ray crystallography (3). They are arranged radially into small compact clusters which, when sectioned, resemble a spherical structure (4).

Physical laws governing crystallization determine the form assumed by a chemical substance. The conditions existent in the solution, e.g., temperature, viscosity, pressure, concentration, space, all combine to produce a particular crystalline shape. The most commonly recognized form of MSUM has been acicular ("needle"). However, it is known that divergence of needles in growth may lead

141

to round, spherically symmetrical, radiating crystal aggregates (spherulites). This growth form is particularly favored in media with high viscosity(5). The aggregate of crystals producing the spherulite is rational crystallographically, i.e., a "polarization cross" is produced under the microscope. Thus, the arms of the cross which are parallel to the gamma plane of the first order red quartz compensator would be yellow in a negatively birefringent crystal, and blue when perpendicular.

Our study of this phenomenon was stimulated by the discovery of circular "beachball" objects, 5-10µ in size, in the synovial fluid of a patient with a gouty clinical syndrome and classical "needles" present in the examined fluid. These round objects displayed the same negative birefringent quality of MSUM when rotated with a first order red quartz compensator under the microscope. After seeing these objects in five similar patients' synovial fluids (occasionally within leukocytes) and in two of six stored and frozen synovial fluids of former patients with gout, we decided to try to identify the forms. Exposure to uricase resulted in digestion of these negatively birefringent objects, while no effect was observed upon similar, but positively birefringent, round objects also observed in some synovial fluids.

We then undertook to synthesize MSUM using a previously described method (6) of dissolving 4 grams of uric acid in 800 ml of distilled water containing 25 ml of 1 N NaOH. The solution was heated to $60^{\circ}C$, pH adjusted to 8.9, and allowed to cool at $23^{\circ}C$ and at $4^{\circ}C$. MSUM crystals precipitated out of both solutions, but at $23^{\circ}C$ an acicular form was predominant and,at $4^{\circ}C$, a spherulite form was predominant. Both types of crystals were strongly birefringent (See Figure 1) and displayed characteristics of negative birefringence with addition of the red plate compensator to the polarizing microscope. The structure of the spherulites was confirmed by scanning electron microscopy. The surface of the spherulites had many needle-like projections attached (Figure 2).

The identity of the spherulites was confirmed using X-ray diffraction, weight of hydration, and sodium analysis. They were found to be identical to the acicular form of MSUM.

In summary, we have observed MSUM crystals _in vivo_ and _in vitro_ as spherulites. The recognition of this form of the crystal should bear the same diagnostic and therapeutic implications as recognition of the classical needles in clinical gout.

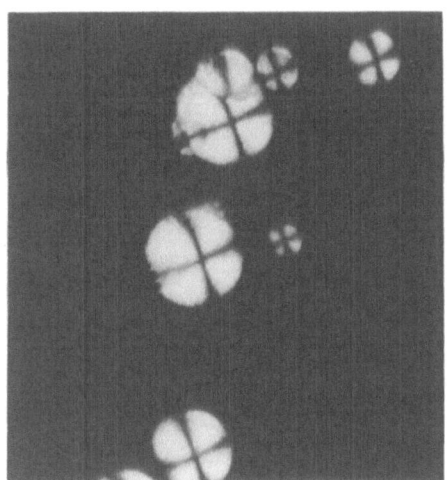 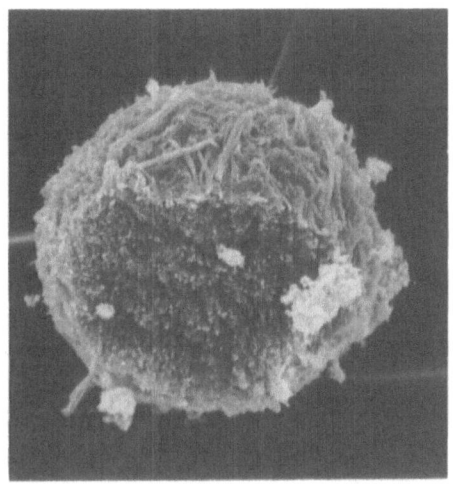

Figure 1. Spherulites of MSUM crystallized at 4°C demonstrating birefringence (polorazing microscope X 400).

Figure 2. Photograph of MSUM in spherulite form (scanning electron microscope X 1500).

REFERENCES

1. *Wallace SL, Robinson H, Masi AT, Decker JL, McCarty DJ, Yü T: Arthritis Rheum 20: 895, 1977.*

2. *McCarty DJ: Arthritis and Allied Conditions, Ninth Edition, pp 60-62, 1979 (Editor: McCarty DJ).*

3. *Howell RR, Eanes ED, Seegmiller JE: Arthritis Rheum 6: 97, 1963.*

4. *Sokoloff L: Metabolism 6: 230, 1957.*

5. *Spry A: Metamorphic Textures, First Edition, pp 152-155, 1969.*

6. *Denko CW, Whitehouse MW: J Rheumatol 3: 54, 1976.*

TUMOURAL HYPOURICEMIA

A. Lesmes, M. Díaz-Curiel, J.M. Castrillo

Unidad Metabolica, Fundación Jimenez Díaz
Universidad Autonoma de Madrid
Av. de los Reyes Catolicos, 2
Madrid -3

INTRODUCTION

A few decades ago, hypouricemia was an unusual laboratory finding, and it usually indicated that the patient was taking drugs that would induce this irregularity by different mechanisms,[1]. Only on a few occasions did it lead to the discovery of cases of xanthinuria,[2].

Further publications,[3,4], and the widespread use of autoanalyzers, which permitted thorough biochemical screenings of large sectors of the population have made it possible to know the incidence of hypouricemia and to lengthen the list of causes.

In a previous study,[5], our group found 126 determinations out of 14,865 in which the level of uric serum acid was below 3mg/100ml. These 126 determinations correspond to 114 patients, 16 (14%) of which had a tumour.

Up to the moment, however, we ignore the incidence of hypouricemia on a cancer affected population, as well as its underlying mechanisms. These questions have made us undertake this study.

MATERIALS AND METHODS

We took at random 15,000 clinical records, corresponding to 1975, from our hospital's main file, proceeding from the medical, surgical, gynecology and pediatrics departments, not including clinical records from the psychiatric department.

We studied the results of the autoanalyzer SMA 12/60 tests, performed after an overnight fast and without any medication. We analyzed the clinical and pathological diagnosis and separated those belonging to patients that showed,pathologically, a neoplasia and whose plasmatic uric acid was below 3 mg/100 ml.,[5].

Uric acid was determined using a phosphotungstate test, and creatine was determined using Jaffe's picric acid, adapted to an autoanalyzer Techicon SMA 12/60.

We studied the fractional excretion of uric acid (Cu/Ccr) on 13 tumoural patients from serum for uric and creatinine and a 24 hour urine sample, en a free diet.

The pyrazinamide test was performed according to Steel's method, [6]. The statistical study was based on a T-test for proportional difference.

RESULTS

Out of 15,000 records, we found 1,093 patients with a tumour, which represents 7.2% of the hospital's population. Of these patients, 58.7% (642 cases) were malignant neoplasias, and 41.3% (541 cases) were benign.

In 46 (4.2%) of the 1,093 tumours, we discovered on a basal situation, that the level of serum uric acid was below 3 mg/100 ml.

When we compared this proportion with that of the general hospital population,published in our previous study,[5], we found a value of $t=6.822$, $p < 0.001$, statistically significant.

5.7% (37 cases) of the 642 malignant tumours had hypouricemia, compared with only 1.9% (9 cases) of the 541 benign tumours, $t = 3.2739$, $0.01 > p > 0.001$.

The average level of serum uric acid in the general tumour hypouricemia population was 2.28 ± 0.41 mg/ml. The average values and the standard deviation in this parameter were 2.26 ± 0.44 mg/ml and 2.57 ± 0.21 mg/ml for the group of malignant and benign tumours, respectively. The comparison of these values gave $t = -1.0867$, $p < 0.05$, nonsignificant.

In 13 cases, the average value for fractional uric acid excretion was 0.276 ± 0.114 (SD). All cases had glomerular filtration above 70 ml/min of endogenous creatinine clearance.

Finally, in one male patient with a poorly differentiated lung carcinoma and in one female patient with a peritoneal carcinomatosis

of ovaric origin, we performed a pyrazinamide test, and compared
the basal values of fractional uric acid excretion and its elimination in urine, in relation to the creatinine (Uu/Ucr), before and
after administration of the drug. The results are shown in table 1.

DISCUSSION

Our results confirm that tumours are a main cause of hypouricemia (14%), and also that this biochemical disorder is more frequent among the tumour patients than among the general hospital
population.

This incidence is also significantly higher among the malignant
than the benign tumours, although both groups show no significant
difference in the levels of plasma uric acid.

Based on these results and on those mentioned previously in
the literature,[5,7,8], the mechanism of hypouricemia seems to be a
renal tubular loss.

In two cases studied which showed a tubular secretion blockage
of uric acid through the administration of pyrazinamide, the normal
renal tubular loss was restored. Therefore the original disorder is
due to an increased tubular secretion of uric acid in tumours.

The cause of this pathogenical mechanism is till unknown, but
the hypothesis of an increased production of pyrimidines in the
tumour as a possible cause of hyperuricemia,[9, 10], is attractive.

Nevertheless, further studies are needed until we can isolate
the substance which produces tumoural hypouricemia, discovering the
cause of its production in only a small number of cases.

Table 1. Pyrazinamide Test in Tumoural Hypouricemia

	Cu/Ccr		%(controls 83-96%)
Case no.	Basal	Pyrazinamide	
1	0.160	0.030	81
2	0.155	0.012	91.7

	Uu/Ucr		%(controls 83-96%)
Case no.	Basal	Pyrazinamide	
1	0.778	0.139	82
2	0.792	0.072	90.0

REFERENCES

1. D.S. Young, D.W. Thomas, R.B. Friedman and L.C. Pestaner, Effects of drugs on clinical laboratory tests, Clin. Chem., 18:1041, 1972
2. A. Rapado, H.J. Castro-Mendoza and L. Cifuentes-Delatte, Genetics of xanthinuria, in "Urinary Calculi", ed. L. Cifuentes-Delatte, A. Rapado and A.S. Hodgkinson, Karger, Basel, 1973.
3. C.M. Ramsdell and W.N. Kelly, Clinical significance of hypouricemia, Ann. Int. Med., 78:239, 1973.
4. D. Lawee, Uric acid: The clinical application of 1,000 unsolicited determinations, Canad. Med. Ass. J. 100:838, 1969.
5. M. Diaz-Curiel, A. Zea-Mendoza, A. Rapado and J. González-Villasante, Significación clínica de la hipouricemia en 14.865 determinaciones del Autoanalizador, Rev. Clin. Esp., 138:365, 1975.
6. T.H. Steele, Urate secretion in man: The pyrazinamide suppresion test, Ann. Int. Med. 79:734, 1973.
7. J.S. Bennet, J. Bond, I. Singer and A.J. Gottlieb, Med. 76:751, 1972.
8. B. Weinstein, F. Irreverre and D.M. Watkin, Lung carcinoma, hypouricemia and Aminoaciduria, Am. J. Med. 39:520, 1965.
9. H.J. Fallou, E. Frei, J. Block and J. Seegmiller, The uricosuric and orotic aciduria induced by 6-azauridine, J. Clin. Invest. 40:1. 906, 1961.
10. J.B. Wyngaarden and W.N. Kelly, Gout and hypouricemia, Grune of Stratton, New York, pg. 411-418, 1976.

HEREDITARY RENAL HYPOURICEMIA WITH HYPERURICOSURIA AND VARIABLY

ABSORPTIVE HYPERCALCIURIA AND UROLITHIASIS - A NEW SYNDROME

Oded Sperling and Andre de Vries

Tel-Aviv University Medical School, Dept. of Clinical
Biochemistry and Rogoff-Wellcome Medical Research
Institute, Beilinson Medical Center and Teva Pharmaceu-
tical Industries, Petah-Tikva, Israel

Genetically determined renal hypouricemia is a relatively rare
condition, usually appearing in association with other tubular
defects, such as in the Fanconi (1) and Hartnup syndromes (2) and
Wilson's disease (3). It is only recently that hereditary renal
hypouricemia due to an isolated tubular defect has been described
in man, firts by Greene et al. in 1972 (4). This condition was
described by Praetorius and Kirk (5) already in 1950, but in their
patient no evidence was adduced for genetic transmission. In 1974
we reported a new hereditary syndrome which included hypouricemia,
hypercalciuria and decreased bone density (6). More recently we
encountered an additional four families with renal hypouricemia
in whom the tubular abnormality pertained to urate handling only
(7-9). Until today, to the best of our knowledge, only 8 cases of
hereditary renal hypouricemia due to an isolated tubular defect
have been reported (4,6-11). Five out of these cases (6-9) were
detected in Israel and studied in our laboratory. In reviewing
the reported cases, it became evident that this hereditary isolated
renal hypouricemia is often associated with additional phenomena,
namely hyperuricosuria,hyperabsorptive hypercalciuria and urolithiasis. Thus, this condition appears to represent a new syndrome.

    With the use of the pyrazinamide test (12) three types of
renal hypouricemia have been distinguished according to the
localization of the defect in urate reabsorption - pre-secretory,
post-secretory and combined defective pre-secretory and post-
secretory reabsorption. There is no evidence that renal hypouricemia is caused by increased tubular urate secretion. According to
the present knowledge on the renal urate handling (13) and the
action of pyrazinamide (12,14,15) and probenecid (13,16-18)

(normally probenecid markedly increases urate excretion, while
pyrazinamide almost totally suppresses it),the three types of renal
hypouricemia would be expected to conform to the following responses:
in presecretory tubular urate reabsorption defect an attenuated
response to both pyrazinamide and probenecid; in post-secretory
defect a normal response to pyrazinamide (in bringing the urinary
urate excretion to a level similar to that reached in pyrazinamide-
treated normal subjects) and no response to probenecid; in combined
pre-secretory and post-secretory defect the urate clearance will
exceed glomerular filtration rate (C_{ur}/GFR > 1) and probenecid will
have no effect while the pyrazinamde response will be attenuated,
reducing the C_{ur}/GFR ratio to about 1. It should be stressed,
however, that all these parameters can be expected to be correct
only when these various reabsorption defects are complete. In
accordance with a C_{ur}/GFR ratio <1 and the observed responses to
probenecid and pyrazinamide, a defect in pre-secretory tubular
reabsorption is suggestive in the subjects studied by Greene et
al. (4), Sperling et al.(6), Benjamin et al. (7,8) and Frank et al.
(9), a total of six subjects. A combined pre-and post-secretory
tubular defect for urate reabsorption is suggestive in the two
propositi studied by Khachadurian (10) and Akaoka (11). It should
be mentioned that three additional subjects have been reported
(5,19,20) who should be classified with the combined reabsorption
defect on the basis of the above criteria. However, for none of
these subjects familiality of the condition was demonstrated.

It is an interesting observation that among the propositi with
documented inborn isolated renal hypouricemia in none evidence was
obtained for a post-secretory tubular reabsorption defect. In the
subject with renal hypouricemia compatible with this type of defect,
reported by Gibson et al. (21), evidence of familiality was not
brought forward. Furthermore, this subject had hyperparathyroidism
and thus the possibility of an acquired tubular defect has not been
excluded. A post-secretory urate reabsorption defect probably
existed also in two additional reported patients in whom the
condition may have been acquired; one patient had Wilson's disease
(22) in whom the renal tubules could have been damaged by copper
accumulation, and one had Hodgkin's disease (23) in whom acquired
renal hypouricemia is known to occur. It is possible that this
type of renal hypouricemia may be found to be characteristic for
acquired hypouricemia, whether associated with disease or drug-
induced. Thus, it may well be, that the inborn renal isolated
hypouricemia will turn out to be of only two types, pre-secretory
tubular reabsorption defect and total tubular reabsorption defect.

The hypouricemia as such has, as far as is known, no clinical
significance. Indeed, speculatively, the hypouricemia might be
advantageous in avoiding the risks associated with hyperuricemia,
mainly the various clinical manifestations of urate crystal-

deposition disease, and possibly other non-gouty conditions, hypo-
thesized by some to be promoted by increased body urate. On the
other hand, four out of the eight propositi with inborn isolated
renal hypouricemia (4,7,9) had urinary tract stones, two uric acid
stones (9), one a calcium oxalate stone (4) and one a stone of
unidentified composition (7).Hyperuricosuria, although moderate,
was present in all eight hypouricemic propositi, thus appearing
to be a constant feature of isolated renal hypouricemia. That not
all the hypouricemic hyperuricosuric patients had evidence of
urolithiasis is not surprising since hyperuricosuria is only one
of the determinants in the causation of uric acid lithiasis, the
others being low urine volume and low urinary pH (24).

   The mechanisms of the hyperuricosuria in these subjects with
renal hypouricemia has not been clarified. Principally, in these
subjects, hyperuricosuria might reflect purine overproduction or
diversion of intestinal urate elimination (25) to urinary urate
excretion consequent to the hypouricemia. There is no evidence in
these hypouricemic-hyperuricosuric subjects for purine overproduction.
Akaoka (11) measured ^{15}N-glycine incorporation into urinary uric
acid in one such patient and found it moderately excessive. However,
this excessive incorporation may as well be explained by decreased
intestinal urate disposal. Furthermore, Kawabe (20) foune [15]N-
glycine incorporation to be normal in his hypouricemic uric acid
stone forming patient.

   Hypercalciuria was found to be another associate of the renal
hypouricemia in four of the propositi, one described by Greene et al.
(4), one by Sperling et al. (6) and two reported by Frank et al. (9).
In all these subjects, in whom there was no detectable etiology for
the hypercalciuria, it may be classified as "idiopathic hypercalciuria". In three of these (4,9) the hypercalciuria was proven
to be of the hyperabsorptive type e.g. secondary to increased
intestinal calcium absorption. Thus far in none of these hypouri-
cemic-idiopathic hypercalciuric patients evidence has been obtained
for a primary abnormality in renal calcium handling. Thus, in these
subjects, at this stage of knowledge, the renal tubular defect
leading to hypouricemia may be considered an isolated tubular abnormality until proven otherwise. The apparently frequent association
between the renal defect for urate handling and the intestinal calcium
hyperabsorption is as yet unexplained. It is noteworthy that also
the patient with renal hypouricemia, described by Gibson et al. (21)
had hypercalciuria. In this patient, however, the hypercalciuria
may have been due to hyperparathyroidism, a condition known to affect
renal tubular function. The decreased bone density found by Sperling
et al. (6) to be associated with the hypercalciuria in two renal
hypouricemic siblings has not been observed in reported hypouricemic
families, and remains unexplained.

   In six of the families both males and females were affected, compatible with autosomal inheritance, whereas in the other two families this mode of transmission has not been excluded. In the family described by Sperling et al. (6), the autosomal inheritance was proven to be recessive, while in the families described by Greene et al. (4) and Frank et al. (9) autosomal recessive inheritance was compatible with this distribution. The results obtained with the pyrazinamide and probenecid tests in the various renal hypouricemic families reviewed indicate the heterogeneity of the genetic abnormality in the tubular defect in renal urate handling. Furthermore, they may be interpreted to indicate that renal urate handling in man is controlled by more than one gene.

REFERENCES.
1.   Wallis LA, Engle RI: Am.J. Med. 22:13-23, 1957.
2.   Baron DN, Dent CE, Harris, H., Hard EW & Jepson JB:
     Lancet 271:421-428, 1956.
3.   Bishop, C., Zindahl, WT & Talbott JH: Proc. Soc. Exp. Biol.
     Med. 86: 440-441, 1954.
4.   Greene ML, Marcus, R., Aurbach GD, Kazman ES & Seegmiller JE:
     Am. J. Med. 53: 361-370, 1972.
5.   Praetorius, E. & Kirk, JE: J. Lab. Clin. Med. 35:865-868,1950.
6.   Sperling, O., Weinberger, A., Oliver, I., Liberman U.A. &
     de Vries A: Ann. Intern. Med. 80: 482-487, 1974.
7.   Benjamin, D., Sperling, O., Weinberger, A., Pinkhas, J &
     de Vries, A:Nephron 18: 220-225, 1977.
8.   Benjamin, D., Sperling, O., Weinberger, A. & Pinkhas, J:
     Biomedicine 29: 54-56, 1978.
9.   Frank, M., Many, M., Sperling, O: Brit. J. Urol, in press.
10. Khachadurian, AK & Arslanian, MJ: Ann. Intern. Med. 78:
     547-550,1972.
11. Akaoka, I., Nishizawa T, Yano E, Takeuchi A, Nishida, Y,
     Yoshimura T, & Horiuchi Y: Ann. Clin. Res. 7: 316-524, 1975.
12. Steele TH, Rieselbach RE: Am. J. Med. 43: 868-875, 1967.
13. Rieselbach RE: Adv. Exp. Me. Biol. 76B, 1-22, 1974.
14. Ellard GA, Haslan KM: Tubercle 57: 97-103, 1976.
15. Yu TF, Berger L, Stone DJ, Wolf J, Gutman AB: Proc. Loc.Exp.
     Biol. Med. 96: 264-267, 1957.
16. Diamond HS and Meisel AD: 42 Annual Meeting, Am. Rheum. Ass.
     Abstract 39, p48, 1978.
17. Gutman AB, Yu TF: Ann. J. Med. 23: 600-622, 1957.
18. Sirota JH, Yu TF, Gutman AB: J. Clin. Invest. 31: 692-700,1952.
19. Simkin PA, Skeith DA, Healy LA: Adv. Exp. Med. Biol. 41B:
     723-737, 1974.
20. Kawabe K, Murayama, T, & Akaoka I: J. Urol. 116: 690-692, 1976.
21. Gibson T, Sims HP & Jimenez SA: Ann. Rheum. Dis. 35: 372-376,
     1976.
22. Wilson DM & Goldstein NP: Kidney Intern. 4: 331-335, 1973.
23. Bennett JS, Bond J, Singer I & Gottlieb A: Ann. Intern. Med.
     76: 751-756, 1972.

24. De Vries, A & Sperling 0:   Ciba Foundation Symp. 48 (New series) Elsevier, 1977, pp 179-195.
25. Sorensen, L.B.   Scand. J. Clin. Lab. Invest. 12, 1-43, 1960.

HEREDITARY AND ENVIRONMENTAL FACTORS INFLUENCING ON THE SERUM URIC ACID THROUGHOUT TEN YEARS POPULATION STUDY IN JAPAN

K. Nishioka, M.D.
K. Mikanagi, M.D.

Rheumatology Division, Dept. of Medicine
School of Medicine, Mie University
Medical Center, Thu City, Mie Pref, Japan

Rheumatology Division, Dept. of Medicine
School of Medicine, Jichi Medical School
Tochigi, Japan

Recently, gout has rapidly been increasing in Japan ,[1]. The epidemiological study has been carried out from 1970 to clarify the incidence of gout, environmental and genetic factors influencing on the serum uric acid level on the 3,300 residents that are living in a small island in Japan ,[2-4].

The characteristics of this population study (Toshi-study), mentioned in this paper are the following.
1. Sex and age distribution of the serum uric acid.
2. The relation of gout to hyperuricemia and the risk of developing gouty arthritis for each serum uric acid level.
3. Correlation of the serum uric acid and environmental factors and other laboratory findings.
4. The nature and extent of familial aggregation of the serum uric acid

MATERIAL AND METHOD

Serum uric acid level; Serum uric acid (SUA) was determined by uricase-catalase method, [4], and its values were grouped into classes with 0.5 mg per 100 ml. The items calculated correlation co-efficients between SUA were summarized in Fig. 1.

Correlation analysis of SUA, creatinine and urea-N within

155

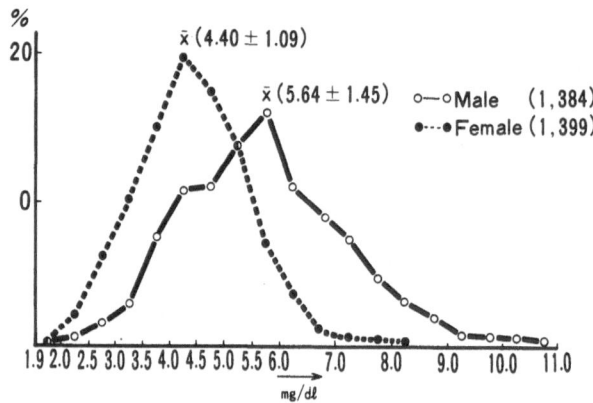

Fig. 1. Distribution of the Serum uric acid.

families were carried out and families were selected and classified
as follows:
Group -1: 29 families with data from "complete" family. (Father,
          mother, first son and first daughter).
Group -2: 70 families (Father, mother and first son).
Group -3: 44 families (Father, mother and first daughter)

Sex and Age Distribution of SUA

The mean, SUA level was 4.40 ± 1.09 mg/dl for females and 5.64
± 1.45 mg/dl for males showed in Fig. 1. The sex and age distribution of the SUA was shown in Fig. 2. The great degree of increase
was seen in the younger males in the range of 10 to 19 years; thereafter, no consistent pattern of difference is to be seen. These
distribution pattern were seemed to the data techmusel study,[5].

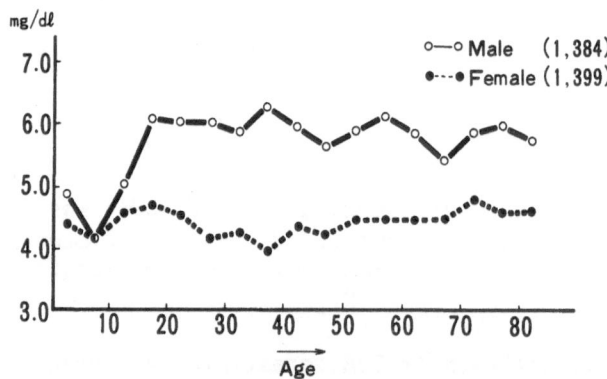

Fig. 2. Sex and age distribution of serum uric acid (Toshi-study).

Table 1. The Prevalence of Gout and Relative Risk.

PREVALANCE OF GOUT IN TOTAL POPULATION		10/2,783	0.4%
RELATIVE RISK	IN MALE	10/1,384	0.7%
	IN ADULT MALE	10/832	1.2%
	IN MALE HYPERURICEMA GROUP	10/67	14.9%
	IN HYPERURICEMIA GROUP ABOVE 10.0MG/DL	10/13	76.9%

Among the female subjects, there was considerable increase in the age above 65 years old, however there was very little change in the age range.

The Prevalence and Incidence of Gout

The prevalence of hyperuricemia is 4.8% in males and 3.4% in females. The 10 patients with definite gouty arthritis were observed throughout this population study series. The prevalence of gout was 0.4% all of the people in this area, however, their relative risk was increased as is shown in Table 1. The prevalence of gout by maximum uric acid category is shown in Fig. 3. Almost all of the hyperuricemia above 10.0 mg/dl had a typical history of gouty attack (76.9%). These data were almost consistent with Framingham population study,[6].

Correlation Analysis on Serum Uric Acid and Other Data

Correlation analysis serum uric acid and various data were

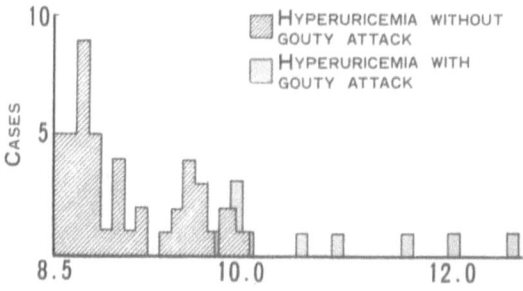

Fig. 3. The prevalence of gouty by maximum uric acid category.

illustrated in Fig. 3. In 689 male subjects, serum creatinine, body weight, height, age, urea-N, blood pressure and alcohol intake were strongly associated with serum uric acid, respectively.

Familiar Study

SUA of 85 fathers were 6.04 ± 1.23 mg/dl, 62 mothers were 4.02 $\pm$ 1.09 mg/dl, 70 subjects of first son were 5.43 ± 1.66 mg/dl and first daughter were 4.60 ± 0.97 mg/dl. Correlation coefficient among each parent and child were summarized in Table 2, classified with each group. These data showed strongly suggested the most consistent relationship found between mothers and their children. These data were consisted with the report by Brunner et al.,[6]. These data also suggested that mothers' serum uric acid level was strongly influencing their children.

SUMMARY

1. The age and sex distribution of the SUA of the Japanes people was the same as of the Caucasians.
2. The prevalence of gout was 0.4% of all the subjects, in adult male it was 1.2% and 76.4% in hyperuricemia above 10.0 mg/dl.
3. The incidence of a year throughout this population study was 0.02% per population.
4. Serum creatinine, body weight, height, age, urea-N, blood pressure and alcohol intake were strongly associated with SUA respectively.
5. Correlation analysis of SUA in familiar members showed the strong correlation between mother and son.

Table 2. Correlation Coefficient Among Each Parent and Child.

	All data	Group 1 (29)	Group 2 (70)	Group 3 (44)
Fathers-mothers	0.074 (156)	0.109	0.114	0.033
Fathers-First sons	0.191* (127)	-0.036	0.032	
Fathers-First daughters	0.164 (87)	0.170		0.139
Mothers-First sons	0.344*** (124)	0.344***	0.369**	
Mothers-First daughters	0.273* (70)	0.382*		0.269

$* \cdots p \leqq 0.05. \quad ** \cdots p \approx 0.01. \quad *** \cdots p \leqq 0.001$

REFERENCE

1. K. Nishioka, K. Mikanagi: Clinical features of 4,000 gouty subjects in Japan. In press.
2. K. Nishioka, K. Mikanagi, M. Kitamura, et al.: Epidemology of
   gout and hyperuricemia in Japan. Rheum., 14: 95-105, 1974 (Japan).
3. K. Nishioka, K. Mikanagi, K. Hironse: Epidemological survey of
   gout and hyperuricemia in Japan. Program Abst., IIIrd Congress
   of SEAPAL, Singapore, 1976, P134.
4. W.M. Mikkelson, H.J. Dodge, H. Valkenburg, et al.: The distribution of serum uric acid values in population unselected as to
   gout and hyperuricemia. Am. J. Med., 39: 242-251, 1965.
5. A.P. Hall, P.E. Barry, T.R. Dawker, et al.: Epidemology of gout
   and hyperuricemia. Am. J. Med., 42: 27-37, 1967.
6. D.Brunner, S. Altma, L. Posner, et al.: Hereditary, Envioment,
   Serum Lipoproteins and Serun Uric Acid. J. Chron. Dis., 23:
   763-773, 1971.

THE NATURAL HISTORY OF URATE OVERPRODUCTION IN SICKLE CELL ANEMIA

Herbert S. Diamond, Allen D. Meisel, and Dorothy Holden

State University of New York Downstate Medical Center
450 Clarkson Avenue, Box #42, Brooklyn, New York 11203

Supported by grants from the General Clinical Research
Center Program of the NIH and by grants from NIH and the
Arthritis Foundation, and from the Irma T. Hirschl
Career Scientist Award (HSD)

In patients who present in adult life with gout, renal stone
or hyperuricemia, urate overproduction is thought to have been
present for many years prior to clinical recognition. These patients are asymptomatic until well into adult life. Thus, there
have been few studies of urate overproduction during its asymptomatic or pre-clinical phase.

In sickle cell anemia, increased red cell turnover is known to
result in uric acid overproduction (1-5) beginning in childhood.
Yet, gout is an uncommon complication of sickle cell disease and
has rarely been noted prior to age 30 (6). Sickle cell disease
patients can readily be identified early in life. To determine
the natural history of serum uric acid and uric acid excretion in
sickle cell disease, ninety-five patients with sickle cell anemia
ranging in age from 17 months to 45 years were studied.

RESULTS

Serum Uric Acid
     Mean serum uric acid was 4.9 ± 0.2 mg/dl before age 10
(Table 1). Hyperuricemia, defined as a serum uric acid greater
than 6.5 mg/dl was present in 4 of 28 children (14%). Hyperuricemia
was frequent in adults with sickle cell disease, being found in 26
of 67 patients studied (39%). Both the mean serum uric acid and
the prevalence of hyperuricemia increased with age to a peak of

6.5 ± 0.3 mg/dl and a frequency of hyperuricemia of 45% in patients
between ages 21-30.

TABLE 1. PREVALENCE OF HYPERURICEMIA IN SICKLE CELL PATIENTS BY AGE

AGE RANGE	NUMBER OF SUBJECTS	PREVALENCE OF HYPERURICEMIA	MEAN PLASMA URATE (MG/DL)
1 - 10	28	14%	4.9 ± 0.2
11 - 20	34	38	6.2 ± 0.3
21 - 30	20	45	6.5 ± 0.5
31 - 40	8	38	5.9 ± 0.5
41 - 50	5	20	5.3 ± 1.4
TOTAL	95	32	

Urinary Uric Acid

Twenty-four hour urinary uric acid excretion was measured after
3 to 5 days on a purine-free diet in 36 adult patients with sickle
cell anemia, ranging in age from 16 to 48 years (Table 2). Mean
urinary uric acid was 574 ± 38.0 mg per 24 hours and exceeded 600
mg per 24 hours in 20 patients (56%). Uric acid excretion greater
than 600 mg per 24 hours was more frequent in younger patients,
occurring in 68% of patients between ages 16 and 25 compared to 36%
of patients over age 25. Uric acid excretion per 24 hours exceeded
600 mg per 24 hours in 67% of patients with a plasma urate below
6.5 mg/dl, but was increased in only 40% of hyperuricemic patients.

Urate Clearance

Urate clearance was inversely correlated with serum uric acid
in both adults and children with sickle cell disease ($r = -0.67$,
$p < 0.001$), suggesting that urate clearance, rather than urate
production was the major determinant of serum uric acid in sickle
cell disease. Mean urate clearance was 5.0 ± 0.7 ml/min in the
hyperuricemic adults with sickle cell disease compared to 13.6 ±
0.7 ml/min in the normouricemic patients ($p < 0.001$).

TABLE 2. URIC ACID EXCRETION 600 MG PER 24 HOURS IN SICKLE CELL
ANEMIA

	NUMBER OF PATIENTS STUDIED	PATIENTS WITH URIC ACID EXCRETION > 600 MG/24 HRS.	
		Number	%
All subjects	36	20	56
Patients: Age <25	22	15	68
Patients: Age >25	14	5	36
Normouricemic Patients	21	14	67
Hyperuricemic Patients	15	6	40

In thirteen adult patients with sickle cell disease and serum urate greater than 6.5 mg/dl, uric acid excretion per 24 hours on a purine free diet was similar to normal controls, and was significantly less than in normouricemic patients with sickle cell disease (Table 3). Urate clearance was 5.0 ml/min in the hyperuricemic patients, significantly less than in both normal controls (8.3 ± 0.6 ml/min; p < 0.001), and in normouricemic patients with sickle cell anemia (13.6 ml/min; p < 0.001). In the normouricemic sickle cell patients, mean urinary acid excretion per 24 hours on a purine free diet was 735 ± 58 mg, and exceeded 600 mg per 24 hours in 10 of 12 patients.

Pyrazinamide suppressible urate clearance was decreased to 4.9 ± 1.2 ml/min in the hyperuricemic patients as compared to 11.8 ± 0.6 ml/min in normouricemic sickle cell patients (p < 0.001) and 7.3 ± 0.5 ml/min in control subjects (p < 0.001) (Table 3). PAH clearance was also decreased to 499 ml/min in the hyperuricemic patients compared to 636 ml/min in the controls. This is consistent with decreased secretion of both urate and PAH in these patients.

TABLE 3. RENAL FUNCTION AND URIC ACID EXCRETION IN HYPERURICEMIC PATIENTS WITH SICKLE CELL ANEMIA COMPARED TO NORMAL SUBJECTS

	HYPERURICEMIC SICKLE CELL PATIENTS	NORMAL SUBJECTS	P
Number of Subjects	13	12	
Plasma Urate (mg/dl)	8.6 ± 0.5	5.3 ± 0.3	0.001
24 Hr. Urinary Uric Acid Excretion (mg/min)	480 ± 45	400 ± 48	N.S.
Uric Acid Clearance (ml/min)	5.0 ± 0.7	8.3 ± 0.6	0.001
GFR (ml/min)	132 ± 15	108 ± 4.5	N.S.
Pyrazinamide Suppressible Urate Clearance (ml/min)	4.9 ± 1.2	7.3 ± 0.5	0.001
Pyrazinamide Non-suppressible Urate Excretion (μg/min)	41.7 ± 8.5	53 ± 6.6	N.S.
PAH Clearance (ml/min)	499 ± 100	634 ± 42	N.S.

Uricosuric response to probenecid would be expected to be intact or enhanced in patients with increased reabsorption of secreted urate and diminished in patients with impaired urate secretion. Peak uricosuric response to probenecid was 18.8 ml/min in the hyperuricemic patients with sickle cell disease. In contrast, the uricosuric response to probenecid was 45.3 ml/min in normouricemic sickle cell patients, and 37.6 ml/min in controls.

DISCUSSION

Although urate overproduction in sickle cell disease begins
in childhood, patients with sickle cell anemia often remain normouricemic into adult life (Figure 1). Normouricemia is maintained
by increased urate secretion and urate clearance. Hyperuricemia
occurs only in those patients who develop diminished urate clearance.
Diminished urate clearance in sickle cell disease is probably
secondary to altered renal tubular function resulting in diminished
urate secretion. The delayed onset of hyperuricemia probably accounts for the low frequency of gout in young adults with urate
overproduction secondary to sickle cell disease.

Moreover, Emerson and Roe (7) have suggested a similar scheme
for the natural history of hyperuricemia in patients with urate
overproduction due to inherited defects in purine metabolism. Thus,
in patients with metabolic abnormalities which cause only moderate
increases in urate overproduction, hyperuricemia and gout may be
limited to those with intrinsic or acquired defects in renal excretion of urate.

STAGE I   SICKLE CELL ANEMIA
                       ⬇
                 INCREASED RED BLOOD CELL TURNOVER
                       ⬇
                 URIC ACID OVERPRODUCTION
                       ⬇
                 HYPERURICOSURIA WITH HIGH URATE
                    CLEARANCE
                       ⬇
                 NORMAL PLASMA URATE

STAGE II   FALL IN URATE CLEARANCE
                       ⬇
                 ASYMPTOMATIC HYPERURICEMIA

STAGE III   SUSTAINED HYPERURICEMIA
                       ⬇
                 CLINICAL GOUT

FIGURE 1.   PROPOSED NATURAL HISTORY OF URATE OVERPRODUCTION IN
            SICKLE CELL ANEMIA

REFERENCES
1. W. J. Crosby, The metabolism of hemoglobin and bile pigment in
   hemolytic disease, Am. J. Med. 18:112 (1955).
2. M. S. Gold, J. C. Williams, M. Spevack, and V. Grann, Sickle
   cell anemia and hyperuricemia, JAMA. 206:1572 (1968).
3. B. R. Walker and F. Alexander, Uric acid excretion in sickle
   cell anemia, JAMA. 215:255 (1971).

4. H. Diamond, E. Sharon, D. Holden, and A. Cacatian, Renal handling
    of uric acid in sickle cell anemia, Adv. Exp. Med. Biol. 41B:759
    (1974).

5. H. Diamond, A. Meisel, D.Holden, E. Sharon, A. Cacatian, and
    R. Virdi, Hyperuricemia in sickle cell anemia, in: "Proceeding,
    First National Symposium on Sickle Cell Anemia," J. Hercules,
    A. N. Schecter, W. A. Eaton, and R. E. Jackson, eds., DHEW
    Publication No. 75-723, Bethesda, Maryland, p. 371 (1974).

6. L. R. Espinoza, I. Spilberg, and C. K. Osterland, Joint mani-
    festations of sickle cell disease, Medicine (Baltimore). 53:295
    (1974).

7. B. T. Emerson and P. G. Roe, An evaluation of the pathogenesis
    of the gouty kidney, Kidney International. 8:65 (1975).

SALVAGE PATHWAY IN ERYTHROCYTES OF PATIENTS WITH PSORIASIS

Gerald Partsch[1], Felix Mayer[2], Rudolf Eberl[1],
Anton Luger[2]

1) Ludwig Boltzmann-Institute of Rheumatology and Bal-
neology, Vienna. 2) Dermatological and Venerological
Dept. of the Hospital of the City of Vienna-Lainz.

INTRODUCTION

Weber[1] found elevated glucose-6-phosphate dehydrogenase activities in serum of patients suffering from psoriasis. In psoriatic
skin lesions this enzyme activity was also enhanced (Weber[2]).
Herdenstam[3] reported that a great part of glucose is metabolized
via the pentose phosphate cycle. Ribose-5-phosphate is the source
for 5-phosphoribosyl-1-pyrophosphate (PRPP) which is of great importance for the purine salvage pathway. In this paper the incorporation of adenine, guanine and hypoxanthine via the salvage
pathway into intact erythrocytes of psoriatic patients was studied
in contrast to healthy people.

MATERIAL AND METHOD

lo ml heparinized blood (Vacuplast vials containing Na-hepa-
rin, Messrs. Greiner) was centrifuged (2.ooo x g/5 minutes). The
plasma and the buffy coat was aspirated and the remaining erythrocytes 2 times washed either with physiological saline or with
Hank's solution. 5oo μl of the sedimented erythrocytes were taken
for estimation of protein (Lowry[4]). 1 ml of the erythrocytes was
used for labelling experiments with ^{14}C-purine bases (adenine,
guanine and hypoxanthine). The cells were incubated with 1 μCi of
the purine bases for 1 hour at 37° C and then well mixed with 2 ml
6% perchloric acid (PCA). After centrifugation and decantation of
the supernatant the sediment was washed with 2 ml PCA. The collected supernatants were neutralized with 2 N KOH against neutral-
red in an ice bath. The solution containing nucleotides was mixed
with prepared charcoal to bind nucleotides. The charcoal was

167

washed 3 times with 15 ml aqua dest. After extraction of the
nucleotides with 4 ml 5o % ethyl alcohol - 2 % concentrated ammonia
the solution was brought to dryness. The remaining substances were
taken up with 5o μl aqua dest. 5 μl of the solution were used for
chromatography. Two different solutions were applied for development of nucleotides:

1. saturated ammonium sulphate : 1 M sodium citrate :
   isopropanol (8o : 18 : 2)
2. n-butanol : aceton : acetic acid : 5 % ammonia : H_2O
   (45 : 15 : 1o : 1o : 2o)

For chromatography cellulose F254 (Merck Co.) was used. The radioactive spots were scanned and after cutting out burned in a sample
oxidizer (Packard B 3o6). Radioactivity was measured in a liquid
scintillation counter (Packard 265o).

RESULTS AND DISCUSSION

In Table 1 the results of the incorporation of adenine, guanine and hypoxanthine into the red cells of psoriatic patients
expressed as nucleotides are shown. Two incubation media were used
for all experiments: physiological saline and Hank's solution.
When adenine was taken for incorporation experiments in psoriatic
erythrocytes no effect of the incubation medium was differentiable.
A limiting factor for discussion of the results is that the standard deviation of all the average values is relatively high so it
is difficult to find out if the incubation medium has any effect
on nucleotide formation. Three blood samples of psoriatic patients
were analyzed in Hank's solution as well as saline. These samples

Table 1. Total radioactivity of ^{14}C-purine base incorporation expressed as nucleotides into
erythrocytes of patients with psoriasis
and controls (DPM/mg protein)

Adenine	N	Hank's Sol.	N	NaCl
Psoriasis	8	1o36,4 + 49o	7	1o29,8 + 554
Control	5	7o4,o $\pm$ 267	5	594,4 $\pm$ 239
Guanine				
Psoriasis	8	751,5 + 326	7	5o7,3 + 2o5
Control	5	584,6 $\pm$ 11o	5	6o1,o $\pm$ 263
Hypoxanthine				
Psoriasis	8	663,5 + 441	7	485,6 + 295
Control	5	6oo,4 $\pm$ 294	5	596,2 $\pm$ 469

showed a distinct enhancement of the adenine incorporation and
interconversion to nucleotides when cells were incubated in Hank's
solution.

In contrast to psoriatic cells controls showed a slightly enhanced incorporation of adenine when Hank's solution was used as
medium. Hershko[5] showed that the uptake of purine bases by incorporation into ribonucleotides was markedly stimulated by increased
P_i in erythrocytes of rabbits. This was supposed as a stimulating
effect on 5-phosphoribosyl-1-pyrophosphate (PRPP) content in the
presence of glucose. But P_i concentration in Hank's solution is
not as high that a stimulating effect may occur. Increased pentose
phosphate cycle in psoriatics (Herdenstam[4]) possibly raises the
PRPP content in erythrocytes. There could not be found any evidence
for this suggestion in the literature. Generally the incorporation
of adenine was higher in psoriatic cells than in erythrocytes of
healthy people without regard to the incubation medium.

The total ^{14}C-guanine incorporation in psoriatic erythrocytes
was lower than with adenine. The average value of 751 DPM/mg protein is approximately two third of that of adenine. The incubation
medium possibly influenced the incorporation of guanine into psoriatic erythrocytes but did not affect nucleotide formation in
controls. The values obtained after incorporation of hypoxanthine
were similar to those with guanine. Radioactivity was slightly
lower in psoriasis with saline. In the control erythrocytes incorporation of hypoxanthine was the same between saline and Hank's
solution.

The thin layer chromatography of the extracted nucleotides

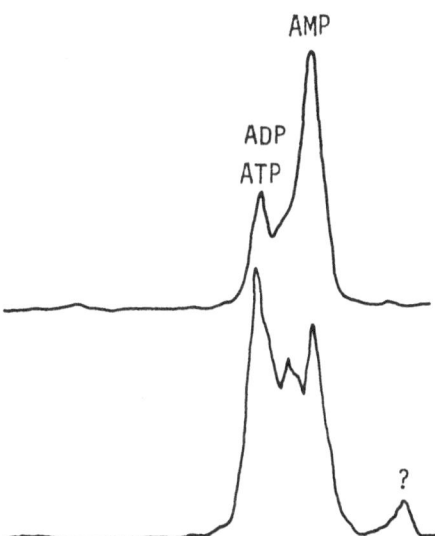

AMP

ADP
ATP

?

Picture 1.   Scanning profiles of synthesized nucleotides after
             incorporation of ^{14}C-adenine in erythrocytes of
             psoriatic patients.

showed very non-uniform results. Only those peaks were cut and
measured in the liquid scintillation counter which were detected
after scanning of the chromatography. Picture 1 shows two scans of
nucleotide preparations which indicate that in psoriatic patients
as well as controls some nucleotides could not be detected following incorporation of purine bases under some conditions. Using
saline as incubation medium, adenine was found in AMP, ADP and ATP
of all psoriatic erythrocytes and in four controls. In one control
sample only a very low radioactivity was detectable in the ATP peak.
In psoriatic erythrocytes incubated in Hank's solution only ATP was
detectable. The absence of AMP in the last cited patient is unclear
because a very high radioactivity in the ADP and ATP peak was measurable. In the controls AMP and ADP were absent in one sample but
in three others no ADP was detectable. Because ATP was measurable
in all erythrocytes the average values were calculated. The DPM
formed ATP/mg protein in psoriatic erythrocytes are distinctly
increased in psoriatic erythrocytes without regard to the incubation medium. ATP production is 100% increased in psoriatic erythrocytes when intact cells were incubated in Hank's solution. In
the controls the difference in ATP synthesis between both incubation media is much higher. The interpretation is difficult because
the biological variance may be of importance. Guanine was converted to GMP and the nucleoside di- and tri-phosphate only in erythrocytes of one psoriatic patient. Other guanine nucleotides were
measurable only in rare cases. In all samples with one exception
hypoxanthine was detectable.

After incorporation of ^{14}C-hypoxanthine no measurable amounts
of IMP were found. Erythrocytes are unable to convert IMP to IDP
and ITP. Radioactivity (ATP) was found only in 5 out of 8 psoriatic
erythrocytes incubated in Hank's solution and in all control
samples. 3 out of 8 psoriatic erythrocytes incubated in saline
showed radioactivity in the ATP peak. There was no radioactivity
in the control samples. This result is unclear because hypoxanthine
is converted to IMP by the hypoxanthine-phosphorylase.

Dean[6] reported that in physiological buffer systems the major
metabolites in intact erythrocytes are ribosides of hypoxanthine
and guanine but IMP and GMP were also found. A possible explanation for the inability to find nucleosides after incorporation of
guanine and hypoxanthine in this system may depend on the incubation procedure.

Summarizing we assume that in psoriatic erythrocytes the
salvage pathway of adenine is enhanced in contrast to controls.
Guanine and hypoxanthine are incorporated to a lower extent than
adenine into psoriatic erythrocytes. In control samples the incorporation of all the three bases was nearly equal. The interconversion of nucleotides varies in each erythrocyte sample.

REFERENCES

1. G. Weber, Über das Vorkommen der Glucose-6-Phosphat-Dehydrogenase im Blutserum von Psoriasis vulgaris-Kranken, Arch. klin. exp. Derm. 215:6o3 (1963).
2. G. Weber and G. W. Korting, Glucose-6-Phosphate Dehydrogenase in Human Skin, J. invest. Derm. 42:167 (1964).
3. C. G. Herdenstam, On the in vitro metabolism of labelled glucose in normal and psoriatic skin slices, Acta derm.-venerol. 42:47 (1962).
4. O. H. Lowry, N. J. Rosenbrough, A. L. Farr and R. J. Randall, Protein measurement with the Folin phenol reagent, J. biol. Chem. 193:265 (1951).
5. A. Hershko, A. Razin, T. Shoshani and J. Mager, Turnover of Purine Nucleotides in Rabbit Erythrocytes. II. Studies in vitro, Biochim. Biophys. Acta 149:59 (1967).
6. B. M. Dean, H. A. Simmonds and A. Cadenhead, A Comparative study of Purine Metabolism by Human and Pig Erythrocytes in vitro, Biochem. Pharm. 22:3189 (1973).

SERUM 5-NUCLEOTIDASE IN PROGRESSIVE MUSCULAR DYSTROPHY

Frieder Làhoda and Klaus Baier

Professor in Neurology; Research Ass.
Neurological Clinic, University of Munich

The information from the literature and personal observations cited hitherto indicate that a connection
between purine metabolism and progressive muscular
dystrophies may possibly exist. This is particularly
valid for the metabolid end-product of purine, uric
acid, which is frequently found to be increased in
the blood-serum in the form of a symptomatic hyperuricaemia. The breakdown of nucleic acids or polynucleotides begins with their depolymerization to
mononucleotides, which are subsequently split up into
their components. The resulting purine bases, adenine
and xanthine, are oxidized via the oxypurines hypoxanthine and xanthine to uric acid, and in this last
stage the enzyme xanthine oxidase plays an important
part. As the observations of CHALMERS et al. showed,
the congenitab absence of xanthine oxidase may, after
a long period, lead to a myopathy. Conversely, the
inhibition of the enzyme with allopurinol in cases of
muscular dystrophy with raised uric acid levels leads
to a temporary and short-lived subjective improvement.

Finally, the increased breakdown of the energy-rich
adenine nucleotides through the dephosphorylating
activity of the proliferating interstitial connective
tissue, the therapeutic administration of nucleoside-
nucleotide mixtures and the abserved activity of
5-nucleotidase in the dystrophic muscle tissue indicate that quite early stages in purine metabolism may
play a part in the pathophysiology of progressive
muscular dystrophy.

The studies of BECKETT an BOURNE in 6 patients of
various ages with progressive muscular dystrophy
revealed increased activities of 5-nucleotidase in
the affected muscle. 5-nucleotidase hydrolyses ade-
nylic acid, inosinic acid and uridylic acid and uridylic acid to the corresponding nucleosides and also
attacks desoxyribonucleosides.

In 1959 BECKETT, BOURNE and GOLARZ observed in the
proliferating endo- and peromysium in progressive
muscular dystrophy histochemically an increased
5-nucleotidase activity compared with normal human
muscle tissue.

Results and discussion:

The 5-nucleotidase in the blood serum was estimated
in 47 patients with progressive muscular dystrophy
by means of the method by PERSIJN. The control group
consisted of 21 volunteers in the age of 17 to 61
vears, who did not suffer from liver or renal diseases. Apart from 3 only slightly increased values the
5-nucleotidase values were within the normal range
in patients with muscular dystrophy.(Precision of the
method: mean coefficient of variation of 5,4 with the
extreme values of 0 - 13,7)

The results correspond with those obtained by

CHOWDHURY in 1962. It has still to be clarified, why
the 5-nucleotidase is not increased in the serum,
although it is fourfold increased in the muscle protein corresponding to the observations by BOURNE.

Abstract:

The 5-nucleotidase in the blood serum was estimated
in 47 patients with progressive muscular dystrophy
by means of the method by PERSIJN. The control group
consisted of 21 volunteers in the age of 17 to 61
years. Apart from 3 only slightly increased values
the 5-nucleotidase values were within the normal range in patients with muscular dystrophy. The results
correspond with those obtained by CHOWDHURY in 1962.

References:

1. BECKETT E.B. and   5'Nucleotidase in normal and
   BOURNE G.H.:   diseased human skeletal mus-
                           cle
                           Journal of Neuropathology
                           and experimental Neurology
                           17 : 199 (1958)

2. BOURNE G.H. and   Human muscular dystrophy as
   GOLARZ M.N.   an aberration of the connec-
                           tive tissue.
                           Nature London. 183, 1741 -
                           1743 (1959)

3. CHALMERS R.A.,   Xanthinuria with Myopathy.
   JOHSON M.,   Quart. J. Of Med., N.S.
   PALLIS C. and   XXXVIII No. 152, 493-512
   WATTS R.W.E.:   (1969)

4. CHOWDHURY S.R.,   Serum enzyme studies in mus-
   PEARSON C.M. and   cular dystrophy III. Serum
   FOWLER W.W.:   malic Dehydrogenase. 5!Nuc-
                           leotidase and Adenosine-
                           triphosphatase
                           Proceedings of the Society
                           for Experimental
                           Biology and Medecine (New
                           York) 109 : 227 (1962)

5. LAHODA F.:   Der erhöhte Harnsäurespie-
                                gel bei der Dystrophia mus-
                                culorum progressiva
                                ERB. Mediz. Monatsschrift
                                26, 12, 558 - 561 (1972)

6. PERSIJN J.P.,   A new method for the Deter-
   VAN DER SLIK W.,   mination of Serum Nucleoti-
   TIMMER C.J. and   dase, IV. Evaluation of the
   REIJNTJES C.M.:   conditions for assays using
                                adenosine deaminase.
                                Zeitschrift für klinische
                                Chemie und klinische Bio-
                                chemie, 8. Jg.: 398 - 402
                                (1970)

PURINE METABOLISM IN DUCHENNE MUSCULAR DYSTROPHY

C.H.M.M. de Bruyn[*], S. Kulakowski[o], C.A. van Bennekom[*],
P. Renoirte[o] and M.M. Müller[**]
[*]Department of Human Genetics, Faculty of Medicine,
University of Nijmegen, Nijmegen, The Netherlands.
[o]Institute "Les Petites Abeilles", Vlezenbeek-Brussels,
Belgium.
[**]II. Department of Medicine, University Hospital Vienna,
Austria.

INTRODUCTION

Duchenne muscular dystrophy (DMD) is the most widely known
of the muscular dystrophies. It is inherited as an X-linked
recessive trait (1). It becomes manifest only in males and it is
transmitted by asymptomatic females, although there are exceptions
(2). The primary manifestation is progressive muscle weakness, but
there are more organs that are progressively affected, e.g. heart
and brain. The primary gene defect is unknown but there is increasing
evidence to suggest an abnormal sarcolemmal membrane (1). Leakage
of ATP from muscles as found in the case of DMD patients (3) might
lead to some defect in muscle purine metabolism. It was speculated
that irreversible loss of purines, caused by the breakdown of ATP by
enzymes in the blood, might be reduced by blocking the last enzymatic
step in purine catabolism in man and by concommitant enhancement of
purine reutilisation. This last step involves the xanthine oxidase
reaction and allopurinol is widely used as an inhibitor of this
reaction in the treatment of hyperuricemia. DMD patients receiving
allopurinol were reported to improve clinically (4,5,6).

Impairment of adenine metabolism in erythrocytes from DMD
patients was reported by Solomons et al. (7). These workers found a
decreased incorporation of radioactive adenine into nucleotides with
DMD erythrocytes. It was suggested that impaired membrane transport
of adenine might account for these findings.

177

In the present study the effect of allopurinol was investigated during a period of 6 months in 14 DMD patients. In contrast to previous studies (4,5,6), where muscle function was measured manually, in the present study more objective quantitative dynamometric measurements with several muscles were done. In addition, some aspects of purine metabolism in erythrocytes from these patients were studied.

MATERIALS AND METHODS

Clinical Studies

The effect of allopurinol was investigated in 14 DMD patients (13 males, 1 female) in different clinical stages (I, II and III) of the disease. These stages had not been considered in previous studies (4,5). Clinical data of the patients are presented elsewhere (8). Before administration of the drug in all patients quantitative dynamometric measurements were done with three muscles (M. Quadriceps, M. Biceps and M. Adductor), using a commercially available dynamometer (Chatillon). The method we used was an adaptation by Honoré and Lesenne (unpublished) of the method described by Hannard (9). Care was taken to perform all measurements under identical conditions in all patients. The usual readaptation program was continued throughout this study. All measurements were carried out in three-fold before starting the oral administration of allopurinol (10 mg Zyloric/kg body weight/24 hours) and after six months.

Biochemical Studies

Routine biochemical parameters, such as serum CPK and serum uric acid, were tested before starting the administration of allopurinol, after 2, 4 and 6 months and 4 months after terminating the allopurinol administration.

Purine metabolism in intact erythrocytes was studied using 8-^{14}C-adenine, 8-^{14}C-adenosine or 8-^{14}C-hypoxanthine (spec. act. 54-59 mCi/mmol; Radiochemical Centre Amersham) at concentrations of 30 µM, 60 µM and 30 µM, respectively; incubation at 37 ^{o}C for 15 min. The methods for isolation and identification of intra- and extracellular labeled purine metabolites have been described previously (10).

Four enzymes involved in purine nucleotide synthesis were assayed in erythrocyte lysates: hypoxanthine phosphoribosyltransferase (HPRT) adenine phosphoribosyltransferase (APRT), adenosine kinase (AK) and adenylate kinase (Aden. K.). The radiochemical assays for these activity determinations are described elsewhere (11,12,13, and 8, respectively).

Adenine transport across the erythrocyte membrane was studied using the method described by Müller and Falkner (14).

RESULTS AND DISCUSSION

    Comparison of the dynamometric measurements before administration of allopurinol and after 6 months of administration showed several effects of the drug. Concerning M. Quadriceps, 5 out of the 6 patients in stage I and II showed either improvement or stabilisation of muscular force (table 1). Although 5 out of 8 patients in stage III showed stabilisation, this might not have clinical significance, since most of the muscular force had already been lost (8). For stage I the findings with M. Biceps and M. Adductor were comparable with those seen in the case of M. Quadriceps. In stages II and III the effects were less pronounced. This is in agreement with the characteristic sequence of muscle detioration in DMD.

    In contrast to other reports (5,6) no clear clinical signs of improvement were noticed in the present study. During the first two months of administration of allopurinol serum CPK in all patients decreased. The earlier the stages, the more pronounced was the decrease. After two months, however, a gradual adaptation became apparent.

    Uric acid levels in patients from stage I and II were in the normal range before administration of allopurinol, whereas in stage III the levels were always increased as compared to normal. Under allopurinol the uric acid levels in these patients normalised as expected.

Table 1. Dynamometric Evaluation of Muscular Force after 6 Months Oral Administration of Allopurinol (10 mg/kg body weight/ 24 hr).

	Stage I (n=4)			Stage II (n=2)			Stage III (n=8)		
	IM[o]	ST[o]	NE[o]	IM	ST	NE	IM	ST	NE
Quadriceps	2	1	1	1	1	-	2	3	3
Biceps	2	1	1	-	2	-	1	6	-[*]
Adductor	2	1	1	-	2	-	-	4	3[*]

[o]IM; improved, ST; stable, NE; no effect.

[*]One patient not available for evaluation. Start of administration: oktober 1977, termination: april 1978.

It was speculated that allopurinol might prevent purine loss and enhance purine salvage, thus creating better conditions for nucleotide synthesis. Therefore, a first attempt was made to study adenine nucleotide synthesis in easily available material such as erythrocytes.

The carrier mediated transport of ^{14}C-adenine into eryhtrocytes from DMD patients in all three stages of the disease did not differ from that in normal controls. Using a rapid sampling technique (14) the time curves for the uptake of ^{14}C-adenine by mutant and normal cells were comparable (data not shown). Kinetic studies on the transport system did not show abnormalities in the DMD erythrocytes (8). It has been suggested that in DMD erythrocytes adenine uptake was reduced; this was concluded from studies in which the incorporation of adenine following the uptake process was measured (7). Our data show that the uptake process in DMD red cells is not impaired and, therefore, the primary genetic defect is not reflected as far as adenine transport is concerned.

The incorporation of ^{14}C-adenine into ^{14}C-AMP, ^{14}C-ADP and ^{14}C-ATP by DMD and control erythrocytes was also comparable. This can be seen in table 2 where the ratios of labeled intracellular ATP and ADP are given. Administration of allopurinol did not change these ratios over a period of 6 months (table 2). Four months after terminating the administration the ATP/ADP ratio still remained the same. No differences between the clinical stages were observed. Our findings are not in agreement with the results of other workers, who reported decreased incorporation of ^{14}C-adenine into DMD erythrocytes (7). The main difference between their methods and ours was the incubation medium: in contrast to autologous plasma we used a phosphate buffered system. However, when DMD and control erythrocytes were incubated in autologous plasma we did not notice differences. The total incorporation in the case of autologous plasma was much lower as compared to the phosphate buffered system and the ratios

Table 2. ^{14}C-ATP/^{14}C-ADP Ratios in Erythrocytes Following Incubation of Intact Cells With 30 µM 8-^{14}C-adenine for 15 min. at 37°C.

	n	mean	s.d.	range
Controls	9	8.1	1.9	5.5 - 10.2
DMD Patients				
no allopurinol	9	8.2	1.5	6.6 - 11.5
2 mo. on allopurinol	13	7.2	0.9	5.9 - 9.3
4 mo. on allopurinol	12	7.8	1.5	6.1 - 10.6
6 mo. on allopurinol	13	8.5	1.5	6.1 - 12.6
4 mo. after termination	12	8.4	2.0	5.5 - 12.5

Table 3. Activities of Some Purine Nucleotide Synthetising Enzymes in Erythrocyte Lysates.

| | Enzyme activity in 10^{-9}moles/mg prot.hr. | | | |
	H-PRT	A-PRT	AK	Aden K
DMD patients (n = 15)	60.5 ± 8.9[*]	11.1 ± 1.5[*]	$7.0 + 0.9$[*]	33.7 ± 4.9[*]
Controls (n = 5)	64.5 ± 6.8	10.7 ± 2.0	6.8 ± 1.2	34.0 ± 1.5

[*] mean $\pm$ s.d.

H-PRT: hypoxanthine phosphoribosyl transferase; A-PRT: adenine phosphoribosyl transferase; AK: adenosine kinase; Aden K: adenylate kinase.

of labeled ATP/ADP were much more subject to fluctuation.

^{14}C-adenosine and ^{14}C-hypoxanthine incorporation by intact DMD and control erythrocytes did not show significant differences either (data not shown). The same was found with the specific activities of several purine nucleotide synthetising enzymes in erythrocyte lysates (table 3). No evidence for increased leakiness of nucleotides in DMD was obtained. Therefore, the erythrocyte does not seem to constitute a suitable model system as far as purine metabolism is concerned. In other cell types (e.g. cultured muscle cells and fibroblasts) the primary defect might be reflected in purine metabolism.

Although neither before nor during or after terminating allopurinol administration abnormalities were noted in the parameters of red cell purine metabolism tested, and also no clear clinical effect of allopurinol was observed, the muscular force had stabilised or even improved, but only in early stages of the disease (table 1). Stabilisation of muscular force over a period of 6 months can already be considered as a positive effect, since in untreated or non-responsive DMD patients a progressive loss of muscular force is always seen. Because the CPK levels in serum were diminished after 2 months on allopurinol it might be speculated that less CPK had been leaking from the muscles. This might implicate either an ameliorated membrane function or a decreased synthesis of CPK. After 2 months, CPK levels rose again in most cases, implicating the transitory effect of the drug.

It seems that allopurinol exerts its effects presumably on the affected muscle fibres rather than by stimulating the unaffected fibres. However, before this speculation can be based on experimental findings, neurophysiological studies should be carried out in a

more objective way, and the decrease of CPK in serum should be proven
to be related to altered nucleotide metabolism and membrane function.

REFERENCES

1. J.N. Walton, Disorders of voluntary muscle, Churchill Livingstone,
    Edinburgh and London (1974).
2. W.H.S. Thomson, J.C. Sweetin and T.E. Hilditch, Clin. Chim.
    Acta 63:353 (1975).
3. L. Stengel-Rutkowski and W. Barthelmai, Klin. Wochenschr. 51:
    957 (1973).
4. W.H.S. Thomson and I. Smith, Lancet i:805 (1976).
5. W.H.S. Thomson and I. Smith, Metabolism 27:151 (1978).
6. W.H.S. Thomson and I. Smith, New Engl. J. Med. 229:101 (1978).
7. C.C. Solomons, S.P. Ringal, E.I. Nwke and H. Suga, Nature
    (London) 268:55 (1977).
8. S. Kulakowski, P. Renoirte and C. de Bruyn, ms. in preparation.
9. C. Hannard, Rev. Assoc. Franc. Myopath. 71(suppl.):9 (1977).
10. C.H.M.M. de Bruyn and T.L. Oei, Adv. Exp. Med. Biol. 76B:139
    (1977).
11. C.H.M.M. de Bruyn, T.L. Oei and P. Hösli, Biochem. Biophys.
    Res. Commun. 68:483 (1976).
12. M.P. Uitendaal, C.H.M.M. de Bruyn, T.L. Oei and P. Hösli, Biochem.
    Genet. 16:1187 (1978).
13. F.L. Meyskens and H.E. Williams, Biochim. Biophys. Acta 240:
    170 (1971).
14. M.M. Müller and G. Falkner, Adv. Exp. Med. Biol. 76B:131 (1977).

METABOLISM OF ADENINE AND ADENOSINE IN ERYTHROCYTES OF PATIENTS WITH MYOTONIC MUSCULAR DYSTROPHY (MMD)

Mathias M. Müller, Michael Frass and
Bruno Mamoli
2nd Dept.of Medicine,Dept.of Medical Chemistry and Dept. of Neurology
University of Vienna, Austria

INTRODUCTION

Myotonic muscular dystrophy (MMD) is an inherited disease with an autosomal dominant trait with high penetrance and a variable clinical presentation. The symptoms may include myotonia, weakness and dystrophy of skeletal muscles, cardiac arrhythmias, dysphagia and abdominal pain, testicular atrophy, hyperinsulinemia in response to a glucose load, cataract and mental retardation. Although the inborn error of metabolism is unknown, it is assumed that the cellular plasma membranes in many organs are impaired. The basis of the altered membrane properties might be a decreased phosphorylation of membrane proteins by the enzyme protein kinase which was demonstrated in muscle biopsies (1) and erythrocytes of patients with MMD (2). Previous studies showed abnormal adenine metabolism of erythrocytes in MMD with a reduced adenine incorporation into ATP and ADP and an increased formation of AMP (3). The authors discussed an impaired adenine uptake responsible for the reduced formation of ATP.

To obtain more information on membrane functions and metabolic routes of purines involved in the synthesis of ATP we investigated the uptake and subsequent metabolism of adenine and adenosine of intact MMD erythrocytes.

MATERIALS AND METHODS

Our experiments were performed with 5 unrelated

183

patients with MMD and 11 control persons with no neuromuscular disease. The clinical diagnosis of MMD was
based on clinical status, EMG examinations and muscle
biopsies. 5 ml heparinized venous blood was obtained.
Immediately after blood collection the erythrocytes
were spun down for 10 minutes at 2000 rpm and washed
4-times with Krebs-Ringer solution (pH 7.4). For transport studies packed erythrocytes were resuspended in
Krebs-Ringer solution containing 16.7 mmol/l glucose.

To study the uptake process of adenine and adenosine 300 ul suspended erythrocytes were incubated at
20° C with ^{14}C-adenine (58 mCi/mmol) or ^{14}C-adenosine
(61 mCi/mmol) to give a final concentration of 30
umol/l. Incubations were performed for 7 seconds.
The separation of erythrocytes from the incubation
medium was performed by means of a rapid filtration
technique (4). Radioactivities of the incubation medium and of perchloric acid extracts of the erythrocytes were measured.

Intracellular metabolism of adenine and adenosine
was studied using 300 ul erythrocytes suspended in
2 volumes of Krebs-Ringer solution were layered
on top of a silicon oil layer (AR 70, Wacker Chemie,
Munich, GFR). Incubation at 37° C for 30 min was
started by the addition of ^{14}C-adenine or ^{14}C-adeno-
sine at concentrations of 33 umol/l and 45 umol/l
respectively. Erythrocytes were separated by centrifugation from the incubation medium as described above.

In aliquots of the perchloric acid extract the incorporation of adenine and adenosine was determined
by high voltage electrophoresis at 2000 V using a
citrate buffer (25 mmol/l, pH 3.5). Areas corresponding to adenine and adenosine AMP, ADP and ATP were
cut out and their radioactivities determined.

RESULTS AND DISCUSSION

Uptake of adenine and adenosine:
Previous studies showed a linearity of purine uptake up to 15 seconds at experimental conditions used.
The initial uptake of adenine and adenosine into erythrocytes of controls and patients are shown in table 1.
The uptake of adenosine was approximately twice that
of adenine. Purine bases and purine nucleosides are
transported into erythrocytes by different proteins
of the red cell membrane. Purine bases are taken up
against a concentration gradient (5). In erythrocytes the adenine uptake is influenced by purine
phosphoribosyltransferases (6). Adenosine transport

seems to be a carrier-mediated faciliated diffusion (7,8).

Table 1. Uptake of ^{14}C-adenine and ^{14}C-adenosine into erythrocytes

	Intracellular Adenine Label umol/l	Intracellular Adenosine Label umol/l
CONTROLS (n=7)		
x̄	27.9	48.1
S.D.	6.5	5.3
MYOTONIC DYSTROPHY (n=5)		
x̄	24.5	53.6
S.D.	9.1	9.8

It is speculated that this transport system is either connected to adenosin kinase (9) or to adenosine desaminase (10). It is generally accepted that the uptake of adenine and adenosine is rate limiting for their subsequent metabolization. Membrane alterations described in erythrocytes of MMD patients (11) could influence adenine and adenosine influx and therefore the intracellular formation of nucleotides. However, no difference in uptake of erythrocytes from controls and MMD patients were observed. This indicates, that the translocators for adenine and adenosine are not impaired in MMD erythrocytes, as it was proposed for adenine uptake (3).

The administered adenine or adenosine were converted during the incubation to adenine nucleotides (AMP,ADP, ATP). In all experiments the precursors were in excess and not rate limiting.

Erythrocytes of controls took up on an average of 24.6 ± 2.7 % of ^{14}C-adenine administered. 9.8 ± 2.5 % of the once intracellular adenine were converted to adenine nucleotides, mainly into ATP. The proportion of newly synthesized AMP : ADP : ATP was 1.00 : 3.58 : 4.22 calculated from the mean values obtained with normal erythrocytes. Adenine was translocated into erythrocytes of MMD patients at an relative amount of 24.9 ± 1.8 %. From this 28.3 ± 4.2 % were metabolized to adenine nucleotides. The observed enhanced conversion of adenine to its nucleotids was mainly based on a 3-fold increase of AMP- and ATP-formation.

Table 2. Incorporation of ^{14}C-adenine into adenine-nucleotides

	AMP	ADP	ATP
		Dpm/test	
CONTROLS (n=11)			
x̄	2.285	8.187	9.653
S.D.	1.428	2.599	3.366
MYOTONIC DYSTROPHY (n=5)			
x̄	7.994	13.9o8	3o.416
S.D.	2.124	1.9o9	8.318

In spite of this the comparison of labelled adenine nucleotides with each other reveals that the amount of high energy nucleotides (ADP, ATP) is decreased in MMD red blood cells compared to normal erythrocytes: the proportion of AMP : ADP : ATP formed was 1.00 : 1.74 : 3.80.

Using normal erythrocytes incorporation experiments with ^{14}-C-adenosine showed an uptake twice that of adenine during 30 min. 54 $\pm$ 6.4 % of administered adenosine were taken up by red blood cells of controls. From this an amount of 29.o $\pm$ 6.7 % was metabolized to adenine nucleotides. The highest amount of ^{14}C-label was observed in the ATP-fraction followed by ADP (AMP : ADP : ATP = 1.00 : 2.65 : 4.76).

In contrast to adenine no differences concerning the

Table 3. Incorporation of ^{14}C-adenosine into adenine-nucleotides

	AMP	ADP	ATP
		Dpm/test	
CONTROLS (n=11)			
x̄	9.739	25.814	46.37o
S.D.	5.244	5.641	11.768
MYOTONIC DYSTROPHY (n=5)			
x̄	1o.797	26.436	45.930
S.D.	1.o78	6.379	5.371

metabolism of adenosine between normal and MMD erythrocytes were observed.

Adenine is incorporated into adenine nucleotides by
the action of APRT with PRPP as cosubstrate. Adenosine
may be phosphorylated under ATP consumption by AK to
AMP or degradated by ADA and PNP to inosine and hypoxanthine which would be salvaged to IMP and metabolized
via adenylasuccinate to AMP. Under the experimental conditions used and under physiological concentrations of
intracellular adenosine the second possibility seems
rather unreasonable since ADA needs relative high adenosine concentrations for saturation in contrast to AK (12),
not being available. Furthermore no labelled IMP and
hypoxanthine were detected.

In normal erythrocytes it is obvious that adenosine
uptake and metabolism is twice that of adenine, which
corresponds well to data presented by others (13).
Under physiological conditions most adenine nucleotides
are formed from adenosine via AK reaction. In favour
of this assumption is the fact that for AMP formation
via the AK reaction onyl 1 ATP is needed, however
2 molecules ATP are necessary for the salvage of adenine.

In MMD erythrocytes a greater demand for adenine
nucleotides seems to exist inspite of normal ATP-levels.

Table 4. AT-concentration of washed erythrocytes.

	ATP mg/dl
CONTROLS (n=23)	
$\bar{x}$	23.9
S.D.	4.7
MYOTONIC DYSTROPHY (n=5)	
$\bar{x}$	24.3
S.D.	5.2

This greater demand is documented by an increased incorporation of adenine into adenine nucleotides. The
increased formation of adenine nucleotides from adenine and the normal concentration of ATP indicate an
enhanced turnover of adenine nucleotides. This could be
the consequence of increased synthesis of membrane
phospholipids (14) or of increased consumption for ATP

for stabilizing the cell membrane (15). The data presented do not favour the assumption of a primary defect
of purine metabolism.

References:

1. Roses A.D., Appel S.H. (1974), Nature 250, 245-247.

2. Roses A.D., Appel S.H. (1973), Proc.Soc.Acad.Sci.
   USA (7o), 1155 - 1159.

3. Solomons C.C., Ringel S.P., Nwuke E.I., Suga H.
   (1977), Nature 268, 55-56.

4. Klingenberg M., Pfaff E. (1967) Methods of Enzymology 10, 680 - 684.

5. Müller M.M., Kraupp M., Falkner G., De Bruyn C.H.M.M.
   (1978) Monogr. hum. Genet., 10, 116 - 121.

6. Müller M.M., Kraupp M., De Bruyn C.H.M.M., (1979)
   Human Heredity 29, 118 - 123.

7. Schrader J., Berne R.M., Rubio R., Amer. (1972)
   J. Physiol.

8. Kraupp M., Chiba P., Müller M.M., (1979), J.Clin.
   Chem.Clin.Biochem., 17, 172

9. Shimizu H., Tanaka S., Kodama T., (1972)J. Neurochem. 19, 687 - 698

1o.Agarwal R.P., Parks, R.E., (1975) Biochem.Pharmacol.
   24, 547 - 550.

11.Plishker G.A., Gittelman H.J., Appel S.H., (1978)
   Science 200, 323 - 325

12.Meyskens F., Williams H.E., (1971) BBA 2340, 170 - 179

13.Lerner M.H.,Rubinstein D.,(197o) BBA 224, 301 - 310

14.Kornberg A.,Priar W.E.,(1953) JBC,200, 204 - 206

15.Wilkinson J.H., Robinson J.M. (1974) Nature 249,
   663 - 664.

CLINICAL AND ENZYMOLOGICAL STUDIES IN A CHILD WITH TYPE I GLYCOGEN STORAGE DISEASE ASSOCIATED WITH PARTIAL DEFICIENCY OF HEPATIC GLUCOSE-6-PHOSPHATASE

George Nuki* and John Parker

Department of Medicine, Welsh National School of

Medicine, Cardiff, U.K.

Type I glycogen storage disease (GSD - I),von Gierke's disease, is usually characterised by hepatomegaly, failure to thrive, metabolic acidosis, hyperuricaemia and a complete absence of hepatic glucose-6-phosphatase (EC 3.1.3.9) activity (1, 2). A number of patients with similar clinical and biochemical features but with normal in vitro hepatic glucose-6-phosphatase activity have been described (3-11), and these are sometimes referred to as glycogen storage disease, type IB.

In this paper we describe a child with GSD - I in whom a modified clinical picture, notable for an absence of hypoglycaemia, was associated with significant residual hepatic glucose-6-phosphatase activity.

Severe hyperuricaemia in this patient was associated with a persistent lactic acidaemia and a reduction in uric acid clearance. Hepatic phosphoribosyl pyrophosphate (PP-ribose-P) availability for nucleotide synthesis did not appear to be increased.

CASE REPORT

The patient was the second child of a non-consanguinous marriage with healthy parents and an elder sister. Birth weight was 2.78kg. following full term normal gestation. She presented at the age of 14 weeks with a two week history of intermittent vomiting but had apparently required frequent bottle feeds since birth. She was

----

*Present Address: Rheumatic Diseases Unit, Northern General Hospital, Ferry Road, Edinburgh EH5 2DQ.

described as having a doll-like appearance with chubby cheeks and a
protuberant abdomen. The liver was enlarged 6cm. below the right
costal margin and there was some clinical evidence of muscular
hypotonia and retardation of growth.

Initial investigations showed:- Haemaglobin 11.2g/dl, white
blood count 14.3 x 10^9/l (polymorphs 4%, lymphocytes 88%, eosino-
phils 4%, basophils 4%) serum bilirubin 11.9micro.mol/l, gamma
glutamyl transpeptidase 66u/l. Alkaline phosphatase 35KA units/100ml.
Total serum proteins 86g/l, albumin 42g/l, globulins 44g/l. Plasma
sodium, potassium, bicarbonate and urea normal. Capilliary blood
gases normal (pH 7.41, pCO$_2$ 37mm/Hg standard bicarbonate 23.5mmol/l).
Plasma lactate 2.9mmol/l. No ketonaemia or ketonuria. Serum uric
acid 0.85-0.94mmol/l. Serum triglycerides 7.5mmol/l. Serum cholesterol 15.5mmol/l. Serum calcium 3.4mmol/l. Fasting plasma glucose
3.7-4.9mmol/l. There was no glucose response to glucagon or
galactose (Figs. 1 and 2).

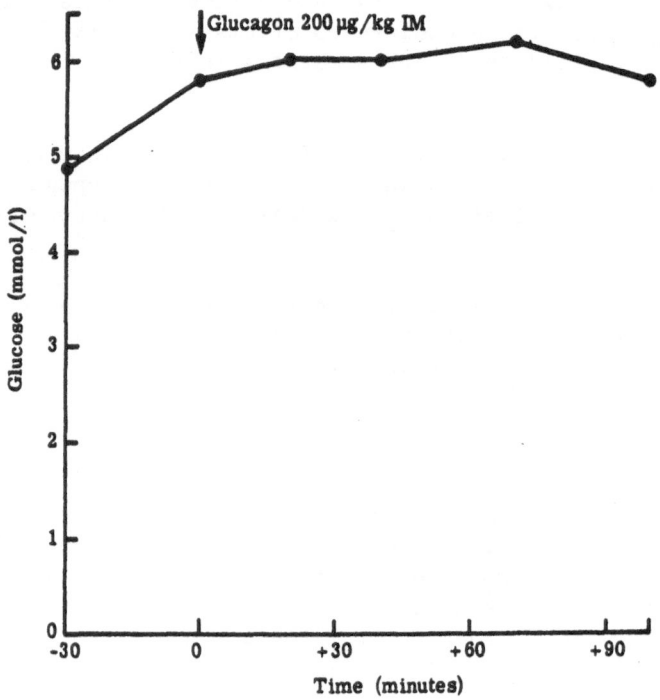

Fig. 1. Glucose response to I.M. glucagon.

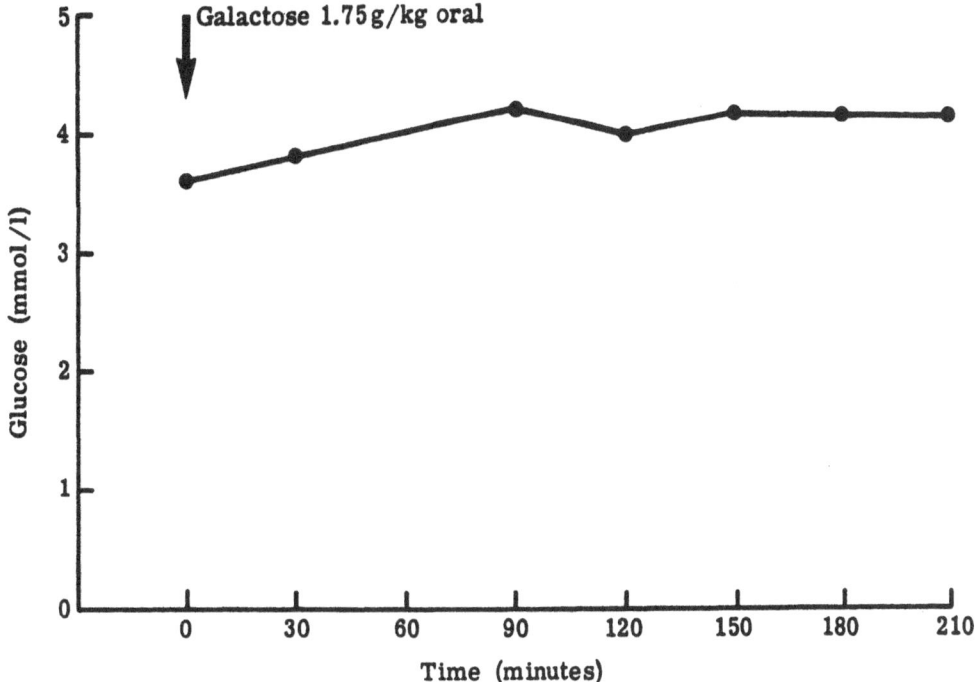

Fig. 2.   Glucose response to oral galactose

The physical appearance and hepatomegaly associated with lactic
acidosis, hyperlipidaemia, hyperuricaemia and failure to mount a
glucose response to glucagon or galactose suggested a diagnosis of
GSD - I although the complete absence of hypoglycaemia was certainly
atypical.

Liver biopsy was not initially undertaken and a diagnosis of
type III glycogen storage disease (GSD - III) was erroneously
entertained as a result of leucocyte assays for debranching enzyme
activity (amylo - 1, 6 - glucosidase EC 3.2.1.33) which failed to
show the normal glucose release from a phosphorylase limit dextrin
which was demonstrable in leucocyte extracts from both parents and
normal controls. The erroneous diagnosis of GSD - III seemed to
be further confirmed by detection of apparently normal levels of
glucose-6-phosphatase in the platelets of the child and her parents
using glucose-1-phosphate as a control substrate (13-15). Subsequent
studies using alpha and beta glycerophosphate as control substrates
and pH inactivated controls strongly suggested that platelet phosphatase activity measurable at pH 6.8 is non-specific (16).

Despite a high protein diet (17) and regular high caloric feeds
by day and night the child failed to grow normally and there was

gross retardation of motor development. She was first able to sit
unsupported at age 14 months and walk unaided at 26 months although
she was speaking in sentences at 18 months. Formal psychological
testing at age 25 months on the Griffiths scales showed a G.Q. of
78 and an apparent mental age of 19.5 months but analysis of per-

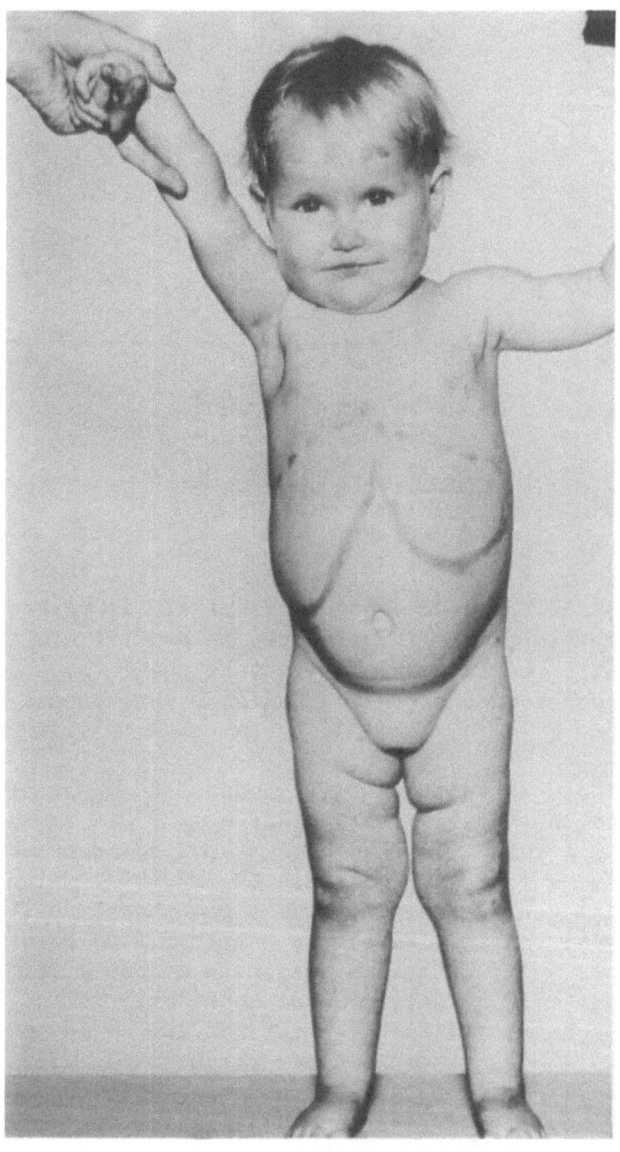

Figure 3   K.W. age 3 years showing marked hepatosplenomegaly and
           growth retardation.

formance in various skills showed her to be 14 months behind in
locomotor tasks and normal in all other areas.

She was readmitted to hospital at age 3 years following a
urinary tract infection and an episode of haematemesis and melaena.
There was marked hepatosplenomegaly (Fig. 3) and her height (74cm)
and weight 10kg) were well below the third percentile.

Investigations showed:- Haemaglobin 9.6g/1, WBC 21.5 x 10^9/1.
Bleeding and clotting screen - normal. Plasma electrolytes and
capilliary blood gas es normal. Plasma urea 8.6mmol/1, serum creatinine 60micro.mol/1. Serum uric acid 1.13mmol/1. Creatinine clearance 43ml/min/1.73m^{2}BSA. Uric acid clearance 3.2ml/min/1.73m^{2}BSA.
Urine uric acid/creatinine ratio 1.7. MSU - pus cells and coliform
organisms but no uric acid crystals detected. Serum lactate 6.4mmol/1,
serum pyruvate 285micro.mol/1. Serum triglycerides 12.7mmol/1. Serum
cholesterol 10.5mmol/1. No ketonaemia or ketonuria. Plasma glucose
after 8 hours fasting - 3.5mmol/1. Liver function tests: bilirubin
5.1micro.mol/1, gamma GT 990u/1. Alkaline phosphatase 60 KAunits/
100ml. Straight x-ray abdomen confirmed gross hepatosplenomegaly.
Intravenous pyelogram - normal. X-ray right hand and wrist - bone
age 2 years. Normal growth hormone response following oral bovril
stimulus.

Erythrocyte PP-ribose-P and hypoxanthine-guanine-phosphoribosyltransferase (HGPRT) levels were normal (Table 1) and serum uric
acid levels were normal in father (0.41mmol/1) and mother (0.33
mmol/1).

She was treated with a course of septrin (trimethoprim and
sulphamethoxazole) and oral iron and later phenytoin sodium 15mg.
t.d.s. was started following reports of successful reduction in
hepatomegaly and lactic acidaemia in patients with hepatic glycogen
storage diseases using enzyme inducing hydantoinates (19). There
was a marked decrease in liver size over the following six months

TABLE 1

ERYTHROCYTE PP-RIBOSE-P CONCENTRATIONS AND HGPRT
ACTIVITY

	PP-RIBOSE-P * (nmol/mg protein)	HGPRT (nmol IMP/hr/mg protein)
PATIENT K.W.	0.062	87.9
NORMAL ($\pm$ SD)	0.078 ($\pm$ 0.039)	106.7 ($\pm$ 12.7)

* method of Sperling et al (18).

with an improvement in liver function tests (gamma GT 144u/l,
alkaline phosphatase 32 KA units/l00ml), serum lactate (1.37mmol/l).
serum triglycerides (3.35mmol/l) and serum uric acid (0.68mmol/l).
Growth, however, continued to be markedly impaired and at age 5
years both her height (85cm) and weight (12kg) were about 5 standard deviations below the mean.

In view of the continuing uncertainty regarding the diagnosis,
and because of the potential therapeutic and prognostic implications, an open diagnostic liver biopsy was undertaken.

LIVER BIOPSY RESULTS

The major portion of the liver tissue obtained at open biopsy
was snap frozen in liquid nitrogen within 10 seconds of operative
removal; fresh slices 1mm. thick were incubated in tissue culture
medium (MEM with 15% foetal calf serum).

Glucose-6-Phosphatase was measured in the snap frozen liver
tissue using a phosphate release assay with pH inactivated controls
to correct for non specific phosphatase activity (20). Repeated
assay demonstrated that a true glucose-6-phosphatase activity was
present with 10-15% of normal catalytic activity (Table 11).
Additional evidence that this residual enzyme activity was true
glucose-6-phosphatase came from the demonstration of PPi-glucose
phosphotransferase activity (21) in the liver tissue. This activity peculiar to true glucose-6-phosphatase was present at approximately 18% of normal levels (Table 11).

Further studies (Fig. 4) showed that the apparent K_m of the
crude liver enzyme from this patient was 8.0mM (K_m normal liver
8.3mM) but that the V_{max} was depressed to 1.06micro.mol inorganic
Pi/min/g. wet wt. (normal = 6.7).

The time course for inactivation of the crude liver enzyme
at pH 5 and 37°C was similar to that observed in normal liver
(Fig. 5).

Further tissue enzyme assays and a glycogen determination were
undertaken on an aliquot of the snap frozen liver biopsy tissue by
Rosemary Eagle in Professor Ryman's laboratory, Charing Cross Hospital, London. In addition to confirming defective glucose-6-phos-
phatase, these demonstrated increased glycogen content and normal
debranching activity (Table III).

PP-ribose-P availability was assayed in fresh liver slices
maintained in tissue culture medium (22) and did not appear to be
increased when compared with controls (Table 1V).

TABLE 11

LIVER GLUCOSE-6-PHOSPHATASE

	GLUCOSE-6-PHOSPHATASE (micro.mol inorganic Pi/min/g.w.w.)	PPi GLUCOSE PHOSPHOTRANSFERASE (micro.mol G6P/min/g.w.w.)
PATIENT K.W.	0.28	0.80
NORMAL CONTROL	2.13	4.26
% NORMAL	13	18

G-6-P + H$_2$O $\longrightarrow$ Glucose + Pi   PPi + Glucose $\longrightarrow$ G-6-P + Pi

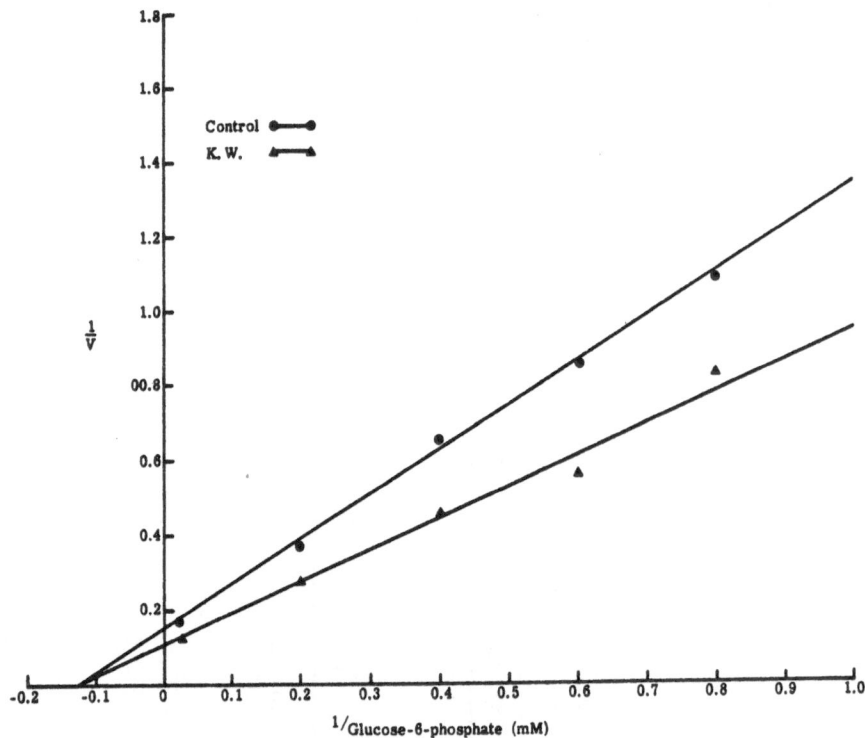

Fig. 4.   Lineweaver - Burk plots for G-6-P in unpurified liver homogenates.

TABLE 111

GLYCOGEN CONCENTRATION & OTHER ENZYME
ACTIVITIES IN THE LIVER TISSUE*

	PATIENT K.W.	NORMAL RANGE
GLYCOGEN (mg./100mg. tissue	8.42	< 4
DEBRANCHING ACTIVITY (a) (micro.mol glucose released from phosphorylase limit/dextrin/min/g.tissue)	0.59	0.2-0.6
ACID GLUCOSIDASE ACTIVITY (micro.mol maltose hydrolysed/min/g. tissue)	0.53	1-2

* R. Eagle in Professor Ryman's laboratory (20).
(a) Combination of glycosyl transferase and alpha 1, 6, glucosidase
    activity.

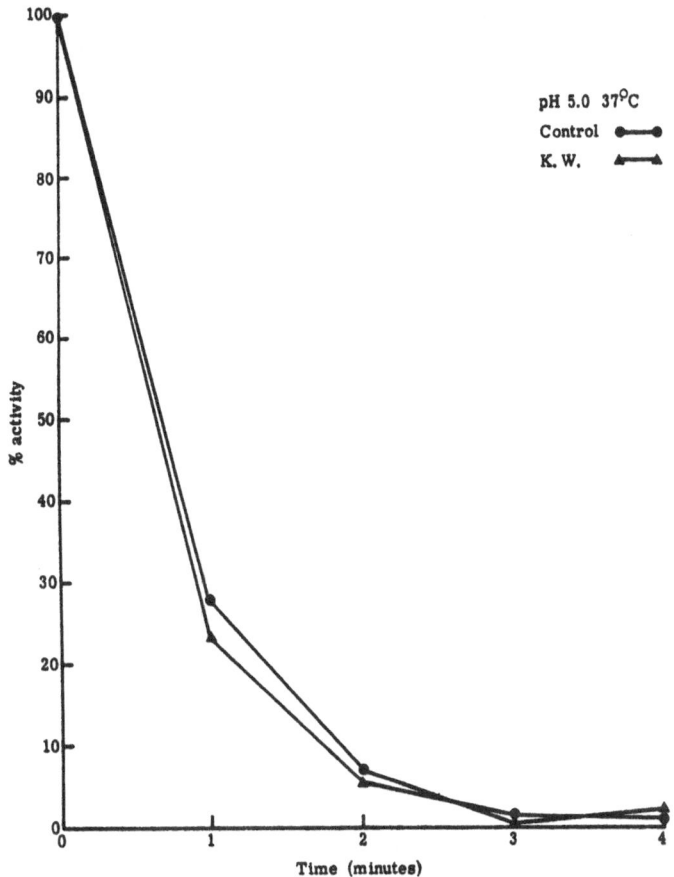

Fig. 5. pH inactivation curves for hepatic G-6-Phosphatase
in normal (100% = 2.13micro.mol inorganic Pi/min/g.
wet wt.) and K.W. (100% = 0.28micro.mol. inorganic
Pi/min/g.wet wt.)

TABLE 1V

AVAILABILITY OF PP-RIBOSE-P FOR NUCLEOTIDE
SYNTHESIS IN FRESH LIVER TISSUE

	PP-RIBOSE-P AVAILABILITY (nmol/g. w.w.)
PATIENT K.W.	54.2
NORMAL CONTROL	65.1

Direct measurement of PP-ribose-P content of snap frozen liver tissue was attempted (23) but not found possible despite freezing the liver biopsy tissue within 10 seconds of operative excision.

Patient Progress

After the establishment of a definitive tissue diagnosis of GSD - I with partial deficiency of hepatic glucose-6-phosphatase a regime of more intensive glucose feeding was commenced. The regimen, modified from Stacey et al (24), consisted of hourly supplements of 4gm. oral glucose by day and a continuous nocturnal infusion of 0.4g. glucose/Kg body weight/hour administered via a nasogastric tube with an IVAC pump. The infusion mixture contained glucose 25%, milk 25% and water 50%.

Treatment was followed by a dramatic increase in both height and weight growth velocity with a satisfactory degree of catch up growth (Fig. 6) although the lactic acidaemia (3.4mmol/l) and hyper-uricaemia (0.67mmol/l) persisted. In view of the longterm risks of

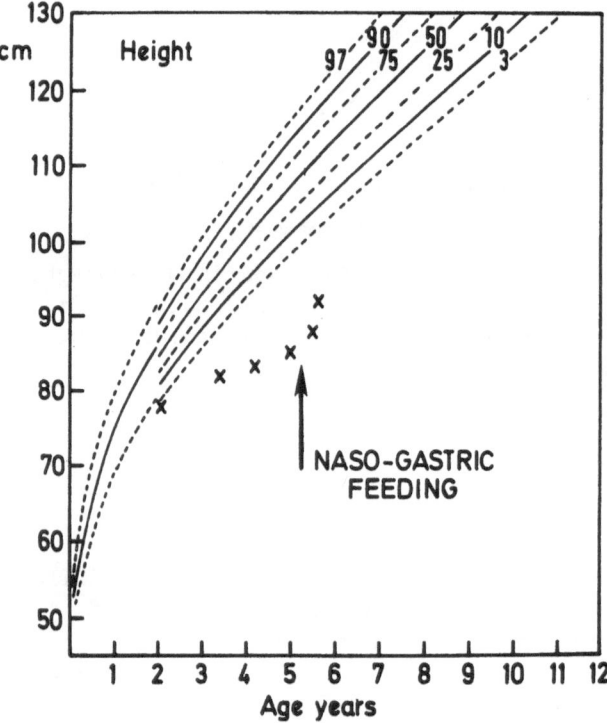

Fig. 6. Growth chart* (standing height) for K.W. showing severe retardation followed by growth spurt and catch up growth after starting intensive regime of glucose feeding.
* Chart prepared by J.M. Tanner & R.H. Whitehouse, Institute of Child Health, University of London.

gouty arthritis and renal complications in patients with GSD - I
and uncontrolled hyperuricaemia (1, 25) prophylactic treatment with
allopurinol was commenced.

DISCUSSION

The characteristic clinical and laboratory features of GSD - I
are usually associated with complete or virtual absence of hepatic
glucose-6-phosphatase (1, 2). The modified clinical presentation
in the patient here described, which was notable for an absence of
hypoglycaemia, may be attributable to the finding of significant
residual catalytic glucose-6-phosphatase activity in the liver
biopsy. Partial deficiency of hepatic glucose-6-phosphatase has
previously been described in an adult female with tophaceous gout
in whom childhood manifestations of glucose-6-phosphatase deficiency were mild and unrecognised (26). In both patients there was
evidence of hypertriglyceridaemia, lactic acidaemia and hyperuricaemia as well as an impaired glucose response to intravenous
galactose although basal plasma glucose levels were normal.

Intermediate levels of jejunal glucose-6-phosphatase have been
reported in heterozygous parents of children with GSD - I whose
biopsies showed complete enzyme deficiency (27) and similar findings
have been reported in platelets (13, 14, 15, 28). Our own studies
have not confirmed these findings and suggest that platelet phosphatase activity detectable at pH 6.8 is non-specific.

Intermediate levels of hepatic glucose-6-phosphatase have been
recorded on other rare occasions (5, 29) but the characteristics of
the presumed mutant enzyme with residual catalytic activity have not
previously been investigated.

Howell (30) has suggested that the purine over-production
observed in some patients with GSD - I (31, 32, 33) may result from
increased synthesis of PP-ribose-P as a consequence of increased
conversion of glucose-6-phosphate to 6, phosphogluconate and pentose
phosphate pathway metabolites. Although technical considerations
precluded the accurate measurement of PP-ribose-P concentration in
the snap frozen liver biopsy material in this patient PP-ribose-P
availability in fresh liver tissue maintained in culture medium did
not appear to be increased. It must be stressed however that there
was no clinical evidence of increased uric acid production or
excretion in this patient; in whom the severe hyperuricaemia could
well be mainly, or entirely, a consequence of reduced uric acid
clearance associated with the persistent lactic acidaemia.

The metabolic basis of the severe growth retardation seen in
children with GSD - I is not clear. Although hypoglycaemia itself
is sometimes implicated (24), the growth of this child with partial
hepatic glucose-6-phosphatase deficiency and a notable absence of

hypoglycaemia was markedly retarded. We have shown,as have others
(24, 34, 35), that remarkable growth acceleration can follow intensive supplemental feeding with glucose even though the lactic acid
osis is not abolished.

As might be expected the persistent lactic acidaemia is associated with persistent hyperuricaemia unless a xanthine oxidase inhibitor is administered.

SUMMARY AND CONCLUSIONS

1.   GSD - I is described in a child with partial deficiency of
      hepatic glucose-6-phosphatase.
2.   Growth retardation and hepatosplenomegaly were major clinical
      features.
3.   Hyperlipidaemia, lactic acidaemia, hyperuricaemia and reduced
      uric acid clearance were major biochemical findings.
4.   Although the glucose response to glucagon and galactose was
      impaired, there was a striking absence of hypoglycaemia which
      may be attributable to residual catalytic activity of the enayme.
5.   Preliminary studies of the crude liver enzyme showed it to
      have a normal pH inactivation profile and apparent Km with a
      reduced Vmax.
6.   No evidence of increased PP-ribose-P availability in fresh
      liver tissue was detected.
7.   Continuous glucose feeding resulted in accelerated growth with-
      out complete correction of lactic acidosis or hyperuricaemia.
8.   GSD - I with partial deficiency of hepatic glucose-6-phosph-
      atase should be considered in patients with gout or hyper-
      uricaemia associated with hypertriglyceridaemia and lactic
      acidaemia even in the absence of hypoglycaemia.

ACKNOWLEDGEMENTS

     The authors are grateful to the Department of Medical Bio-
chemistry, W.N.S.M. for help with clinical biochemical studies;
Miss R. Eagle and Professor B. Ryman for help with studies of liver
enzymes and glycogen; Dr. A. Palit and Professor O. Gray for allowing us to study their patient and Professor R. Mahler for useful
discussions. Financial support for this work came from a block
grant from the Arthritis and Rheumatism Council.

REFERENCES

1.   Howell, R.R. (1978). The Glycogen Storage Diseases In the
      Metabolic Basis of Inherited Disease, McGraw-Hill, New York,
      4th Ed. pp. 137-159. Ed: Stanbury J.B., Wyngaarden J.B. and
      Frederickson D.S.
2.   Mahler, R. (1976). Clinics in Endocr. Metab., 5, 579-598.
3.   Senior, B. and Loridon, L. (1968). New Eng. J. Med. 279,
      958-965.

4. Senior, B. and Loridon, L. (1968) New Eng. J. Med. 279, 965-970

5. Sokal, J.E., Lowe, C.V., Sarcione, E.J., Mosovich, L.L. and Doray, B.H. (1961) J. Clin. Invest. 40, 364-374

6. Briggs, J.N. and Haworth, J.C. (1964) Am. J. Med. 36, 443-449.

7. Francois, R., Hermier, M. and Ruitton-Ugliengo A. (1965), Paediatrie, 20, 37-50.

8. Spencer-Peet, J., Norman, M.E., Lake B.S., McNamara, J., and Patrick A.D. (1971). Quart. J. Med. 40, 95-114.

9. Grenet, P., Badoual, J., Lestradet, H., Marinetti, J., Sonna, N. and Voyer, J. (1972). Ann. Paediat. 19, 499-506.

10. Badoual, J., Lestradet, H., Tichet, J., Sonna, N. and Grenet, P. (1972). Ann. Paediat. 19, 507-514.

11. Moses, S.W. (1973). Adv. Exp. Med. Biol. 41A, 353-360.

12. Chalmers, R.A., Ryman, B.E. and Watts R.W.E. (1978). Acta Paediatr. Scand., 67, 201-207.

13. Soyama, K., Shimada, M., Kusonoki, T. and Nakamura, T. (1973) Clin. Chim. Acta. 44, 327-331.

14. Negishi, H., Morishita, Y., Kodama, S. and Matsuo, T. (1974) Clin. Chim. Acta, 53, 175-181.

15. Stormont, D., Davies, C., and Emmerson, B.T. (1976). Clin. Chim. Acta. 71, 303-308.

16. Parker, J.R. and Nuki G. (unpublished findings).

17. Fernandes, J., and van der Kamer (1968). Paediatrics, 41, 935-944.

18. Sperling, O., Eiam, G., Persky-Brosh, S. and De Vries, A. (1972). J. Lab. Clin. Med. 79, 1021-1026.

19. Jubiz, W. and Rallison, M.L. (1974). Arch. Int. Med. 134, 418-421.

20. Ryman, B.E. (1976) in The Principles and Practice of Diagnostic Enzymology. Ed. Wilkinson J.H., Edward Arnold, London.

21. Nordlie, R.C. and Arion, W.J. (1966). in Methods in Enzymology. Vol. 9, p. 619.

22. Boer, P., Lipstein, P., de Vries, A. and Sperling, O. (1976). Biochim. Biophys. Acta. 432, 10-17.

23. Lalanne, M. and Henderson, J.F. (1974) Anal. Biochem. 62, 121-133

24. Stacey, T.E., McNab, A. and Strang, L.B. (1979) (in press).

25. Holling, H.E. (1963) Am. Int. Med. 58, 645-663.

26. Stamm, W.E. andWebb, D.I. (1975). Arch. Int. Med. 135, 1107-1109.

27. Field, J.B., Epstein, S. and Egan J. (1965) J. Clin. Invest. 44, 1240-1247.

28. Linneweh, F., Lohr, G.W., Waller, H.D. and Gross, R. (1963). Klin. Wochenschr. 41, 352-354.

29. Cori, G.T. and Cori, C.F. (1952). J. Biochem. 199, 661-667.

30. Howell, R.R. (1965) Arthritis Rheum. 8, 780-785.

31. Alepa, F.P., Howell, R.R., Klinenberg, J.R. and Seegmiller, J.E. (1967). Am. J. Med. 42, 58-66.

32. Jacovcic, S. and Sorenson, L.B. (1967) Arthritis Rheum, 10, 129-134.

33. Kelley, W.N., Rosenbloom, F.M., Seegmiller, J.E. and Howell, R.R.

34. Burr, I.M., O'Neill, J.A., Karzon, D.T., Howard, L.J. and Greene, H.L. (1974). J. Paediatr. 85, 792-795.
35. Greene, H.L., Slonim, A.E., O'Neill, J.A. and Burr, I.M. (1976). New Eng. J. Med., 294-423-425.

DIFFERENTIAL ABSORPTION OF PURINE NUCLEOTIDES, NUCLEOSIDES AND BASES

C. F. Potter, A. Cadenhead, H. A. Simmonds and
J. S. Cameron
Clinical Science Laboratories, Guy's Hospital, London;
Department of Applied Nutrition, Rowett Research
Institute, Aberdeen

INTRODUCTION

Substantial contributions to our knowledge of purine metabolism and the origin and elimination of uric acid, have been made from studies in man and animals. Studies early in the century[1] established that the body could be sustained on a diet free of purine (endogenous metabolism) and the role of dietary (exogenous) purine in the aetiology of the gouty attack. The effect of protein in increasing the urinary excretion (clearance) of dietary derived uric acid, and the opposing effect of starvation, on endogenous uric acid clearance was also demonstrated[1]. Recent studies have confirmed these findings and indicate that age onset classical gout is predominantly a disease of plenty[2,3].

Knowledge of the exact form in which dietary purine is absorbed is thus important, though scanty. Studies in both man[2] and animals[4] have demonstrated competitive absorption between purines (guanine[4] and RNA[2]) and purine analogues (allopurinol - a xanthine oxidase inhibitor used in gout). We have now extended these studies[4] to include other purine bases, nucleosides and nucleotides, DNA, RNA and single cell protein nucleic acid. For the above reasons, we also looked at the absorption of the compounds in competition with allopurinol.

METHODS

The breed of pigs (four animals) and design of the experiment (basal low purine diet, fixed water intake, acclimatisation in metabolic cages following insertion of cannulas for blood sampling etc) have been described elsewhere[4] After four days on the basal diet, the supplement (11 mmol/day) was given for a further four days, allopurinol (11 mmol/day) was then added to the diet for the next four days, followed by a return to basal diet. The following supplements were given:

Polynucleotides: DNA, RNA, single cell protein nucleic acid (Pruteen)
Mononucleotides: 5'AMP, 5'GMP
Nucleosides: adenosine (AR), guanosine (GR)
Bases: adenine (A), guanine (hydrochloride) (G)

Urinary pH, creatinine, total nitrogen, urea, 24 h purine and pyrimidine excretion were determined as previously described[4].

RESULTS

Urinary purine excretion

Allantoin (purine end product pig, equivalent to uric acid) represented approximately 90% of the total purine end product on the basal diet, and during the periods of supplementation with each purine additive in all but two instances: In the case of adenine and adenosine (but interestingly not AMP) hypoxanthine represented

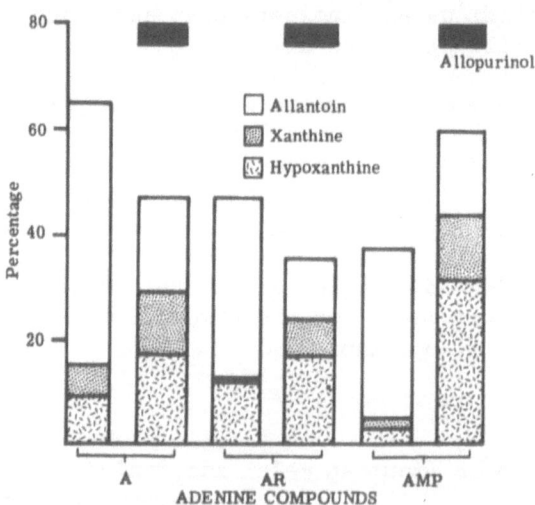

% DOSE OF PURINE SUPPLEMENT RECOVERED IN THE URINE

Fig. 1. Effect of different adenine compounds with and without allopurinol on the excretion of allantoin, xanthine and hypoxanthine (the contribution made by uric acid was insignificant and was not included in the figure).

up to 30% of the increment above the basal diet (Figs. 1-3). Only one purine, guanosine, appeared to be extremely well absorbed (86%) (Fig. 2). On all other occasions the percentage of diet supplement recovered in urinary metabolites varied from 30-60%, the nucleic acids (Fig. 3), DNA, RNA and Pruteen, showing the lowest levels (30-40%).

Effect of allopurinol

Allopurinol reduced the dietary derived increment in total purine excretion (xanthine plus hypoxanthine plus uric acid plus allantoin) except in the case of the mononucleotides AMP and GMP where the reverse effect was noted (Figs. 1-3). Again guanosine

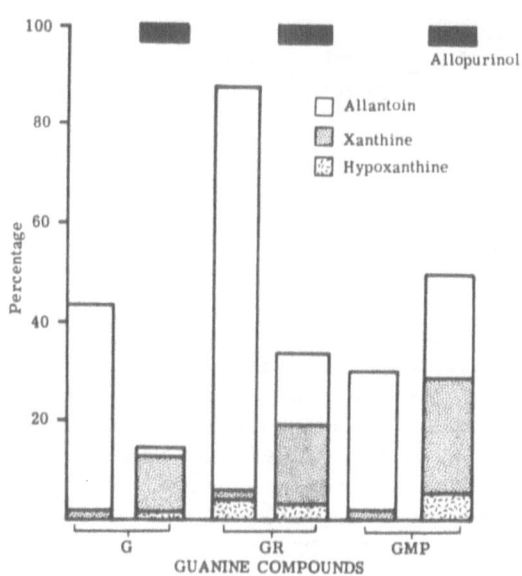

Fig. 2. Effect of different guanine compounds with and without allopurinol on the excretion of allantoin, xanthine and hypoxanthine (the contribution made by uric acid was insignificant and was not included in the figure).

% DOSE OF PURINE SUPPLEMENT RECOVERED IN THE URINE

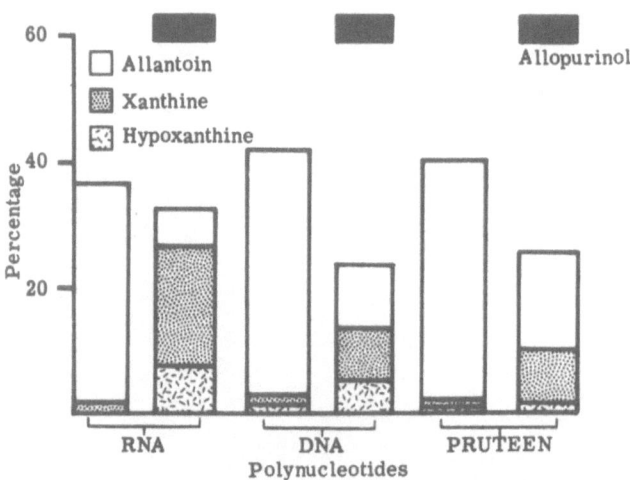

Fig. 3. Effect of the polynucleotides DNA, RNA and Pruteen with
         and without allopurinol on the excretion of allantoin,
         xanthine and hypoxanthine (the contribution made by uric
         acid was insignificant and was not included in the total).

showed the greatest effect (>60%), adenosine and adenine the least
(up to 25%) reduction.

     A difference was noted in the effect of allopurinol on the
increment in precursor xanthine and hypoxanthine with the different
additives. With all three adenine compounds the increment was
most marked in hypoxanthine (Hx:X ratio 2.5:1), while with the
guanine compounds the results resembled more closely those during
allopurinol therapy in man, the greater increment being in xanthine
(Hx:X ratio 0.2:1). With the nucleic acids the ratio was approximately 0.5:1. Allantoin excretion was reduced below control levels
without exception, representing only 40-50% of the total with the
guanine compounds, Pruteen and DNA, as compared with an even lower
value (25-30%) for the adenine compounds or RNA (Figs. 1-3).

Allopurinol metabolites

     The absorption (>75%) and metabolism (>60% oxipurinol) was
similar to that in previous studies[4], with the exception that with
the adenine compounds (particularly adenine) where high hypoxanthine levels were attained, little or no allopurinol riboside
was excreted (Fig. 1); a situation noted in the Lesch Nyhan

syndrome and attributed to hypoxanthine being a better competitive substrate for nucleoside phosphorylase[4].

DISCUSSION

Discrepancies in the literature regarding the effect of specific dietary purine additives on uric acid (or allantoin) excretion levels are almost certainly due to the form administered (e.g. insoluble guanine as distinct from soluble guanine hydrochloride) as well as the species used. Some studies have based assessment on the change in 24 hour urine uric acid alone, or have looked at plasma uric acids, but only uric acid/creatinine ratios in urine[5]. Moreover, the situation in man is complicated by the low clearance of uric acid: a dietary additive may thus raise both plasma and urinary uric acid levels, making it difficult to evaluate the extent of purine absorption from urinary levels. By contrast, allantoin is cleared rapidly so that urinary levels should be an accurate representation of absorption. This and other features[6] make the pig an excellent model for studies of purine metabolism relative to man[4]. Several noteworthy observations have resulted from the present studies:

First, hypoxanthine represented at least 30% of the total increment in urinary purine during supplementation with adenine and adenosine, but with no other purine. (Hypoxanthine in man is normally less than 2% of the urinary purine total, except in the Lesch Nyhan syndrome.) This suggests that these purines are normally unlikely contributors to urinary purines and, more importantly, that studies which looked only at uric acid (allantoin) would have underestimated the effect of diet on urine levels by this amount. The high recovery with guanosine was unique and again has relevance to man in that guanosine is the principal purine in beer; a finding which could thus explain the acute gouty attack which frequently follows a drinking bout. Oral guanosine in man (7 mmol) has produced a sharp increment in plasma urate (2 mg%; 0.12 mmol) but did not change urine levels[7]. Our results of the contribution of dietary DNA, RNA and Pruteen to urinary purine levels are similar to those in man[3] but those with AMP and GMP are much lower.

The results with allopurinol and the different additives (except AMP and GMP) reaffirm previous work by ourselves and others[4,8] that allopurinol will reduce the urinary increment derived from dietary purine by up to 60%: an important additional advantage of allopurinol therapy in the gouty, particularly gouty beer drinkers. This beneficial effect of allopurinol was originally attributed solely to inhibition of de novo purine production. In a companion paper (Van Acker et al, this symposium) we confirm a substantial (and similar - up to 60%) reduction by allopurinol of the absorption of dietary purine in man. The greater increment

produced by allopurinol was also in xanthine. This finding, cont-
rasting with the marked hypoxanthine increment on allopurinol with
all three adenine compounds in this study, is additional evidence
for a minor contribution of the latter to urinary uric acid in man.
(This also confirms little salvage of hypoxanthine of dietary
(exogenous) origin and the futility of efforts to control endogenous
purine production by this means.)

These results, together with similar studies in man, suggest
that the balance of dietary purine absorption must favour guanine
compounds in the nucleoside or base form. However, it will be
important to extend these studies to competitive experiments
between adenine and guanine compounds as well as between purines
and pyrimidines to establish the relevance of these experiments in
the pig to our understanding of purine metabolism in man.

REFERENCES

1. L. B. Mendel and E. W. Brown, The rate of elimination of uric
    acid in man, J.A.M.A., XLIX:896 (1907).
2. W. Gröbner, M. Berlin and N. Zöllner, Der underschiedliche
    Einfluss von Allopurinol auf die endogene und exogene
    Uratquote, Verh. Dt. Ger. Inn. Med., 76:847 (1970).
3. J. Bowering, D. H. Calloway, S. Margen and N. A. Kaufmann,
    Dietary protein level and uric acid metabolism in normal man,
    J. Nutrition, 100:249 (1970).
4. H. A. Simmonds, P. J. Hatfield and J. S. Cameron, Metabolic
    studies of purine metabolism in the pig during the oral
    administration of guanine and allopurinol, Biochem. Pharmacol.,
    22:2537 (1973).
5. A. J. Clifford, J. A. Riumallo, V. R. Young and N. S. Scrimshaw,
    Effect of oral purines on serum and urinary uric acid of
    normal, hyperuricemic and gouty humans, J. Nutr., 106:428
    (1975).
6. J. S. Cameron, H. A. Simmonds, P. J. Hatfield, A. S. Jones and
    A. Cadenhead, The pig as an animal model for purine metabolism
    studies, in: "Purine Metabolism in Man", 41B, Plenum Press,
    New York (1974).
7. R. Kotz, H. Metzerroth and M. M. Müller, Stoffwechselbelastung
    mit Guanosin bei Gesunden und bei Patienten mit Arthritis
    Urica, Zeitschrift fur Rheumatologie, 34:108 (1974).
8. A. Griebsch and N. Zöllner, Effect of ribomononucleotides given
    orally on uric acid production in man, in: "Purine Metabolism
    in Man", 41B, Plenum Press, New York, pp443 (1974).

INFLUENCE OF DIETARY PROTEIN ON SERUM AND URINARY URIC ACID

W. Löffler, W. Gröbner, N. Zöllner

Medical Policlinic, University of Munich

Pettenkoferstraße 8a, Munich

The relationship between dietary protein intake and uric acid metabolism has been studied by many workers, such as Taylor and Rose (1914), Leopold et al. (1925), Bien et al. (1953), Waslien et al. (1968), and Bowering et al. (1969). However, this relationship was still unsettled as far as serum uric acid concentration is concerned. We therefore investigated the influence of a high dietary protein intake as compared to an average intake on serum and urinary uric acid.

METHODS

Five healthy students received an isoenergetic purine-free liquid formula diet for a total of 45 days (three periods of 15 days each). The composition of the diet during each period is shown in Table 1. The sources were commercially available skimmed milk powder, sunflower-oil and a glucose polymerisate.

Table 1. Composition of the isoenergetic purine-free liquid formula diets used in the study. The table shows energy percent values.

	I	II	III
Protein	12	24	36
Fat	3o	24	18
Carbohydrates	58	52	46

Vitamins were given orally and the intake of sodium and potassium was adjusted to that of diet III by adding NaCl and KCl to diet I and II. Daily fluid intake was unrestricted but subjects were told to drink enough to provide a urine volume of at least 15oo ml per day.

An equilibration period of 9 days was allowed and the data used for calculations were those measured from day 1o to 15 of each period. Uric acid was determined enzymatically, urinary oxipurines were measured according to Klinenberg et al. (1967). The concentrations of creatinine, sodium, potassium, and calcium were determined by auto-analyzer methods.

RESULTS

The daily average uric acid values ($\bar{x} \pm$ range) are shown in Fig. 1. There is a fall of serum uric acid and a rise of renal uric acid excretion, when dietary protein intake is increased. The average values for each period are shown in Table 2.

Table 2. Average values ($\bar{x} \pm$ SD) of serum and urinary uric acid in each period (n = 3o in each period).

	I	II	III
Serum uric acid (mg/1oo ml)	3,72 ± 0,55	2,84 ± 0,6o	2,51 ± 0,44
Urinary uric acid	287 ± 48	356 ± 73	386 ± 84

There were clear differences between the three periods, when evaluated by sequential analysis. However, these differences were not statistically significant when averages (data from day 1o to day 15 for all subjects) were compared. In Fig. 2 average values for each period are shown. Assuming a linear relationship this would mean a decrease of serum uric acid concentration of o,o86 mg/1oo ml and an increase of urinary uric acid of 7 mg/day for every 1o g of protein (based on a 24oo kcals diet) added on cost of fat and carbohydrates. Uric acid clearance was 5,4 ± o,94, 8,8 ± 1,9, and 1o,9 ± 1,7 ml/min, respectively, for the three diets studied.

There was also a small rise in urinary oxipurine excretion with increasing protein intake. The average values were 16,1 (diet I), 22,1 (diet II), and 27,8 (diet III) mg/day, respectively.

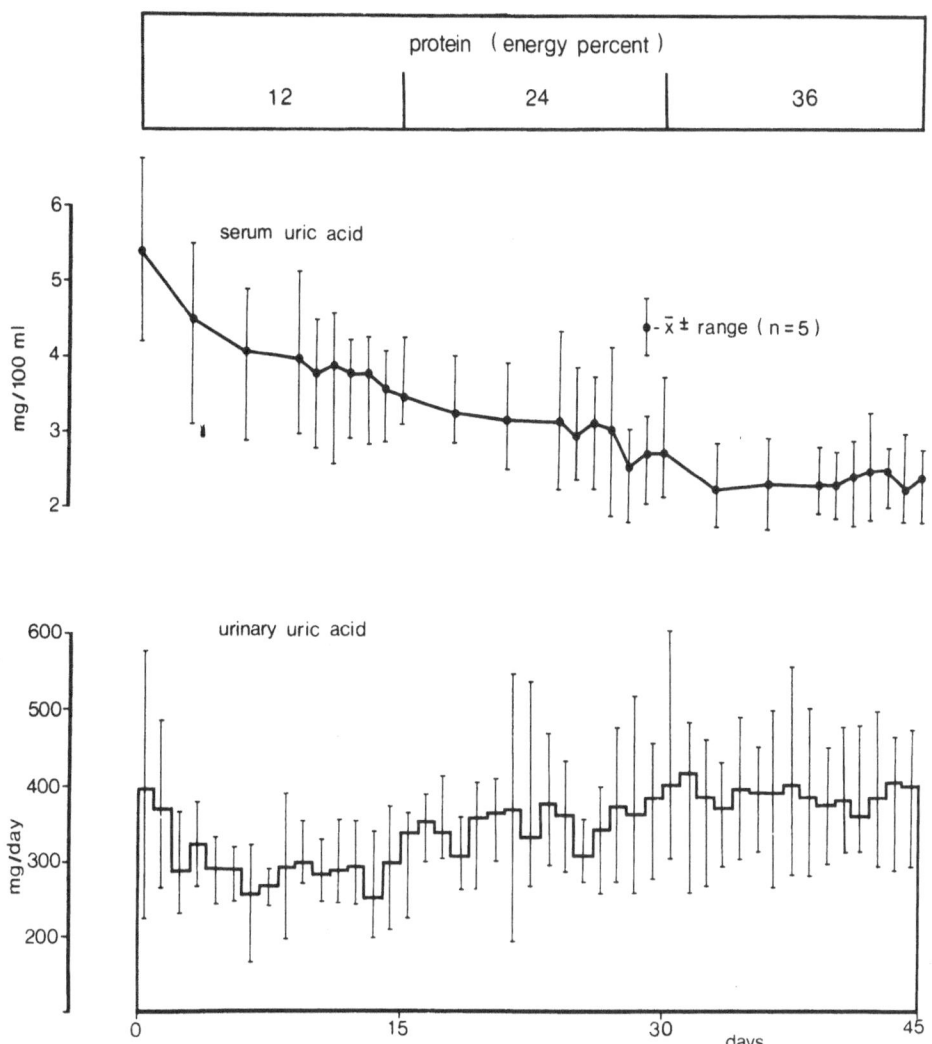

Fig. 1. Influence of dietary protein on serum and urinary uric
          acid in five normal subjects ($\bar{x} \pm$ range).

DISCUSSION

In most publications there is agreement concerning the influence of dietary protein on urinary uric acid excretion. However,
there are controversial results concerning serum uric acid levels
(Waslien et al., 1968; Bowering et al., 1969). Our data demonstrate an increase of urinary uric acid and a fall of serum uric
acid with increasing protein ingestion. The enhanced uric acid
clearance might be due to an uricosuric effect of amino acids.
Matzkies and Berg (1977) have shown that intravenous administra-

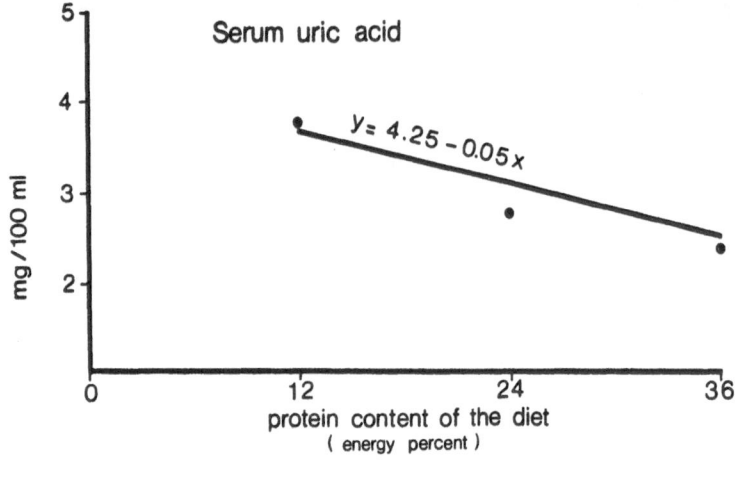

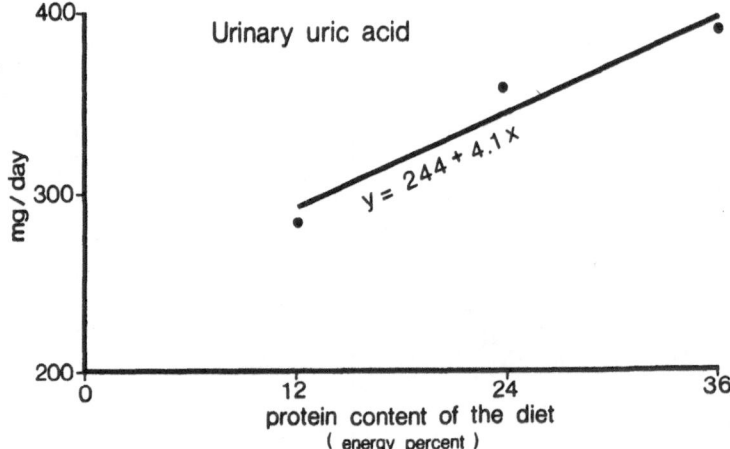

Fig. 2. Influence of dietary protein on serum and urinary uric
acid. The figure shows average values for each of the
three diets under investigation and the relationship,
which can be calculated from the underlying data.

tion of a mixture of amino acids produces an uricosuric effect.
However, an additional mechanism is very likely to exist because
uric acid excretion during protein supplementation does not reflect
the classical uricosuric response seen after administration of
uricosuric drugs. An increased endogenous uric acid production
has to be considered as a possible cause. This hypothesis would
be consistent with increased incorporation of labelled glycine
into urinary uric acid under protein supplementation as demonstrated by Bien et al. (1953). They found a surplus of 62 mg

urinary uric acid per day, when changing from a 48 to a 84 grams protein diet in the one normal subject studied.

For practical purposes we conclude from our data that there is no need of restricting the purine-free protein content of the gouty patients' diet, because there is a slight decrease in serum uric acid and the increase of urinary uric acid is very small.

References

Bien, E. J., Yü, T. F., Benedict, J. D., Gutman, A. E., Stetten, D. W. Jr., 1953, The relation of dietary nitrogen consumption to the rate of uric acid synthesis in normal and gouty man. J. Clin. Invest. 32:778.

Bowering, J., Calloway, D. H., Margen, S., Kaufmann, N. A., Dietary protein level and uric acid metabolism in normal man. 1969, J. Nutr. 1oo:249.

Klinenberg, J. R., Goldfinger, S., Bradley, K. H., Seegmiller, J. E., 1967, An enzymatic spectrophotometric method for the determination of xanthine and hypoxanthine. Clin. Chem. 13: 834.

Leopold, J. S., Bernhard, A., Jacobi, H. G., 1925, Uric acid metabolism of children. Amer. J. Dis. Childr. 29:191.

Matzkies, F., Berg, G., 1977, The uricosuric action of amino acids in man. Adv. Exp. Med. Biol. B 76:36.

Taylor, A. E., Rose, W. C., 1914, The influence of protein uptake upon the formation of uric acid. J. Biol. Chem. 18:519.

Waslien, C. J., Calloway, D. H., Margen, S., 1968, Uric acid production of men fed graded amounts of egg protein and yeast nucleic acid. Amer. J. Clin. Nutr. 21:892.

EFFECT OF HYPOXANTHINE IN MEAT ON SERUM URIC ACID AND URINARY

URIC ACID EXCRETION

W. K. Spann, W. Gröbner, N. Zöllner

Medical Policlinic, University of Munich

Pettenkoferstraße 8a, Munich

Isoenergetic quantities of meat and vegetables contain about
the same concentration of RNA purine nitrogen (Zöllner and Griebsch,
1973). Clinical experience however shows, that meat eating leads
more easily to gout than a vegetarian diet.
The purine mononucleotide concentration in meat is higher than in
vegetables. Predominantly degradation products of ATP exist in
high concentration. Besides RNA und DNA other uric acid precour-
sors in meat may be responsible for the increase of uric acid. We
determined therefore the hypoxanthine content of pork meat and
evaluted the influence of hypoxanthine on serum uric acid and urinary uric acid excretion in a nutrition experiment.

METHODS

The concentration of hypoxanthine in meat was determined by
a modification of the method of Pottast and Hamm (1969). Meat
was homogenized, and protein was precipitated with perchloric
acid. Hypoxanthine was separated from other purine compounds by
thin layer chromatography and quantitated by elution and measuring
the UV absorbance.

To evaluate the effect of hypoxanthine on serum and urinary
uric acid 5 healthy volunteers received an isoenergetic liquid
formula diet free of purines, to which measured amounts of hypoxanthine (Serva Chemie, Heidelberg) were added. The protein content of the diet was 15 energy % and the proportion of fat to
carbohydrates was roughly 1:2. Every experimental period was continued over 1o days. Hypoxanthine laod was o,5 g/7o kg body

weight and 1 g/7o kg body weight respectively. Serum uric acid
was determined every second day, the concentration of uric acid
and creatinine in the 24 hour urine daily. Uric acid was determined
enzymatically, creatinine colorimetrically.

RESULTS AND DISCUSSION

    The hypoxanthine concentration in meat increases with storage
time and temperature (Fig. 1). Edible pork had a maximum hypoxanthine concentration of 3o mg/1oo g.

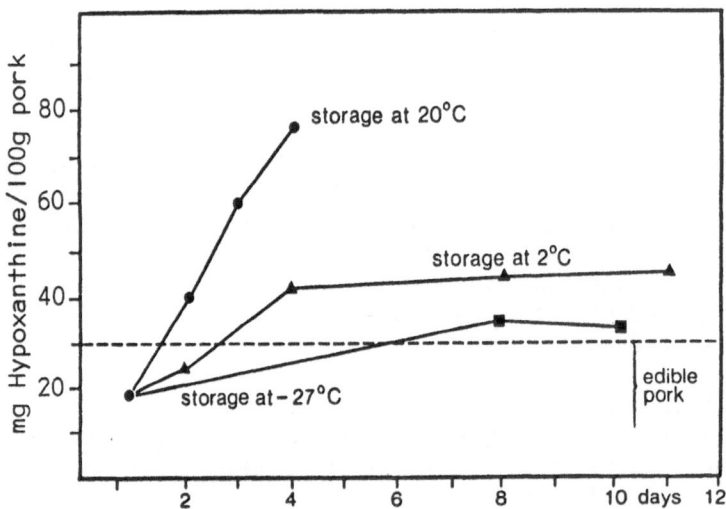

Fig. 1: Hypoxanthine content of pork during storage at tempe-
          ratures of 2o°C, 2°C and -27°C.

    Under a purine free isoenergetic diet serum uric acid and
urinary uric acid excretion reaches a minimum within 1o days and
remains constant thereafter. When hypoxanthine is added to the
basal diet, the plasma concentration and urinary excretion of
uric acid increase until they reach a new level after about a
week. When hypoxanthine supplementation is doubled, there is
a further rise (Fig. 2).

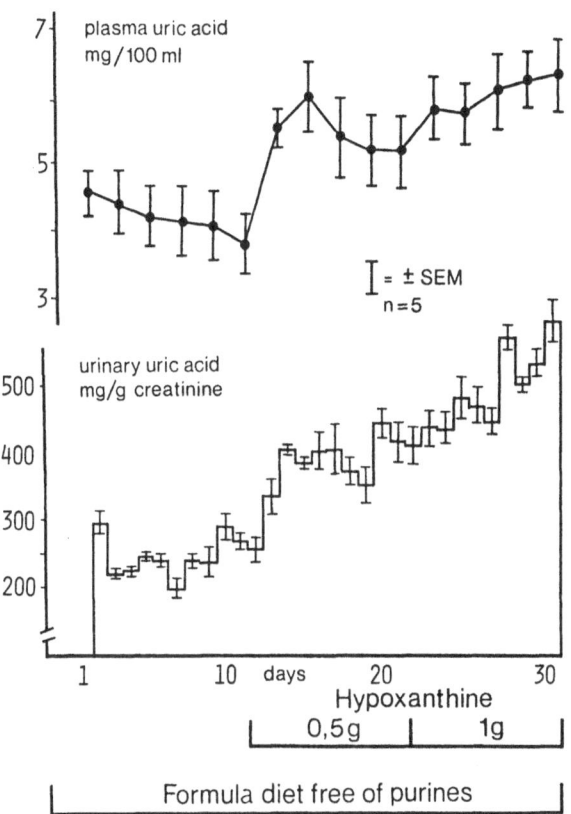

Fig. 2: Plasma levels (above) and renal excretion (below) of uric acid during administration of amounts of o,5 and 1 g hypoxanthine per 7o kg body weight.

o,o1 mol hypoxanthine produces an increase of serum uric acid
of 3,5 mg/1oo ml. In comparison to equimolar doses of AMP, GMP
and RNA-purines (Zöllner et al., 1972) hypoxanthine causes a
minor, in comparison to DNA a greater effect on serum uric acid
(Fig. 3) and urinary uric acid excretion (Fig. 4).

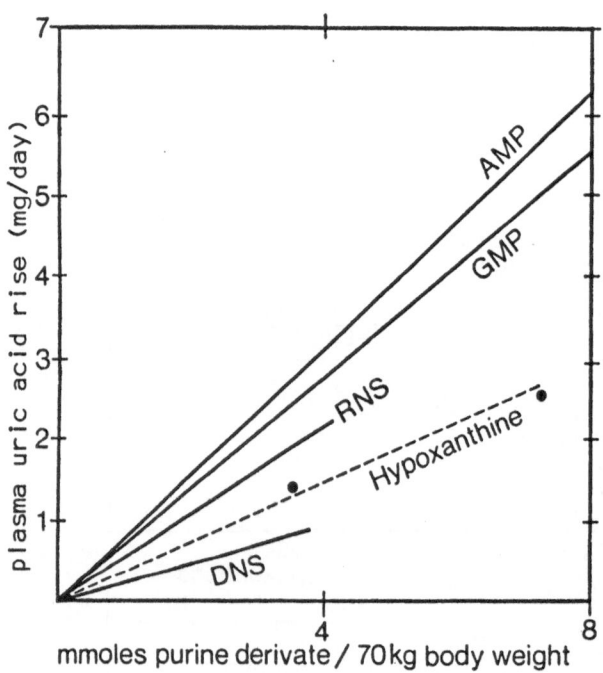

Fig. 3: Rise of serum uric acid level in comparison to purine
          load from different sources on molar basis.

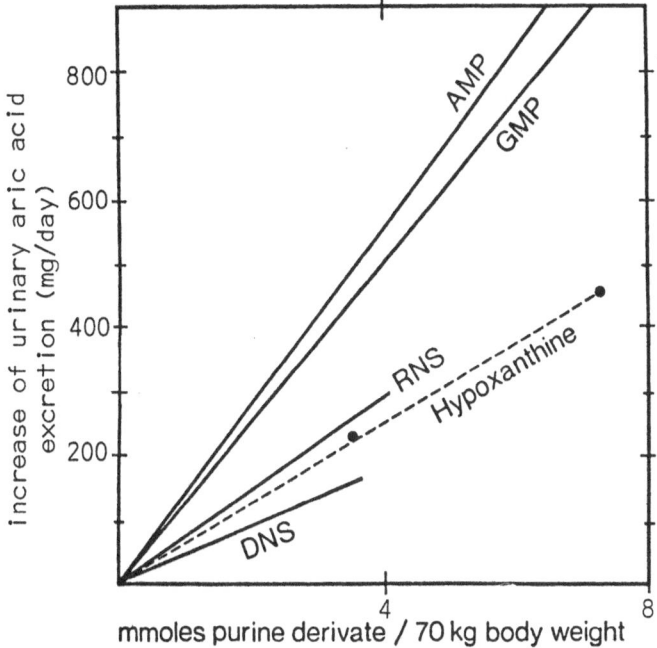

Fig. 4: Rise of urinary uric acid excretion incomparison to purine load from different sources on molar basis.

If we combine the result of food analysis and nutritional experiment we can approximate the rise of serum and urinary uric acid induced by hypoxanthine. If a healthy man of 7o kg body weight eats 2oo g of pork with a hypoxanthine content of 3o mg/1oo g, the increase of serum uric acid by this purine base amounts to about o,2 mg/1oo ml. The increase of serum uric acid in gouty subjects after purine administration is twice as high.

ACKNOWLEDGEMENT

This work was supported by the Deutsche Forschungsgemeinschaft.

REFERENCES

Zöllner, N., Griebsch, A., Uric Acid Lithiasis, Urinary Calculi. Int. Symp. Renal Stone Res., Madrid 1973, 84.
Pottast, K., Hamm, R., 1969, J. Chromatog., 42:558.
Zöllner, N., Griebsch, A., Gröbner, W., 1972, Ernährungsumschau 3:79.

ON THE MECHANISM OF THE PARADOXICAL EFFECT OF SALICYLATE ON URATE

EXCRETION

Herbert S. Diamond, Gary Sterba, Krishnasha Jayadeven,
and Allen D. Meisel

State University of New York Downstate Medical Center
450 Clarkson Avenue, Box #42, Brooklyn, New York 11203

Supported by grants from the Kroc Foundation, the New
York Chapter of the Arthritis Foundation, the General
Clinical Research Center Program of the National In-
stitutes of Health, and from the Irma T. Hirschl Career
Scientist Award (HSD)

In 1959 Yu and Gutman reported that aspirin in a dose of 2g/day
or less resulted in decreased urate excretion while at doses
greater than 3g/day, urate excretion was increased (1). They also
found that alkalinization of the urine with bicarbonate increased
both salicylate and urate excretion. These results were interpreted
as showing that low concentrations of salicylate in tubular urine
inhibited tubular secretion of urate and high concentrations inhibited both tubular secretion of urate and reabsorption of urate.
Since the overall result of tubular transport of urate is net reabsorption, uricosuria appeared when tubular salicylate concentrations were sufficiently increased.

This interpretation fails to consider the complex sequence by
which ingested salicylate is conjugated and salicylate conjugates
are preferentially excreted so that while the principal compound in
plasma is salicylate, over 80 percent of excretion is as conjugates.
These conjugates include salicylurate, salicyl acyl glucuronide,
abbreviated as (SAG) and salicy phenolic glucuronide, abbreviated as
(SPG).

To assess the effects of salicylate and its metabolites on
renal tubular transport of urate, 7 volunteer subjects with normal
renal function were studied as in-patients on a clinical research
unit. Subjects were maintained on a normal protein, constant purine

221

isocaloric diet throughout the study. All medications affecting
urate excretion were discontinued 4 to 7 days prior to the study,
as appropriate. Three days were allowed for dietary adjustment.
Salicylate and salicylate metabolites were measured by spectro-
fluorometry (2).

RESULTS

     Salicylate was administered in varying doses to 7 subjects.
As salicylate dose was increased from 1.2g per day to 4.8g per day,
serum salicylate increased from 1.9 ± 1.0 mg/dL to 18.7 ± 2.5 mg/dL
and urate clearance (Cur) increased from 3.0 ± 1.0 ml/min to 8.6 ±
2.6 ml/min (p < 0.05). Serum salicylate concentration and urate
clearance were positively correlated with r = 0.59, and p < 0.005.

     Infusion of sodium bicarbonate alone has a minimal effect on
urate excretion. Six subjects were treated with aspirin for 4-5
days in doses ranging from 3.6g to 4.8g per day sufficient to maintain urate clearance near control values. When sodium bicarbonate
was administered as 134 meq intravenously over 10-15 minutes
followed by infusion of 44.6 meq over the next hour, urate clearance
increased from 5.4 ± 1.5 ml/min to 15.4 ± 4.2 ml/min (p < 0.02)
(Table 1). Serum salicylate concentration did not change during
bicarbonate infusion. However, salicylate excretion increased from
1715 ± 1564 μM/L to 3499 ± 1800 μM/L (p < 0.05). Excretion of
salicylurate, salicyl phenolic glucuronide and salicyl acyl glucuronide were unchanged during bicarbonate infusion. In the bicarbonate infusion studies, clearance was positively correlated with
salicylate excretion and varied independent of serum salicylate concentration.

     Para aminohippuric acid (PAH) and urate appear to be secreted
by different organic acid transport systems in the human kidney.
PAH was administered intravenously to 6 subjects as a competitive
inhibitor of organic acid secretion by the PAH carrier. PAH was
given as a priming dose of 50 ml of 20 percent PAH in 50 ml of 25%
mannitol, followed by a continuous intravenous infusion of 20% PAH
in mannitol at a rate of 1-2 ml/min. When PAH was administered
without salicylate, urate clearance increased from a control value
of 8.1 ± 0.8 ml/min to 11.5 ± 0.9 ml/min at plasma PAH concentrations of 30-80 mg/dL (p < 0.02). In contrast, in 6 subjects given
3.6-4.8 gms aspirin per day for 4 days prior to the PAH infusion,
urate clearance decreased from 9.6 ± 2.6 ml/min on aspirin alone to
4.0 ± 1.8 ml/min after PAH plus aspirin (p < 0.05) (Figure 1). The
decrease in urate clearance was associated with a concomitant decrease in salicylate excretion from 3485 ± 1663 μM/min after aspirin
alone to 2468 ± 1316 μM/min after aspirin plus PAH (Table 2).
Salicylurate excretion was also decreased from 4370 ± 1026 μM/min
after aspirin to 3490 ± 920 μM/min after aspirin plus PAH. There

was no change in salicyl phenolic glucuronide or salicyl acyl
glucuronide excretion.

TABLE 1. EFFECT OF SODIUM BICARBONATE INFUSION (NaHCO$_3$) ON URATE
CLEARANCE AND ASPIRIN METABOLITE EXCRETION AFTER 4 GRAMS ASPIRIN
(HSA)

	n	Cur (ml/min)	SALICYLATE (μM/L)	SALICYLURATE (μM/L)
HSA	6	5.4±1.5	1715±1564	2091±741
HSA after NaHCO$_3$	6	15.4±4.2*	3499±1800**	2214±990

*p < 0.02; **p < 0.05

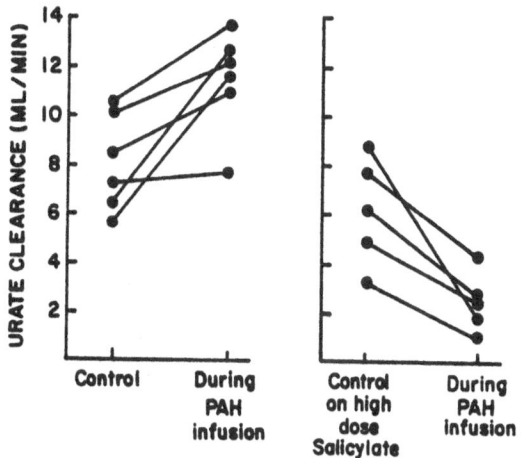

FIGURE 1. EFFECT OF PAH INFUSION UNDER CONTROL CONDITIONS AND
          AFTER 4-4.5 GRAMS OF ASPIRIN PER DAY FOR 4 DAYS

TABLE 2. EFFECT OF PAH ON URATE CLEARANCE AND ASPIRIN METABOLITE
          EXCRETION AFTER 4 GRAMS ASPIRIN (HSA)

	n	Cur (ml/min)	SALICYLATE (μM/L)	SALICYLURATE (μM/L)
HSA	6	8.6±2.6	3485±1663	4370±1026
HSA after PAH	6	4.0±1.8*	2468±1316*	3490± 920*

*p < 0.05

Following oral administration, pyrazinamide is metabolized to the organic acid pyrazinoic acid. At the concentration attained in man following a single 3g oral dose, pyrazinoic acid inhibits urate excretion but does not affect PAH excretion. Inhibition of urate excretion appears to result from inhibition of the urate secretory system without significant inhibition of urate reabsorption. Pyrazinamide 3g orally was administered to 3 subjects after 4-5 days of aspirin in a dose ranging from 3.6 to 4.8g per day. Urate clearance decreased from 11.5 ± 5.0 ml/min on aspirin alone to 5.5 ± 5.2 ml/min when pyrazinamide was administered with aspirin ($p < 0.05$) (Table 3). Salicyl urate excretion fell from 5249 ± 1459 meq/L after aspirin alone to 3305 ± 1190 meq/L when pyrazinamide was added. There was no significant change in excretion of salicylate, salicyl acyl glucuronide or salicyl phenolic glucuronide.

DISCUSSION

Interpretation of these results depends in part upon present concepts of renal transport of organic acids in man.

The most probable interpretation of present knowledge of urate transport supports the following sequence: 1) urate is completely or nearly completely filtered at the glomerulus; 2) filtered urate is largely, if not completely reabsorbed in the proximal tubule; 3) additional urate enters the tubule by secretion through a transport system independent of the PAH secretory system; and 4) there is substantial reabsorption of secreted urate co-extensive with and/or distal to the urate secretory site.

Salicylate appears to be secreted by the PAH secretory system and not by the urate system since salicylate secretion was inhibited by PAH but not by pyrazinamide. Salicylate is reabsorbed by passive non-ionic diffusion in acidic tubular fluid. The uricosuric effect of aspirin is correlated best with urine salicylate concentration and not with serum concentration or urine concentration of salicylate metabolites indicating that salicylate accounts for the uricosuric effect.

TABLE 3.   EFFECT OF PYRAZINAMIDE (PZA) ON URATE CLEARANCE AND
           ASPIRIN METABOLITE EXCRETION AFTER 4 GRAMS ASPIRIN (HSA)

	n	Cur (ml/min)	SALICYLATE (μM/L)	SALICYLURATE (μM/L)
HSA	3	11.5 ± 5.0	2707 ± 1045	5248 ± 1459
HSA after PZA	3	$5.5 \pm 5.2^*$	1679 ± 714	$3385 \pm 1190^{**}$

*$p < 0.05$
**$0.05 < p < 0.10$

Salicylurate secretion is inhibited by pyrazinamide and to a lesser extent by PAH, suggesting that salicylurate is secreted by both the PAH and urate secretory systems but predominately by the urate system. Salicylurate, therefore, competes with urate for secretion, and renal retention of urate occurs when salicylurate is the major excretory product of aspirin. Thus salicylurate appears to be the inhibitor of urate secretion.

When aspirin is administered in low doses, salicylate elimination is predominately by conjugation to form salicylurate. Serum salicylate concentration is low and there is little salicylate in acid urine. Urate reabsorption is therefore minimally affected. The major excretory product is salicylurate. Urate secretion is therefore inhibited. The net effect is decreased urate clearance.

When aspirin intake is sufficiently increased, hepatic metabolism of salicylate to salicylurate is saturated. Salicylate concentration increases in serum and salicylate becomes a major excretory product. Since the net tubular urate transport is reabsorption, when urine salicylate concentration is great enough, salicylate inhibition of urate reabsorption overrides salicylurate inhibition of secretion and uricosuria results.

REFERENCES
1. T. F. Yu and A. B. Gutman, Study of the paradoxical effects of salicylate in low, intermediate and high dosage on the renal mechanisms for excretion of urate in man, J. Clin. Invest. 38:1298 (1959).
2. D. Schachter and J. G. Manis, Salicylate and salicyl conjugates: Fluorometric estimation of biosynthesis and renal excretion in man, J. Clin. Invest. 37:800 (1958).

The Uricosuric Action of Protein in Man

F. Matzkies *, G. Berg, H. Mädl

Abteilung für Stoffwechsel und Ernährung (Vorsteher:
Prof.Dr.Dr.h.c.G.Berg) in der Medizinischen Klinik
mit Poliklinik der Universität Erlangen-Nürnberg
8520 Erlangen, Krankenhausstr. 12
* new adress: 8740 Bad Neustadt/Saale, Kurparkklinik

The influence of dietetic measures on uric acid metabolism was
repeatedly investigated (1,2,3,6). In our studies we were able
to show that intravenously administered amino acids display a
definite uricosuric effect (5). Our aim was then to investigate whether the uric acid metabolism can also be affected under oral application of purine free protein by modifying the
dietary protein portion.

Material and methods

8 normal subjects, aged 22 to 40 years (4 women and 4 men),
ingested a variable, balanced formula diet free of purine
(Berodiät V) with an energy supply of 1920 calories (8000 kJ)
over a period of 14 days. The applied low-sodium formula diet
is divided into 3 components. The diet consists of a protein
component, an energy component and a mineral-vitamin component. The formula diet was administered in such a way that in
one period 59 g protein, 256 g carbohydrate and 72 g fat were
given, in a second period 113 g protein, 224 g carbohydrate
and 53 g fat, and in a third period 167 g protein, 232 g carbohydrate and 34 g fat.

4 days after the beginning of the dietary treatment with the
formula diet, the protein portion was modified in periods of
3 days each. During this period, venous blood was daily
collected in the morning for the determination of uric acid
concentration. Simultaneously, the daily uric acid excretion
was determined and uric acid clearance was calculated.

Results
Uric acid in serum.
An increase in protein supply to 113 and 167 g protein/day results in a decrease of the serum concentration which, however,
could not be established statistically (table 1).

Uric acid excretion.
With the formula diet free of purine, uric acid excretion was
found to be between 279 and 330 mg/24 hrs. when 59 g of protein were administered. With increased protein intake uric
acid excretion rose to 406-525 mg/24 hrs.. With further increase of protein intake uric acid excretion also increased.
The increased renal excretion of uric acid during increase of
protein intake was statistically significant ($p < 0.05$).

Uric acid clearance.
When 59 g of protein were given, uric acid clearance was between 4.2 and 5.0 ml/min.. If protein intake was increased, a
continuous significant increase in uric acid clearance up to
values between 7.8 and 8.9 ml/min. resulted.

Discussion
In a previous study we were able to show a significant drop of
uric acid levels from 6.8 to 5.0 mg/100 ml within a period of
12 hours during parenteral amino acid application. During the
same observation period, an elevated uric acid excretion of
845 (730-961) mg/12 hrs. was recorded (5). Because of these
results we were induced to reinvestigate the effect of protein
on the uric acid metabolism. The results suggest that serum
uric acid concentration tends to decrease, whereas simultaneously the renal elimination is enhanced and uric acid
clearance is significantly increased. The overall increase in
excretion, however, is only minimal. When the protein supply
was raised from 60 to 113 g, uric acid excretion rose by an
average of 166 mg. With further increase of protein intake up
to 167 g/day, a further increase by an average of 40 mg/24hrs.
occurred. The increase in uric acid excretion of the individual subjects varied between 88 and 373 mg/24 hrs. compared
with the previous values.

Our findings are in accordance with the results of HARDING,
ALLIN and EAGLES. These authors observed a remarkable rise in
uric acid excretion when a protein diet was applied. In their
study, however, the uric acid increase had been induced by a
diet rich in fat (3). WASLIEN, CALLOWAY and MARGEN found a
uric acid excretion of 337 - 48 mg/24 hrs. during administration of 22 g egg albumin with a corresponding serum value of
approx. 5.2 mg/24 hrs.. When the protein supply was raised to

Table 1

Uric acid excretion, uric acid concentration in serum and uric acid clearance in healthy adults.
n = 8, t = 9 days.

Mean value and standard deviation or median and decils are indicated.
The underlined value significantly differs from the initial value.

days	59 g protein			113 g protein			167 g protein		
	1	2	3	1	2	3	1	2	3
uric acid in serum (mg/100ml)	5.0 ±1.6	5.7 ±1.5	5.3 ±1.1	4.5 ±1.3	5.4 ±2.1	5.1 ±1.7	4.2 ±1.1	4.6 ±1.2	4.7 ±1.4
uric acid excretion (mg/24h)	330 ±83	350 ±111 (225-428)	279	406 ±54	444 ±104	525 ±111	472 ±71	546 ±145	478 ±67
uric acid clearance (ml/min)	5.0 ±1.4	4.5 ±1.6 (2.7-11.2)	4.2	6.7 ±2.3	6.5 ±3.1	7.8 ±3.6	8.4 ±3.2	8.9 ±3.7	7.8 ±2.4

75 g/day, an increase in uric acid excretion up to 392 +
66 mg/24 hrs. and a corresponding decrease of serum uric acid
levels to 4.7 + 0.6 mg followed. The authors also discussed an
elevated renal elimination during protein supply (6). WILSON,
BISHOP and TALBOTT varied the protein, fat and carbohydrate
consumption in 4 subjects; they observed a significant increase in uric acid excretion in 3 subjects during the pro-
tein-rich diet (7).

The mean uric acid excretion of 330 mg during application of a
purine-free formula diet is in very good accordance with studies carried out by ZÖLLNER (4,8). ZÖLLNER and co-workers also
found a uric acid excretion of 330 mg/24 hrs..

An elevated renal excretion of uric acid after the addition of
protein to a low-protein diet was also observed by BIEN and
BOWERING (1,2). Primarily, it was concluded from these studies
that protein induces the de novo synthesis of uric acid. BIEN
postulated that 10 g of protein would result in an excretion
surplus of 50 mg uric acid/day. However, we could not confirm
these results. Despite an increased protein portion of
108 g/day, a total excretion surplus of only 206 mg resulted;
which amounts to an increased excretion of 2 mg/g protein.
Therefore, it is unlikely that the de novo uric acid synthesis
is stimulated by an increase in protein supply. On the other
hand, if we would have used isotope-labeled glycine, it would
have been incorporated into uric acid, a fact which does not
contradict our own results (1).

Our results suggest that protein has a low uricosuric effect
which becomes especially evident if a change-over from a low-
protein diet to a diet rich in protein is effected. The observed excretion rates of uric acid during administration of
purine-free formula diets therefore also depend on their protein content (8). In connection with the uricosuric action of
intravenously supplied amino acids, it may be assumed that
orally administered proteins also display a slight uricosuric
effect through an increase of amino acids in blood. Theoretically, it could be assumed that subsequent to an elevated protein supply more amino acids appear in the glomerular filtrate
which must be reabsorbed by the tubules. The necessity to
reabsorb larger amounts of amino acids might result in an inhibition of the tubular reabsorption of uric acid. This hypothesis, however, would have to be supported by animal experiments.

Summary

In 8 subjects, the oral protein supply is modified in periods
of 3 days each. With increasing protein intake, a decrease of
the serum uric acid levels results as well as a significant
increase in uric acid excretion and uric acid clearance. On
the basis of the data obtained, it may be assumed that protein
has a uricosuric action.

References

1. Bien, E.J., T.F. Yü, J.D. Benedict, A.B. Gutman,
                          D.J. Stetten:
                          The relation of dietary nitrogen
                          consumption to the rate of uric acid
                          synthesis in normal and gouty man.
                          J.clin.invest 32(1953), 778
2. Bowering, J., D.H. Calloway, S. Margen, W.A. Kaufmann:
                          Dietary protein level and uric acid
                          metabolism in normal man.
                          J. nutr. 100 (1969), 249
3. Harding, V.J., K.D. Allin, B.A. Eagles:
                          Influence of fat and carbohydrate diets
                          upon the level of blood uric acid.
                          J.Biol.Chem. 74(1927), 631
4. Löffler et al: Ernährung des Gichtkranken
                          Klinikarzt 6 (1977), 865
5. Matzkies, F., G. Berg:
                          The uricosuric action of amino acids
                          J.Clin.Chem.Clin.Biochem. 14(1976), 308
6. Waslien, C.I., D.H. Calloway, S. Margen:
                          Uric acid production of men fed graded
                          amounts of egg protein and yeast nuc-
                          leic acid.
                          Am.J.Clin.Nutr. 21 (1968), 892
7. Wilson, D., Ch. Bishop, J.H. Talbott:
                          A factorial experiment to test the
                          effect of various types of diets on
                          uric acid excretion of normal human
                          subjects.
                          J. appl. Physiol. 4(1952), 560
8. Zöllner, N.:   Die Manifestation der Gicht
                          Beobachtungen über das Zusammenwirken
                          von Ernährung und biochemischer
                          Individualität
                          Internist 18 (1977), 474

HYPOXIC EFFECTS ON PURINE METABOLISM STUDIED WITH HIGH PRESSURE LIQUID CHROMATOGRAPHY

R.A. Harkness, R.J. Simmonds and M.C. O'Connor

Division of Perinatal Medicine
Medical Research Council's Clinical Research Centre
Watford Road, Harrow HA1 3UJ UK.

The purpose of our studies was to obtain estimates of the metabolic effects of hypoxia especially in the perinatal period. Perinatal hypoxia is believed to be probably the most important cause of cerebral palsy and an important factor in the aetiology of mental subnormality (Illingworth, 1979). A cumulative measure of the metabolic effects of hypoxia should therefore be useful in obstetrics and pediatrics.

ATP is the major energy currency of the cell but is intracellular and therefore inaccessible to clinical sampling. Extracellular compounds released by ATP breakdown have therefore been studied, especially the purine base hypoxanthine; this is the major form of performed purine transport between cells (Murray et al., 1970). The increased rate of release of hypoxanthine and other purine compounds during hypoxia has been widely studied experimentally in a variety of animal organs (Arch and Newsholme, 1978) and there is limited evidence for such release in man (Saugstad, 1975).

The concentrations of the purines, pyrimidines and their nucleosides have been estimated in trichloracetic acid extracts using a Waters ALC 200 liquid chromatograph with a model 440 absorbance detector (Waters Ltd., Hartford, Cheshire, UK) operated at 254 and 280 nm with sensitivity of 0.005 absorbance units for a full scale response. 250 x 5 mm columns of ODS Hypersil (Shandon Southern Ltd., Runcorn, Cheshire, UK) were used with a mobile phase of 0.01 mol/l KH_2PO_4 with 1% methanol (v/v) adjusted to pH 6.5. An internal standard of allopurinol was used with quantitative calculations corrected for the concentration of the internal standard, column efficiency and compound retention.

At present a widely used marker of fetal distress and an in-
dicator of a high risk of further damage is meconium staining of
amniotic fluid.   Amniotic fluid from patients before the onset of
labour was therefore studied and the results in Figure 1 show that
hypoxanthine and xanthine concentrations were elevated in most
meconium stained samples above concentrations in unstained amniotic
fluid.   Statistical analyses of the data as logarithims showed
that the mean elevations of hypoxanthine and xanthine were highly
significant (P < .0001) and that the hypoxanthine increased more
than xanthine.   A quantitative biochemical assessment of the
probable metabolic effects of 'hypoxia' therefore agreed with the
accepted qualitative clinical criterion.

Preliminary studies have shown that urinary oxypurine excretion
by newborn infants is increased after episodes of 'hypoxia' thus
extending the work of Manzke et al (1977) and suggesting that at
least part of the origin of the oxypurines in amniotic fluid is
fetal urine.   This is also consistant with our finding that oxypurine concentrations are lower at 16-18 weeks compared to 38-41
weeks gestation.

Obstetricians use a variety of 'risk' factors to decide when
to assist the birth of a child either by forceps or by Caesarean
section.   Mean umbilical arterial plasma oxypurine concentrations
after such assistance were significantly elevated and were thus
consistant with the overall clinical assessments which were largely
qualitative.   In babies delivered by forceps mean hypoxanthine
concentration was significantly raised (P < .001) whereas mean
xanthine concentration was only just raised (P < .05).   Similar
results for babies delivered by Caesarean section showed that the
mean hypoxanthine concentration was just significantly raised (P
< .05) whereas the xanthine concentration was significantly raised
(P < .01).   Since forceps deliveries are performed for short
periods of 'fetal distress' late in the labour and Caesarean
sections for more chronic fetal distress especially early in labour,
it seems possible that in umbilical arterial plasma, elevations of
xanthine concentration reflect a longer period of fetal distress
than that which may cause an elevation of hypoxanthine.

The brain is probably the most sensitive and important organ
in the body especially since damage to brain is largely irreversible.   Cerebrospinal fluid hypoxanthine and xanthine concentrations did not exceed about 5 µmol/l in adults and in children with
no clinical evidence of hypoxia, gross cerebral ischemia or renal
failure.   Inosine was usually undetectable.   In contrast, the
raised hypoxanthine and xanthine concentrations in stillbirths and
in live infants who were hypoxic by two clinical criteria are clear

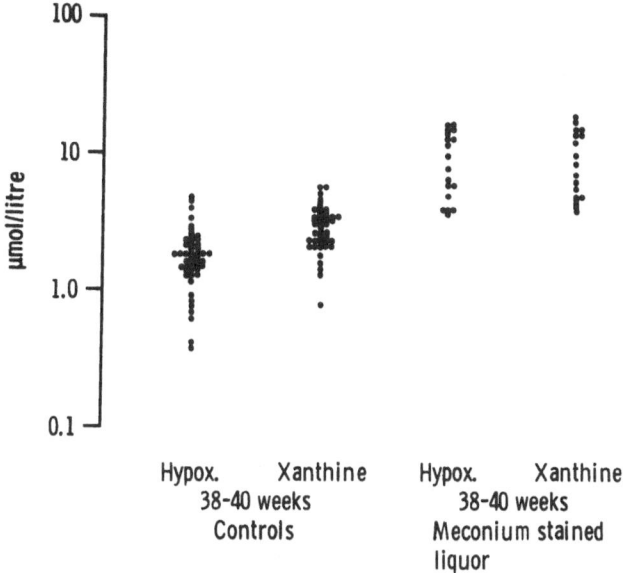

Fig. 1.   Elevated concentrations of hypoxanthine and xanthine in
          meconium stained amniotic fluid obtained before labour
          from women 38-40 weeks pregnant, compared with concen-
          trations in unstained fluid.

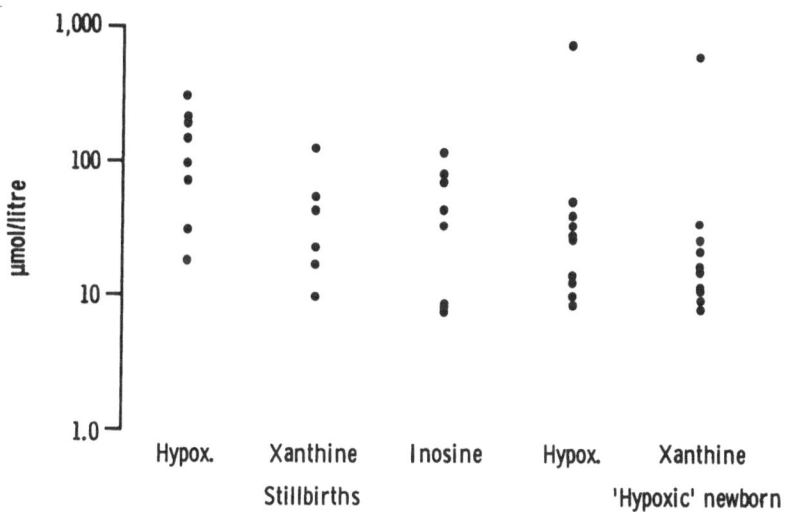

Fig. 2.   Elevated concentrations of hypoxanthine, xanthine and
          inosine in cerebrospinal fluid samples from fresh
          stillbirths and 'hypoxic' newborn as judged by two
          established criteria.

from the results in Figure 2.

From existing evidence it seems justifiable to suggest that the estimation of hypoxanthine and xanthine in extracellular fluids may provide cumulative estimates of the metabolic effects of 'hypoxia'.

References

Arch, J.R.S., and Newsholme, R.A., 1978, Control of metabolism and the hormonal role of adenosine, Essays in Biochem., 14:82.

Illingworth, R.S., 1979, Why blame the obstetrician? A review, Brit. Med. J., 1:79.

Manzke, H., Dorner, K., and Grunitz, J., 1977, Urinary hypoxanthine xanthine and uric acid excretion in newborn infants with perinatal complications, Acta Paed. Scand., 66:713.

Murray, A.W., Elliot, D.C., and Atkinson, M.R., 1970, Nucleotide biosynthesis from performed purines in mammalian cells : Regulatory mechanisms and biological significance, Progr. Nucleic Acid Res. Mol. Biol., 10:87.

Saugstad, O.D., 1975, Hypoxanthine as a measurement of hypoxia, Ped. Res., 9:158.

URATE METABOLISM IN A MONGREL DOG

Peter A. Simkin

Division of Rheumatology, Department of Medicine

University of Washington, Seattle, Washington

For many years, mongrel dogs have been widely used for physiologic and pharmacologic studies of uric acid excretion. Despite
the abundance of excretion data, relatively little is known about
urate metabolism in these animals beyond the fact that there is
an abundance of uricase activity. In order to examine this question, we conducted three short-term turnover studies in a trained
mongrel dog.

This animal had taken part in a number of preliminary metabolic
studies, as well as in previous transport work, and was a willing
participant in the experiments. Without anesthesia or sedation, but
with mild restraint in a Pavlov stand, the dog was equipped with an
indwelling Foley catheter, a continuing infusion of 5% dextrose in
water at 1 ml/minute in one hind leg, and a heparin lock for periodic
sampling of venous blood from the other hind leg.

After a satisfactory urine flow was achieved, 10 µCi of uric
acid, 2-^{14}C was injected to initiate each experiment. Ten samples
of urine were collected at 15-minute intervals thereafter with concomitant blood specimens taken during the midpoint of every other
collection period. After 90 minutes, an injection was made consisting of 720 mg of Pyrazinamide in Experiment 1, 600 mg of Probenecid
in Experiment 2, and 1 g of Sodium Salicylate in Experiment 3.

One ml alloquots of each urine specimen and of TCA precipitated
plasma supernates were fractionated over a polyacrylamide (BIO GEL
P-2) column (1). This process cleanly separates urate from its
major metabolite, allantoin, and isolates both urate and allantoin
from most other urinary constituents. Column fractions underlying
the urate peak were analyzed for urate by the enzymatic spectropho-

tometric method while fractions underlying the allantoin peak were
analyzed colorimetrically for allantoin (2,3). All fractions were
counted by standard liquid scintillation techniques for [14]C. These
studies thus permit quantification of labelled and unlabelled urate
and allantoin in both serum and urine.

From semi-log plots of specific activity against time, we are
able to determine turnover rates of both uric acid and allantoin.
The most conspicuous feature of such plots is the rapidity with
which the isotopic urate equilibrates with the body pool of urate
and with which labelled allantoin is produced and equilibrates with
the allantoin pool. Turnover rates were calculated on the basis
of the final 7 of the 10 overall 15-minute experimental periods.
The slopes of such curves were remarkably linear for urate with
correlation coefficients of 0.999, 0.997, and 0.994 in the three
experiments with respective half-lives of 27.4, 25.7, and 37.6 minutes. When these curves were extrapolated back to time 0, we were
able to calculate values for the apparent urate pool of 192 μM,
246 μM, and 232 μM in the same three experiments. The average
"urate space" was 39.2% of the body weight, which agrees well with
the value of 41% found earlier by Logan et al in their study of a
Springer spaniel (4).

The turnover curves for isotopic allantoin were comparably
linear, with correlation coefficients of 0.992, 0.996, and 0.997 in
the three experiments. Extrapolation of these curves back to time
0 suggested larger allantoin pools of 439, 482, and 331 μM. The
respective half-life values for allantoin were 70.0, 64.0, and 66.9
minutes. These observations of a larger allantoin pool turning over
at a slower rate than urate were confirmed in one additional study
in which labelled allantoin was administered alone. In all three
major experiments, there was no indication that the injected Pyrazinamide, Probenecid, or Sodium Salicylate caused any perturbation
in the turnover results.

The preceding turnover data suggest that the isotopic urate
rapidly equilibrated with body stores, and was then metabolized and
excreted in a uniform manner. Further analysis of the urinary data
indicate, however, that the process is considerably more complex
than it first appears. We chose to evaluate the "uricolytic index"
which has been employed for many years to assess endogenous catabolism of urate (5). In this index, the concentration of allantoin is divided by the sum of the concentrations of urate and allantoin in the same urine specimen. The uricolytic index for endogenous unlabelled urate and allantoin remained quite constant around
a mean value of 95.3%. As might be expected, the uricolytic index
for labelled urate and allantoin began much lower, and rose progressively to full equilibration with the unlabelled material by the
4th 15-minute experimental period. Unexpectedly, however, the
uricolytic index of the labelled material continued to rise throughout the remainder of the experiment, reaching a mean value of 98.4%
in the 10th experimental period. This difference between the

endogenous and the labelled material was highly significant statistically, and reflects a major difference in the way labelled and
unlabelled urate are excreted. I believe that this difference supports the observations of Quebbemann, who found that infusions of
labelled xanthine into isolated renal arteries of dogs resulted in
excretion of labelled uric acid from these animals' kidneys (6).
His findings thus indicate that the dog kidney can synthesize and
directly excrete urinary uric acid. Our observations of differing
uricolytic indices suggest much the same conclusion since urate
synthesized and directly excreted by the kidney might escape the
activity of hepatic uricase, whereas infused uric acid would be more
likely to be catabolized. If this analysis is correct, our observations indicate that at least half of the urinary uric acid in this
animal was directly synthesized and excreted by the kidney.

The data from these studies also permit independent analysis
of the clearance of labelled and of unlabelled urate. The infused
isotopic urate was catabolized so rapidly, however, that reliable
serum concentrations could be measured only through the first 6
collection periods. From the foregoing analysis of uricolytic index
data, one might expect that the clearance of unlabelled urate would
be greater than that of isotopic urate, because of direct renal
synthesis and secretion. We found instead that the clearance of
isotopic urate consistently exceeded that of the endogenous material.
In the first two experimental periods, the clearance of isotopic
urate was close to that of creatinine (45 ml/minute), but by the 3rd
experimental period, it had levelled out to values between 15 and 20
ml/minute. The clearance of endogenous urate, on the other hand,
remained entirely stable at approximately 5 ml/minute. These findings are most disturbing because they appear to violate the basic
premise of all isotopic, physiologic studies: that the isotopic
molecule is and will be recognized as the same as the endogenous
molecule. The findings are consistent, however, with a host of previous investigations in which infusions of endogenous urate in mongrel dogs resulted in accelerated urate clearance compared to baseline values. In our experiments, it is the exogenous urate that is
cleared more rapidly, while the endogenous urate continues to be
cleared at its usual rate. These observations suggest, then, that
the renal tubules in some way recognize the infused exogenous urate
as different from the native, endogenous urate. It is conceivable,
but unlikely, that the reason for this difference is that the ^{14}C
atom at the 2 position is recognized by tubular transport systems.
I think it is more likely that there is a more profound difference
between these molecular species, such as persistent protonation,
stable tautomers, covalent hydration, or the formation of addition
products with other small molecules. This possibility deserves
further exploration in dogs and in man.

In summary:
(1) We have performed short-term turnover studies in a trained
     mongrel dog, and found remarkably linear turnover curves
     for both urate and its principle metabolic product, allan-
     toin.

(2) Infusions of Pyrazinamide, Probenecid, and Sodium Sali-
     cylate had no apparent effect on urate metabolism.

(3) The "uricolytic index" for endogenous urate is consis-
     tently lower than that of infused, isotopic urate. This
     finding suggests renal synthesis and direct excretion of
     unlabelled urate.

(4) The renal clearance of labelled urate is higher than that
     of endogenous urate despite apparent renal synthesis.
     This finding suggests that urate may exist in more than
     one molecular form.

REFERENCES

1. Simkin, P.A. J. Chromatogr. 1970, 47, 103.
2. Liddle, L., Seegmiller, J.E., Laster, L. J. Lab. Clin. Med.
    1959, 54, 903.
3. Vogels, G.D., Van Der Drift, C. Anal. Biochem. 1970, 33, 143.
4. Logan, D.C., Wilson, D.E., Flowers, C.M., Sparks, P.J.,
    Taylor, F.H. Metabolism 1976, 25, 517.
5. Rose, N.M.C. Physiol. Rev. 1923, 3, 567.
6. Quebbemann, A.J., Cumming, J.D., Shideman, J.R. Am. J. Physiol.
    1975, 228, 959.

THE EFFECT OF WEIGHT REDUCTION ON PLASMA AND URINARY LEVELS OF OXYPURINES IN AN OBESE XANTHINURIC PATIENT

C. Auscher[*], C. Pasquier[*], N. Amory[*], G. Gay[**],
A. Aisène[**] and G. Debry

[*] Institut de Rhumatologie, INSERM U. 5 - ERA 337
CNRS, Hôpital Cochin, Paris, France
[**] Département de Nutrition et des Maladies
Métaboliques, Université de Nancy, INSERM U. 59,
Nancy, France

INTRODUCTION

Xanthinuria, a rare hereditary disorder is characterized by a deficiency of xanthine oxidase activity and by an urinary excretion of oxypurines (xanthine and hypoxanthine) which replace uric acid as the end products of purine metabolism.

The mechanism whereby xanthine and hypoxanthine are excreted by the human kidney is unknown. It was only suggested that oxypurines bases are reabsorbed in renal tubules by a mechanism distinct from that responsible for urate reabsorption (1). Renal clearance of oxypurines approaching that of the glomerular filtration rate was found in two patients with xanthinuria (2,3,4) and may be regarded as being the same as in normal patient (1,5,6). It does not seem necessary to postulate a separate tubular defect in renal handling of xanthine as it was suggested by Dickson and Smellie (3) in the original xanthinuric patient of Dent and Philpot (2). Therefore xanthinuria affords the possibility to obtain informations concerning the renal handling of xanthine and hypoxanthine.

A new case of xanthinuria was incidentally discovered through the finding of hypo-uricemia in an obese girl during biochemical investigations prior to the reduction of her overweight of 151 % of the ideal value.

The present study was undertaken in order to investigate the variations of plasma and urinary levels of xanthine and hypoxanthine during hypo-caloric diet.

CASE REPORT

The french young girl was 17 year-old. Birth and childhood were normal. Obesity of gynoid type set up when she was 12 year-old. Then appeared also psycho-motor and schooling difficulties.

Glycoregulation, endocrine and renal functions were normal. She had no nephretic colics but sharp crystals in urine of which 95 % were identified as xanthine. I.Q. was rather low and she was in struggle with her parents. Daily caloric intake when outpatient was difficult to evaluate.

BIOCHEMICAL INVESTIGATION

None xanthine oxidase activity was demonstrated in duodenum biopsies by both histochemical (7) and radiochemical techniques (8).

As complete collection of urine was difficult to obtain, urinary levels of oxypurines were expressed per g of creatinine (Cr). Plasma levels of uric acid and oxypurines were 1.65 and 0.58 mg % of which 73 % was xanthine and urinary levels of uric acid and oxypurines were 8.5 and 348 mg per g of creatinine of which 82 % was xanthine (table I).

Plasma urate lowering drugs given orally : allopurinol 300 mg/day during 7 days, orotic acid 6 g/day during 6 days and probenecid 1 g/day during 3 days did not alter plasma and urinary levels of oxypurines (table I).

Allopurinol and thiopurinol were rapidly converted into their 6-hydroxylated metabolites. The cumulated percentage of dose recovered in urine 2, 6, 10 and 24 h after a single dose of 400 mg given orally were respectively 5 %, 19 %, 31 % and 50 % of oxipurinol and 13, 52, 69 and 76 % of oxithiopurinol. That is to say, in the same magnitude as in control patient (9).

FAMILIAL INVESTIGATION

The pedigree of the family is shown on fig. 1. According to the criteria previously reported (10) no abnormality of oxypurines was detected in the 12 relatives investigated among whom the presumed obligate heterozygotes. Therefore the mode of inheritance was consistent with an autosomal recessive disorder.

Table I

Treatment per day	n	$m \pm se$	Allopurinol 300 mg	Orotic Ac. 6 g	Probenecid 1 g
Plasma[1]					
Uric Ac.	5	1.65 ± 0.2	1.37	1.26	1.32
Oxypurines	5	0.58 ± 0.1	0.62	0.56	0.51
Xanthine %	5	73.3 ± 4.3	-	-	70
Urine[2]					
Uric Ac.	10	8.5 ± 1.6	8.6	9.7	10
Oxypurines	10	348 ± 78	423	286	349
Xanthine %	10	82.5 ± 4.8	79	85	86

(1) mg % ; (2) mg/g creatinine.

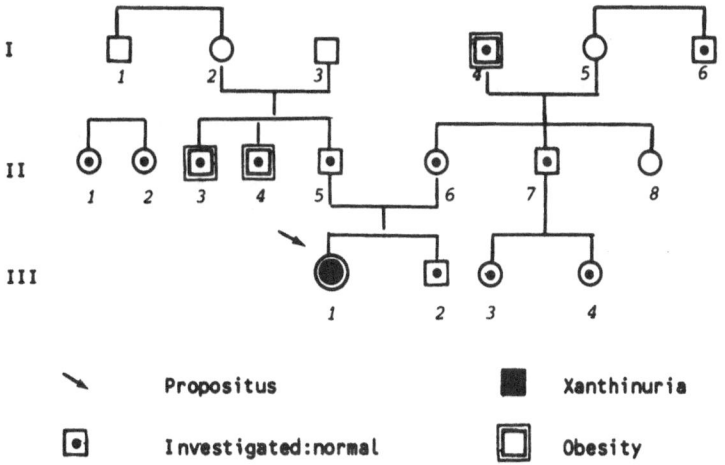

Fig. 1 : Pedigree of the family.

Diabete, obesity and lithiasis were also investigated.
Obesity was the only abnormality discovered in 3 relatives.

HYPOCALORIC DIET AND WEIGHT REDUCTION

The patient was admitted in hospital during hypocaloric
diet from usual down to 500 cal/day, set up gradually as indicated
on fig. 2. After eight days given 500 cal., she was authorized
to go home for a week-end and did not control her diet. When she
came back 500 cal/day were given again for a week. For this reason,
the weight reduction was divided in two periods : in the first
one she lost 7 kg in 29 days and in the second one, 3 kg in 6
days.

As shown on fig. 2, neither plasma nor urinary levels of
oxypurines were altered during these two periods. The slight
increase of oxypurine clearance during the second period might be
correlated to the great variation of diuresis. As the patient was
admitted in an hospital far from our laboratory, plasma lactate
was difficult to carry out during the experiment. It was investigated only at the end of the two periods of weight reduction and
was slightly increased (fig. 2).

Interpretation of these results must be cautious and we
intend to confirm them during a total caloric restriction which
would be performed in an important department of diabetology where
plasma lactate, β-hydroxybutyrate and ketosis would be strictly
controlled. Moreover non-xanthinuric obese patients would be
previously caloric restricted according to the same format, as
control.

Nevertheless the high renal clearance of xanthine and
hypoxanthine in this patient, and the fact that there was no
alteration of plasma and urinary levels of oxypurines when
hypocaloric diet or probenecid were given are additional data
to those previously reported (11) concerning the study of the
renal handling of xanthine and hypoxanthine in xanthinuria.

ACKNOWLEDGMENTS

We would like to acknowledge Mrs F. Loyer for her technical
assistance and Miss J. Chevallier for the preparation of this
manuscript.

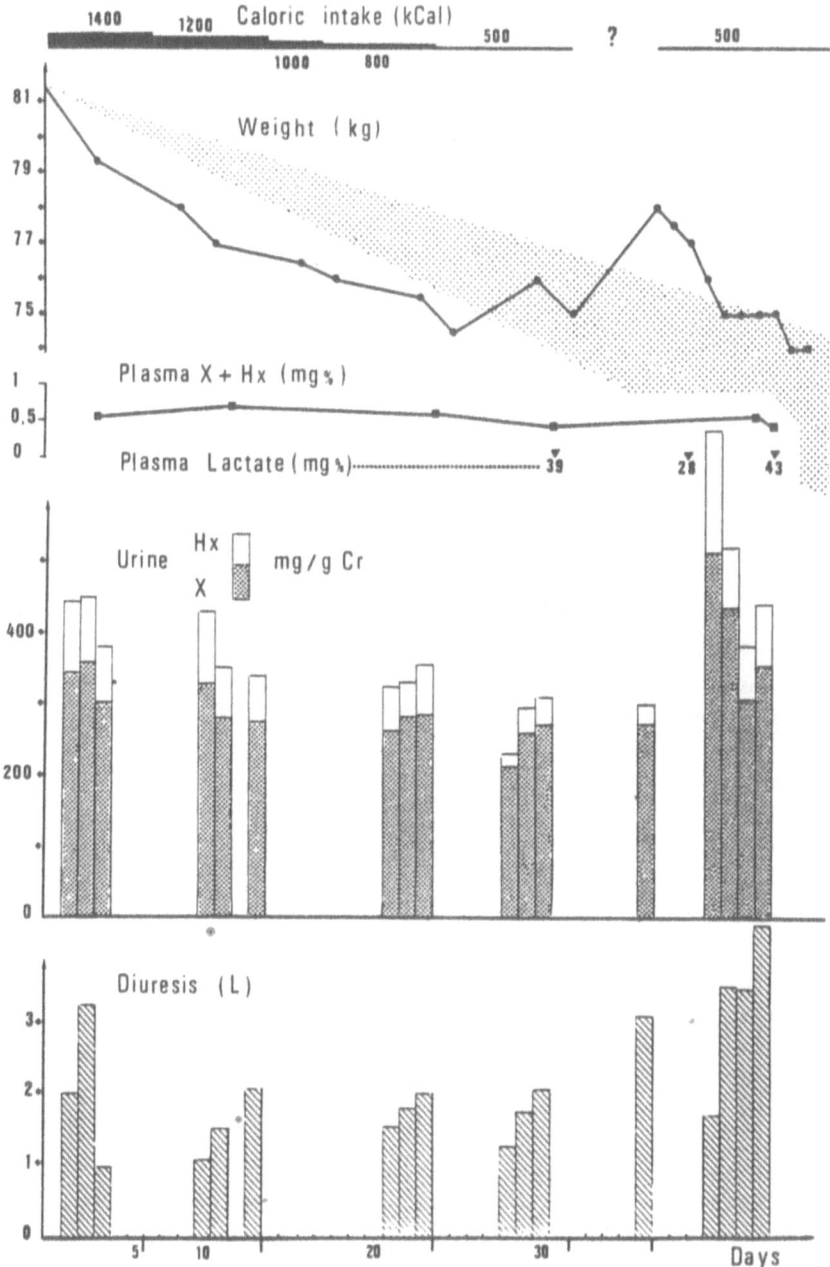

Fig. 2 : weight loss ; plasma and urinary levels of oxypurines
during hypo-caloric diet.

REFERENCES

(1). J.B. Wyngaarden, Hereditary xanthinuria p. 1037, in :
     The metabolic basis of inherited disease, J.B. Stanbury,
     J.B. Wyngaarden, D.S. Fredrickson, fourth ed., Ed.
     Mac Graw Hill, New-York (1978).

(2). C.E. Dent and G.R. Philpot, Xanthinuria,an inborn error
     (or deviation)of metabolism , Lancet I : 182 (1954).

(3). C.J. Dickson and J.M. Smellies, Xanthinuria Brit. Med. J. 2 :
     1217 (1959).

(4). K. Engelman, R.W.E. Watts, J.R. Klinenberg, A. Sjoerdsma and
     J.E. Seegmiller, Clinical, physiological and biochemical
     studies of a patient with xanthinuria and pheochromocytoma,
     Amer. J. Med. 37 : 839 (1964).

(5). S. Goldfinger, J.R. Klinenberg and J.E. Seegmiller, The renal
     excretion of oxypurines, J. Clin. Invest. 44 : 623 (1965).

(6). J.R. Klinenberg, S. Goldfinger, J. Miller and J.E. Seegmiller,
     The effectiveness of xanthine oxidase inhibitor in the
     treatment of gout, Arthr. and Rheum. 6 : 779 (1963).

(7). C. Auscher and N. Amory, The histochemical localisation of
     xanthine oxidase in the rat liver, Biomed. Express 25 : 37
     (1976).

(8). C. Auscher, N. Amory and P. van der Kemp, Xanthine oxidase
     activity in human intestines.Histochemical and radiochemical
     study, this symposium (1979).

(9). C. Auscher, C. Pasquier, N. Mercier and F. Delbarre, Urinary
     excretion of 6-hydroxylated metabolite and oxypurines in a
     xanthinuric man given allopurinol or thiopurinol, Adv. Exp.
     Med. Biol. 41B : 663 (1974).

(10).C. Auscher, C. Pasquier, A. de Gery, R. Weissenbach and
     F. Delbarre, Xanthinuria : Study of a large kindred with
     familial urolithiasis and gout, Biomedicine 27 : 57 (1977).

(11).C. Auscher, C. Pasquier, P. Pehuet and F. Delbarre, Study of
     urinary pyrazinamide metabolites and their action on the
     renal excretion of xanthine and hypoxanthine in a xanthinuric
     patient, Biomedicine 28 : 129 (1978).

XANTHINURIA: THE CAUSE OF HYPOURICEMIA IN HEPATIC DISEASE

H.J. Castro-Mendoza, A. Rapado, C. De La Piedra,
J.M. Castrillo

Biochemistry Department
Metabolic Unit
Jimenez Díaz Foundation
Madrid, Spain

In 1966, the chemical analysis of a vesical calculus extracted from V.Y. (II-7) showed to be composed of 94% xanthine, 4% hypoxanthine and 2% other substances, principally calcium phosphate. As a result, a biochemical study of the purine metabolism in blood and urine was carried out, revealing a new case of xanthinuria with xanthine lithiasis.

A screening test was carried out on 28 members of his family, and three were found to have asymptomatic xanthinuria M.Y. (II-6), R.Y. (II-9) and L.Y. (III-3). The possible heterozygocity was unable to be shown due to the complexity and difficulty of finding an adequate system of detection in determined circumstances, as Harris,[1] and Wats and cols.[2] have stated. These investigations constitute our second case of hereditary xanthinuria,[3].

The biochemical examination of blood and urine corresponding to L.Y. (II-4) did not show any abnormality in his purine metabolism,[4].

Eight years later, 1974, this patient was admitted to the Metabolic Unit of the Jimenez Díaz Foundation in order to undergo a new clinical examination. The main symptoms that were found revealed the existence of a severe hepatic insufficiency, possibly due to an etilic chronic hepatitis with cirrhosis, aside from a previous total gastrectomy, due to an ulcer.

In view of the family background of hereditary xanthinuria, a

new study of his purine metabolism was carried out and another case
of asymptomatic xanthinuria was found, number 8 in our protocol.

In the following tables 1 and 2, the biochemical parameters of
our case are recorded.

We must add to the preceding results,[4], that we were unable to
show any activity of oxidase-xanthina in concentrates of leukocytes
and platelets, following the methods used by Feigelson and col.,[5],
Ultman,[6], Ramia and col.,[7], and Haining and Legan,[8].

The findings in plasma and urine correspond to a clinical
picture of xanthinuria. Others are directly related to a severe
functional deterioration of the hepatic cell, possibly due to ethanol.

DISCUSSION

The above data confirm the biochemical diagnosis of xanthinuria
added or subsequent to a toxic, chronic hepatitis with a cirrhotic
component in the phase of functional insufficiency.

It is difficult to explain the possible biochemical mechanisms,
which led to the clinical expression of xanthinuria, found eight
years after the fist screening in a case that had previously been
studied and where an absolutely normal purine metabolism had been
found.

The probable heterozygocity, even with a certain deffect in
xanthina oxidase allowed relative normality to be maintained in his
purine metabolism, as he belongs to a family with hereditary xanthinuria.

The toxic action derived from his chronic alcoholism could
have produced an impairment of the remaining oxidase xanthina activity reaching a marked hypouricemia due to the following metabolic
possibilities:
a)- The use of xanthina oxidase in ethanol metabolism through the
     catalase system,[9], as a way of liberating excess protons, thus
     trying to maintain the biochemical integrity of the hepatocyte.
b)- In certain circumstances, xanthina oxidase is capable of oxi-
     dizing the acetaldehide, intermediary of ethanol metabolism.
c)- The protein defficiency produced by an abnormal absorption of
     dietetic proteins due to total gastrectomy, along with the poor
     intake of proteins and of other elements provoked a decrease in
     the synthesis of xanthina oxidase as occurs in rats on a diet,
     low or lacking in proteins, according to the experience of
     Mangoni and col.,[10], Vitale and col.,[11], and Rowe and Wyngaarden,
     [12].

Our patient has a heterozygotous trait that with the use of

xanthina oxidase in the metabolism of ethanol and with a shortage of protein intake manifested a biochemical picture of xanthinuria.

Table 1. Plasma

Parameter	1,966		1,974	
Hematocrit	44.2	%	34.6	%
Hemoglobin	15.2	g. "	11.9	g. "
Total proteins	7.400 "	"	7.990 "	"
Albumin	3.620 "	"	2.010 "	"
-Globulin	0.830 "	"	4.187 "	"
Total lipids	0.690 "	"	1.379 "	"
Triglycerides	0.112 "	"	0.480 "	"
Free fatty acids (Nefa)	0.022 "	"	0.056 "	"
Total cholesterol	0.171 "	"	0.220 "	"
Alkaline phosphatase	-----		18.7 u.	
Lactic dehydrogenase	-----		380 "	
Glutamic-oxalacetic transaminase	11.6 u.		302 "	
Glutamic-pyruvic transaminase	14.2 u.		230.4 "	
γ-Glutamiltranspeptidase	23.6 mU/ml		433.2 mU/ml	
Total bilirubin	0.6 mg %		3.2 mg %	
Glucose	84 " "		71 " "	
Lactic acid	------		29.8 " "	
Alkaline reserve	------		16.2 mEq/1	
Urea	26 " "		11.3 mg %	
Creatinine	0.9 " "		0.8 " "	
Uric acid	5.1 " "		0.4 " "	
Oxypurine (xanthine and hypo- xanthine)	0.2 " "		0.6 " "	

Table 2. Urine

Parameter	1.966		1,974	
Creatinine	765	mg/24 hrs.	706	mg/24 hrs.
Uric acid	492	" "	34	" "
Xanthine	4.8	" "	369.7	" "
Hypoxanthine	9.3	" "	67.7	" "

REFERENCES

1. H. Harris, "The principles of human biochemical genetics", pg.243,
    North Holland Pub., London, 1971.
2. R.W.E. Watts, K. Engelman, J.R. Klimberg, J.E. Seegmiller and A.
    Sjoerdsma: Enzyme deffect in a case of xanthinuria. Nature,
    201-395, 1964.

3. H. Castro Mendoza, A. Rapado and L. Cifuentes: Una nueva observación de xantinuria familiar", Rev. Clin. Esp., 124-341, 1972.

4. A. Rapado, H.J. Castro Mendoza, J.M. Castrillo, M. Frutos and L. Cifuentes: "Xanthinuria as a cause of hypouricemia in liver disease", Brit. Med. J., 2, 560, 1975.

5. P. Feigelson, J.M. Ultiman, S. Harris and T. Dashman: "Cellular xanthine oxidase and uricase levels in leukemia and normal mouse leukocytes", Cancer Res., 19, 1230, 1959

6. J.M. Ultman, P. Feigelson and S. Harris: "A method for studying the properties of intracellular anliberated xanthine oxidase", Anal. Bioch. 1, 417, 1960.

7. J. Ramia, J. Bozal and F. Calvet: "Influencia de algunos agentes terapeuticos sobre la biosíntesis del caido úrico", Rev. Esp. Fisiol., 22;85, 1966.

8. J.L. Haining and J.S. Legan: "Fluorometric assay for xanthine oxidase", Anal. Biochem., 21,337, 1967.

9. E. Chance: "Alcohol and aldehyde metabolizing systems". pg. 169, Academic Press, London, 1974.

10. A. Mangoni: "Xanthine oxidase", Bull. Soc. Ital. Biol. Esper., 31, 1397, 1955.

11. J.J. Vitale, P.L. White, L. Sinesterra, D.M. Hegsted and N. Zameheck: "Interrelation of protein and calories during caloric restriction", Fed. Proc., 15, 575, 1956.

12. P. Rowe and J. Wyngaarden: "The mechanism of dietary alteration in rat hepatic xanthine oxidase levels", J. Biol. Chem., 241, 5571, 1966.

XANTHINE-COPROPORPHYRIN III

H.J. Castro-Mendoza

Biochemistry Department
Jimenez Díaz Foundation
Madrid (Spain)

The observation of a weak reddish fluorescence to ultraviolet light at 253 nm wave length in a chromatographic plate destined to identify purine bases contained in a vesical calculus of V.Y. (11-7), member of our second family of hereditary xanthinuria,[1], made us suspect the presence of porphyrins or of pharmacological agents adsorbed by the purines. This presumption brought us to restudy the urethral calculi and a vesical calculus of a xanthinuric patient, F.P. (11-4) belonging to our first family with hereditary xanthinuria,[2].

As control, 47 urinary calculi were studied whose fundamental composition consisted of oxalates, uric acid, calcium phosphates, ammonium-magnesium phosphate and cystine. The result of these investigations was totally negative, and it gave us the impression that it was a specific phenomenon.

METHODS AND RESULTS

a) 50 mg. of each of the different xanthine calculi underwent Schwartz and Wikoff's procedure[3], valid for the isolation of uro, copro, and protoporphyrins in bone, red cells, liver and other tissues. The final result showed that the method was not adequate for the present study.

b) New samples of the same weight were treated with 25 ml. of HCl 1 N; after 5 hours of continuous agitation, the liquid acquired a very pale violet-reddish tone, slightly fluorescent to ultraviolet light at 320 nm wavelength.

c) Under the same conditions, ex⁺raction with HCl 6 N was more complete. The supernatant exhibited a violet-reddish color and the in-
tensity of fluorescence increased from 100 to 150 times as much in
relation to what was obtained in the previous experience.

The supernatants of each calculi, diluted to the concentration
of HCl 0.1 N were examined in the spectrophotometer Unicam SP 800,
and spectroscopic curves of adsorption were obtained which specifically corresponded to coproporphyrins.

Maximum absorptions of coproporphyrins were compared with Rimington's data,[4], and the results appear on Table 1.

With the purpose of confirming these results, porphyrins, isolated and sterified with methanol according to Rimington's method,[4],
had their corresponding spectrums determined as appears on Table 2.

Both spectroscopic series correspond to porphyric rings of the
coproporphyrin type. Thus the presence of degradation products of
Hem or pharmacologic metabolites are excluded.

We stressed the similarity of Rimington's data with that obtained from the study of both types of calculi; vesical and urethral.

With the purpose of defining the isomer type of coproporphyrins
isolated from the calculi, the paper chromatographic method of Falk
and al.,[5], was used. We also compared them with crystallized samples
of coproporphyrin III obtained in our laboratory. The results obtained showed that the Rf of the calculi's porphyrins have an isomer
coproporphyrin III pattern.

A quantitative determination of coproporphyrins in the calculi
appears in Table 3.

The methyllic esters of porphyrins isolated from the calculi
underwent a crystallization process and their microscopic appearance
was analogous to that obtained by other investigators. The porphyrin
methyllic ester crystals exhibited two fusion points, 144°C and
163°C, respectively, and that constitutes a special characteristic
of porphyrin III,[6].

It was permissible to suppose that our patients could deal with
two concomitant diseases: xanthinuria and porphyria. In order to
discard that possibility, even when no symptoms of this second metabolic abnormality were manifest, a 24 hour urine sample of all
the xanthinuric patients from both families was analyzed, and no
porphyrins were found.

A large number of bacteria and fungi are capable of synthe-

Table 1.Dicationic Porphyrins in HCl 1 N.

Porphyrins	λmax(nm)	Soret	II	I	
Copro	λ	399.5	548	590	Rimington
	λ	400	546	590	V.Y.(II-4) vesical calculus
	λ	399.5	545	590	F.P.(II-3) vesical calculus
	λ	400	547	590	F.P.(II-3) urethral calculus

Table 2. Tetra-methyl ester porphyrins in chloroform.

Porphyrins	λmax(nm)	Soret	IV	III	II	I	
Copro	λ	400	498	532	566	621	Rimington
	λ	402	497	531	568	621	V.Y.(II-4) vesical calculus
	λ	401	496	531	567	622	F.P.(II-3) vesical calculus
	λ	401	496	531	566	621	F.P.(II-3) urethral calculus

Table 3.

Patient F.P. (II-3):
    Urethral calculi   44.3 μg./100 mg. sample
    Vesical calculus, nucleus   24.4 μg./100 mg. sample
    Vesical calculus, outer layer   traces

Patient V.Y. (II-4):
    Vesical calculus, nucleus   21.3 μg./100 mg. sample
    Vesical calculus, outer layer   0.9 μg./100 mg. sample

sizing tetrapyrrolic rings in order to form uro, copro and protoporphyrins. Because of that special circumstance urine cultures from our
two patients were made, and again a negative result was obtained.

Finally, the adsorption ability of xanthine and hypoxanthine
was measured, and dicationic solutions of coproporphyrins I and III
within the physiologic pH of our patients' urine showed no incorporation of the porphyrins to the purine bases.

DISCUSSION

Our interest was to interpret the possible mechanisms in the
incorporation of coproporphyrin III in the xanthine calculi.

The presence of uro and coproporphyrins I, biologically inactive, is frequent in calcified structures such as bones and teeth, especially in animals in growing state, probably due to the adsorption of these porphyrins during the Hem synthesis by calcium phosphates under the condition of an excess of ionized calcium,[4].

In the case of F.P. (II-3) vesical calculus, we detected coproporphyrins III in the xanthine nucleus, while we only found traces in the outer layers, constituted exclusively by calcium phosphates,[1].

The fact that xanthine and hypoxanthine lack adsorbing power to coproporphyrins at physiological pH and osmolarity, discards the presumption of a possible phenomenon of adsorption of the porphyrins that are normally contained in the urine or the synthetized tetrapyrrolic end products by bacteria or fungi, as the urines of our patients were always sterile.

Porphyrins are highly reactive compounds due to their free carboxyllic groups, capable of binding to proteins, mainly albumin, through a covalent bond. Lipids also have great capacity to bind with tetrapyrrols and to form stable compounds,[4].

The only possibility is to speculate that xanthine and hypoxanthine are eliminated by the kidney previously combined with coproporphyrin III, and due to their lower solubility are both precipitated together. Also their supposed high molecular weight could explain that both of them constitute the nucleus of the calculus. We have no documented evidence that the formation of this new compound which we provisionally denominated "Coproporphyrin-Xanthine III" is synthetized enzymatically.

The conjugation of coproporphyrin III can only be done among the NH group of xanthine or hypoxanthine and the carboxyl group of tetrapyrrol through a union denominated "bounding ligands".

We have deduced, theoretically, the reaction of 4 mol. of xanthine or hypoxanthine with 1 mol. of coproporphyrin III, with the elimination of 4 mol. of water and a molecular weight of 1.191.

Our attention is drawn by the fact that our other xanthinic patients, a total of ten (including a third family with hereditary xanthinuria presently under study), do not have any symptoms of renal lithiasis, which makes us think of another total or partial metabolic deffect united to xanthinuria and possibly related to Hem metabolism.

REFERENCES

1. H.J. Castro mendoza, L. Cifuentes Delatte and A. Rapado: Una

nueva observación de xantinuria familiar. Rev. Clin. Esp., 124, 341, 1971.
2. L. Cifuentes Delatte and M.J. Castro Mendoza. Xantinuria familiar. Rev. Clin. Esp., 107, 244, 1967.
3. S. Schwartz and H. Wikoff. Methods of Biochemical Analysis (Ed. D. Glick), Interscience Publishers. London, 1963, p. 253.
4. C. Rimington. Porphyrins and Metalloporphyrins (Ed. J. E. Falk), Elsevier Publishing, London, 1964, p.232 and 236.
5. C. Rimington and P.R. Miles. Porphyrins and Metalloporphyrins (Ed. J. E. Falk. Elsevier Publishing, London, 1964, p. 125.
6. J.E. Falk. Porphyrins Biosynthesis (Ed. G.E. Wolstenholme and E.C. Millar). Churchill, London, 1955, p.63.

A CONTROLLED STUDY OF THE EFFECT OF LONG TERM ALLOPURINOL TREATMENT ON RENAL FUNCTION IN GOUT

T. Gibson, H.A. Simmonds, C. Potter, V. Rogers,

Guy's Arthritis Research Unit and Department of Medicine, Guy's Hospital, London, SE1 9RT

An association between gout and kidney dysfunction has been long recognised. It is a reasonable presumption that gout might be the cause of the renal disease.[1] Hypertension, urolithiasis and urinary infection may contribute to this association but it is far from certain whether hyperuricaemia itself may exert a deleterious effect. The available data have been variously interpreted as showing that hyperuricaemia is detrimental[2] or of no consequence.[3]

It might be anticipated that if elevated blood uric acid levels were crucial to the progression of renal dysfunction, treatment with allopurinol should have a beneficial effect in this regard. We have attempted to determine whether such an effect can be demonstrated over a 2 year period.

SUBJECTS AND METHODS

All patients had primary gout of at least one year's duration. None had received regular hypouricaemic therapy prior to their initial investigation. Two had a history of renal stones, 13 had mild hypertension (diastolic B.P.$\geqslant$ 100) controlled by non-diuretic drugs. Renal function was assessed by measurements of blood urea, creatinine, GFR (^{51} Cr. EDTA method)[4] and urine osmolality after 15h. fluid deprivation. After a 4 day low purine diet, each voided sample of urine was timed and collected separately for 24h. After pH estimations the samples were pooled for measurement of protein, uric acid,[5] ammonium and titratable acid.[6] Simultaneous blood samples were obtained for uric acid. Patients were randomly allocated to treatment with colchicine alone or colchicine combined with allopurinol 200mg daily. Renal function was reassessed after 1 and 2 years of treatment. Wherever possible patients were seen

at 2 or 3 month intervals when blood pressure, blood urea, creatinine
and uric acid were measured. Sequential results were compared
statistically by Student's t test for paired data.

RESULTS

57 patients were admitted to the study. Their characteristics are
outlined in Table 1.

TABLE 1   Principal Clinical Features of
                  Gout Patients in the 2 Treatment Groups

	Colchicine	Allopurinol
No. patients	32	25
Sex	32M	24M 1F
Mean Age $\pm$ S.D.	49 $\pm$ 11.9	49.4 $\pm$ 11.9
Mean body wt. $\pm$ S.D. (kg)	80 $\pm$ 9.8	82.9 $\pm$ 14.2
Family History Gout	6 (19%)	12 (48%)
Hypertension	6 (19%)	7 (28%)
Regular Alcohol	23 (72%)	17 (68%)
Mean duration Gout $\pm$ S.D. (yrs.)	5.6 $\pm$ 6	7.1 $\pm$ 7
Tophi	7 (22%)	3 (12%)
Renal calculi	1	1

RESULTS

     Results over the 2 years were obtained for 51 subjects and
these are summarised in Table 2. The most striking findings were
significant declines in GFR and urine concentrating ability at 1
and 2 years in those treated with colchicine alone. There was no
significant change of blood urea or creatinine in either treatment
group. The pattern of daily urine pH in terms of minimum and
maximum values did not alter during the 2 years. In the colchicine
treated group, ammonium and titratable acid excretion did not change
but in those receiving allopurinol there was a significant fall of
ammonium excretion with no significant change of titratable acid.
As anticipated, blood and urine uric acid values were reduced
significantly by allopurinol. Urate clearance (Cur) and fractional

TABLE 2   Mean ± S.D. of renal function tests over 2 years. Figures in parenthesis denote number of paired observations

Time	Colchicine			Allopurinol		
	0	1yr	2yrs	0	1 yr	2yrs
GFR (ml/min/1.73m²)	98±17.5	93±15.4(32)	91±16.5(30)*	89±22	92±23.5(25)	89±24(21)
Urine concentration (mOsm/kg)	826±106	794±135(28)*	764±131(27)***	792±145	792±149(20)	776±152(19)
Ammonium excretion (mmol/24h)	32±12	30±9(21)	30±9(21)	40±13	30±14(16)**	31±10(10)*
Titratable acid excretion (mmol/24h)	21±10	20±9(21)	19±8(21)	23±13	21±13(15)	23±7(18)
No. patients with proteinuria (range g/24h)	6(0.2-0.5)	5(0.2-0.3)	7(0.2-1.6)	6(0.2-4.0)	6(0.2-4.1)	7(0.2-3.4)
B.U.A. (mmol/l)	0.38±0.06	0.37±0.1(32)	0.37±0.09(29)	0.41±0.07	0.27±0.09(25)***	0.28±0.07(21)***
Urate excretion (mmol/24h)	3.2±0.94	3.1±0.86(26)	2.87±1.0(25)	3.05±1.03	2.0±1.0(21)***	1.83±0.73(20)***

*p < 0.05   ** p<0.02   ***p<0.001

urate clearance (Cur/GFR x 100) did not change in the colchicine
group but in the allopurinol patients, mean Cur ($\pm$ S.D.) declined
from 4.71 $\pm$ 1.43 to 3.8 $\pm$ 1.26ml/min/1.73m^2 after 2 years treatment
(t = 4.01; p<0.001). Fractional urate clearance fell from
5.48 $\pm$ 1.63% to 4.41 $\pm$ 1.43% (t = 4.0; p< 0.001). When those treated for 2 years with allopurinol were compared with an equal number
of age-matched normouricaemic subjects (BUA 0.23 $\pm$ 0.08 mmol/l) both
Cur and Cur/GFR x 100 were significantly lower than those of controls
which were 7.0 $\pm$ 2.2 ml/min/1.73m^2 (t = 4.56; p< 0.001) and 7.1$\pm$2.46%
respectively (t = 4.08; p<0.005)

TABLE 3   Initial features of 12 colchicine patients
               whose GFR declined $\geq$ 10ml/min/1.73m^2
               compared with 18 similarly treated subjects

	Reduction GFR >10ml/min/1.73m^2	No or minimal change GFR
No. patients	12	18
Mean age $\pm$ S.D.	46.4 $\pm$ 14	50 $\pm$ 10
Tophi	3 (25%)	3 (22%)
Body wt. $\pm$ S.D. (kg)	79.7 $\pm$ 7.1	81.8 $\pm$ 10.7
Family history gout	2 (17%)	3 (22%)
Regular Alcohol	10 (83%)	12 (67%)*
Hypertension	2 (17%)	3 (17%)
Pre-treatment values: BUA $\pm$ S.D. (mmol/l)	0.37 $\pm$ 0.07	0.38 $\pm$ 0.06
Urate excretion $\pm$ S.D. (mmol/24h)	3.31 $\pm$ 1.12	3.07 $\pm$ 0.87
Urine conc. ability $\pm$ S.D. (mOsm/kg)	833 $\pm$ 84	832 $\pm$ 113
Maximum diurnal urine pH $\pm$ S.D.	6.2 $\pm$ 0.56	6.3 $\pm$ 0.81

* x^2 = 1.66; N.S.

Closer anaylsis of the colchicine treated patients revealed that the significant deterioration of renal function could be ascribed mainly to 12 (40%) patients whose GFR fell more than $10ml/min/1.73m^2$. By contrast, only 2 (9%) of those receiving allopurinol had reductions of GFR which exceeded 10ml/min. ($x^2 = 5.76$; $p < 0.025$). The features of those with > 10ml/min. fall of GFR were compared with the remaining 18 colchicine treated patients whose GFR values fell slightly or were unchanged (Table 3). No obvious differences were apparent.

DISCUSSION

An earlier study, based partly on anecdotal and selected data, implied that renal disease may cause death in 25% of gouty patients.[7] Subsequently it has become evident that although renal dysfunction may occur in gout it may not be severe and it cannot be assumed that renal urate deposition is an exclusive cause.[8,9] There is conflicting evidence about the role of hyperuricaemia,[3,10] and there is no firm evidence that reducing blood uric acid prevents progression of kidney involvement.[1] Our data now suggest that treatment of hyperuricaemia with allopurinol may prevent the slow progression of renal dysfunction in gout and imply that uric acid does exert a harmful effect. Over a 2 year period the decline of GFR was insufficient to result in nitrogen retention and there was no change of the frequency and magnitude of proteinuria. Patients who exhibited most deterioration of GFR could not be distinguished by clinical or laboratory features and in particular they did not manifest either worse urine concentration or more acid urine, features which have been considered premonitory of gouty nephropathy.

The low urine pH observed in gout is associated with a relative impairment of ammonium excretion. It is uncertain whether this reflects overall renal insufficiency or is linked to uric acid metabolism and excretion.[11] The reduction of ammonium excretion in those treated with allopurinol was a phenomenon that has not been previously observed and cannot be readily explained by us.

Diminished urate clearance is a feature of our gouty population.[12] As a causative factor in hyperuricaemia its role has been controversial.[13,14] Restoration of normouricaemia in those patients treated with allopurinol accentuated this defect relative to a control population. This suggests that impaired urate clearance is an important mechanism in the evolution of hyperuricaemia and is not an artefactual observation.

ACKNOWLEDGEMENTS

We would like to acknowledge the help of the Arthritis and Rheumatism Council and the Wellcome Foundation.

REFERENCES

1. Emmerson, B.T.   Gout, uric acid and renal disease.   <u>Med. J. Aus.</u>
    1 : 403 (1976)
2. Steele,T.H.   Asymptomatic hyperuricemia.   Pathogenic or innocent
    bystander?   <u>Arch. Int. Med.</u>   139 : 24 (1979)
3. Berger, L. and Yu, T.S.F.   Renal function in gout <u>IV</u> An analysis
    of 524 gouty subjects including long-term follow-up studies.
    <u>Am. J. Med.</u> 59 : 605 (1975)
4. Garnett, E.S., Parsons, V., Veall, N.   Measurement of glomerular
    filtration rate in man using a ^{51}Cr. Edetic acid complex.
    <u>Lancet</u> 1 : 818 (1967)
5. Simmonds, H.A.   A method of estimation of uric acid in urine and
    other body fluids.   <u>Clinica Chim. Acta</u>   15 : 375 (1967)
6. Chan, J.C.M.   The rapid determination of urinary titratable
    acid and ammonium and evaluation of freezing as a method of
    preservation.   <u>Clin. Biochem.</u>   5 : 94 (1972)
7. Talbott, J.H. and Terplan, K.L.   The kidney in gout.   <u>Medicine</u>
    39 : 405 (1960)
8. Barlow, K.A. and Beilin, L.J.   Renal disease in primary gout.
    <u>Q.J. Med.</u>   37 : 79 (1968)
9. Klinenberg, J.R., Kippen, I., Bluestone, R.   Hyperuricemic
    nephropathy: pathologic reatures and factors influencing urate
    deposition.   <u>Nephron</u> 14: 88 (1975)
10. Klinenberg, J.R., Gonick, H.E., Dornfield, L.   Renal function
    abnormalities in patients with asymptomatic hyperuricemia.
    <u>Arthritis Rheum.</u>   18 : 725 (1975)
11. Gibson, T., Hannan, S.F., Hatfield, P.J., Simmonds, H.A.,
    Cameron, J.S., Potter, C., Crute, C.M.   The effect of acid
    loading on renal excretion of uric acid and ammonium in gout.
    <u>Adv. Exp. Med. and Biol.</u>   76B : 46 (1977)
12. Gibson, T., Simmonds, H.A., Potter, C., Jeyarajah, N.,
    Highton, J. Gout and renal function.   <u>Europ. J. Rheum. Inflam.</u>
    1 : 79 (1978)
13. Yu, T.S.F., Berger, L., Gutman, A.B. Renal function in gout II.
    Effect of uric acid loading on renal excretion of uric acid.
    <u>Am. J. Med.</u>   33 : 829   (1962)
14. Nugent, C.A., MacDiarmid, W.D., Tyler, F.H.   Renal excretion of
    urate in patients with gout.   <u>Arch. Int. Med.</u>   113 : 115 (1964)

STUDIES WITH ALLOPURINOL IN PATIENTS WITH IMPAIRED RENAL FUNCTION

Gertrude B. Elion, Fran M. Benezra, Thomas D. Beardmore
and William N. Kelley
Wellcome Research Laboratories, Research Triangle
Park, NC and Duke University Medical Center,
Durham, NC.

Studies on the metabolism and pharmacokinetics of allopurinol
have shown that the principal metabolic product, oxipurinol
(4,6-dihydroxypyrazolo(3,4-d)pyrimidine), is formed rapidly and
has a prolonged plasma half-life[1]. The plasma half-life of
allopurinol is in the range of 0.5-1 hour, while that of oxipurinol
has been reported to range from 18 to 30 hours[1,2]. Since oxipurinol
binds very tightly to the reduced form of xanthine oxidase[3,4] and
inactivates the enzyme, its inhibitory effects on urate synthesis
are apparent long after allopurinol itself has been cleared from
the body.

Oxipurinol is cleared by the same mechanisms that govern the
clearance of uric acid (glomerular filtration, tubular reabsorption and secretion) but is higher by a factor of about three[2].
Individuals with poor urate clearance generally also have a slow
clearance of oxipurinol. To avoid an accumulation of the drug in
such patients a study was undertaken to determine suitable maintenance regimens of allopurinol.

METHODS

Patients taking 300 mg/day of allopurinol received 100 mg
t.i.d. Blood samples were drawn 2 to 4 hours after ingestion of
a 100 mg tablet. Patients on hemodialysis were given a single
dose of 400 mg after each dialysis. The methodology for the
determination of oxipurinol in plasma and urine used for these
studies was an isotope dilution procedure which has been previously
described[1,2]. The serum and urinary urate concentrations were

determined enzymatically as were the urinary oxypurines, hypoxanthine and xanthine[5].

RESULTS AND DISCUSSION

Two types of patients were studied. One group had hyperuricemia and creatinine clearance values between 10 and 20 ml/min.
The second group had severe renal impairment, C_{Cr} < 3 ml/min.,
and required dialysis.

Plasma levels of oxipurinol are dependent on the dose and
absorption of allopurinol and on the renal clearance of oxipurinol.
Patients with normal renal function taking 300 mg of allopurinol
per day generally have plasma oxipurinol concentrations of 0.5 -
1.0 mg/dl[2,6]. Patients S.S. and E.C., who had C_{Cr} = 20 and 10
ml/min respectively, showed plasma oxipurinol concentrations of
3.0 and 2.9 mg/dl after prolonged treatment with 300 mg allopurinol
per day (Table 1). This is approximately the concentration which
would be achieved by doses of 600 mg/day in those with normal
renal function[2]. It, therefore, seemed possible that daily
300 mg doses might be excessive in such patients and that a lower
dose of allopurinol might achieve effective oxipurinol levels
when renal function was impaired. Two patients, C.C. and S.L.,
with C_{Cr} = 10 ml/min were treated with only 100 mg of allopurinol
per day. They achieved plasma urate concentrations of 6.4 and
6.3 mg/dl, while maintaining an oxipurinol level of 1-2 mg/dl.
Urinary excretion of urate in these individuals ranged between
155-295 mg/day while on treatment with allopurinol; the highest
urinary oxypurine excretion was 79 mg/day (calculated as xanthine).

Table 1

Effect of Daily Administration of Allopurinol in Patients
with Impaired Renal Function

PATIENT	C_{Cr} ML/MIN	DAILY DOSE MG	TIME, DAYS	PLASMA URATE MG/DL	PLASMA OXIPURINOL MG/DL
S.S.	20	300	>90	7.5	3.0
E.C.	10	300	0	10.3	--
			120	4.9	2.9
C.C.	10	100	31	6.4	2.0
			66		1.9
S.L.	9.8	100	22	6.3	1.4
			43		0.9
			64	6.2	1.2

Patients with renal failure, i.e. with C_{Cr} < 3 ml/min, would
be expected to accumulate oxipurinol in the plasma to an even
greater extent than those described above. One such individual,
E.A.C., who had received 300 mg of allopurinol daily for two
weeks, did indeed show an oxipurinol concentration of 5.3 mg/dl
(Table 2). After one week with no further therapy the plasma
oxipurinol fell to 2.7 mg/dl ($t_{\frac{1}{2}}$ of one week) with no rise in
plasma urate. This individual was subsequently put on a regimen
of 400 mg of allopurinol every fourth day. This was sufficient
to maintain his plasma urate within normal values, with plasma
oxipurinol levels of 0.6 - 0.8 mg/dl. Two other patients with
renal failure showed oxipurinol levels of 0.8 - 1.3 mg/dl for
four days after a single 400 mg dose of allopurinol (Table 2).
In case G.A. the plasma urate remained below 7 mg/dl; however, in
patient A.S. the plasma urate rose to 10.2 mg/dl on the fourth
day.

Since neither allopurinol nor oxipurinol is bound to plasma
proteins[1], it was to be expected that dialysis would remove these
drugs from the blood. This was verified by some earlier studies
in which the oxipurinol concentrations were shown to drop approximately 50% during an 8-hour hemodialysis in patients who had
received 400 mg of allopurinol daily[7]. These patients generally
showed oxipurinol levels of 2.2±0.6 mg/dl, as measured by an
enzymatic method. In the present study, both urate and oxipurinol
concentrations were measured before, and after two hours and six
hours of dialysis (Table 3). For patient E.A.C., these analyses
were repeated during a second dialysis. In each case, the fall
in oxipurinol levels closely paralleled the fall in urate levels
(Table 3). The mean plasma urate before dialysis was 6.7 mg/dl;
it fell a mean of 57% during the 6 hour period of dialysis. The
mean oxipurinol concentration before dialysis was 0.85 mg/dl; it
fell an average of 43% during the 6 hours.

These results have confirmed the fact that oxipurinol
clearance is greatly reduced in patients with poor renal function.
The incidence of side-effects appears to be greater in patients
with impaired renal function and it is possible that sustained,
elevated levels of oxipurinol may be involved. It is hoped that
these studies will provide guidelines for the adequate but not
excessive treatment of such patients. In those with C_{Cr} values
of 10-20 ml/min, a dose of allopurinol of 100-200 mg/day appears
sufficient to maintain normal urate levels. In those with severe
renal insufficiency, the effects of a single 400 mg dose can be
retained for 4 days. Since dialysis removes oxipurinol from
blood at about the same rate as it removes uric acid, dosing with
allopurinol should follow dialysis.

Table 2

Persistence of Effect of Allopurinol in Patients with
Renal Failure.

PATIENT	C_{Cr} ML/MIN	DAILY DOSE MG	TIME, DAYS	PLASMA URATE	OXIPURINOL MG/DL
E.A.C.	2.6	300 x 14	15	6.7	5.3
			22	6.6	2.7
		400 x 1	1	4.2	0.8
			4	6.1	0.6
A.S.	≦3	400 x 1	1	6.8	1.1
			4	10.2	0.8
G.A.	<3	400 x 1	1	3.8	1.3
			4	6.6	1.3

Table 3

Effect of Dialysis on Plasma Concentrations of Urate
and Oxipurinol.

PATIENT	HOURS OF DIALYSIS	PLASMA URATE	OXIPURINOL MG/DL
E.A.C.	0	6.1	0.57
	2	4.4	0.47
	6	3.2	0.37
	0	5.4	0.57
	2	3.3	0.43
	6	2.3	0.31
A.S.	0	10.2	0.83
	2	7.4	0.67
	6	4.8	0.46
G.A.	0	5.0	1.32
	2	3.2	0.77
	6		0.54

REFERENCES

1.   G.B. Elion, A. Kovensky, G.H. Hitchings, E. Metz and R.W.
      Rundles, Metabolic studies of allopurinol, an inhibitor of
      xanthine oxidase, Biochem. Pharmacol. 15:863 (1966).
2.   G.B. Elion, T.-F. Yü, A.B. Gutman and G.H. Hitchings, Renal
      clearance of oxipurinol, the chief metabolite of allopurinol,
      Amer. J. Med. 45:69 (1968).
3.   V.Massey, H. Komai, G. Palmer and G.B. Elion, On the mecha-
      nism of inactivation of xanthine oxidase by allopurinol and
      other pyrazolo(3,4-d)pyrimidines, J. Biol. Chem. 245:2837
      (1970).
4.   T. Spector and D.G. Johns, Stoichiometric inhibition of
      reduced xanthine oxidase by hydroxypyrazolo(3,4-d)pyrimidines,
      J. Biol. Chem. 245:5079 (1978).
5.   R.W. Rundles, J.B. Wyngaarden, G.H. Hitchings, G.B. Elion
      and H.R. Silberman, Effects of a xanthine oxidase inhibitor
      on thiopurine metabolism, hyperuricemia, and gout, Trans.
      Assoc. Amer. Physicians 76:126 (1963).
6.   G.P. Rodnan, J.A. Robin, S.F. Tolchin and G.B. Elion,
      Allopurinol and gouty hyperuricemia. Efficacy of a single
      daily dose, J. Amer. Med. Assoc. 231:1143 (1975).
7.   C.P. Hayes, Jr., E.N. Metz, R.R. Robinson, and R.W. Rundles,
      Use of allopurinol (HPP) to control hyperuricemia in patients
      on chronic intermittent dialysis. Trans. Amer. Soc. Artif.
      Intern. Organs 11:247 (1965).

PHARMACOLOGICAL EFFECTS OF 1,3,5-TRIAZINES AND THEIR EXCRETION

CHARACTERISTICS IN THE RAT

Max Hropot, Fritz Sörgel, Bela v. Kerékjárt6,

Hans J. Lang, Roman Muschaweck

Hoechst AG, Postfach 800320
D-6230 Frankfurt(M) 80, W.-Germany

INTRODUCTION

Hyperuricemia is a metabolic and/or excretory disorder of purines associated with humans only. Thus the availability of experimental animal models for the study of human hyperuricemia has been very limited. In most mammalian species uric acid is converted to allantoin by uricase. Uricase is lacking in man and the apes. In these species uric acid rather than water-soluble allantoin is the end-product of purine metabolism (Fanelli[3]). To obtain an animal model for studies on hyperuricemia, hepatic uricase must be blocked with a selective inhibitor. Recently, use of the rat as a hyperuricemic animal model was described and subsequently used in different areas of research. Fridovich[4] and Johnson[1] have shown that certain s-triazines are potent competitive inhibitors of uricase. Particularly oxonic acid and amide of oxonic acid have been described as effective inhibitors of uricase activity in vitro and in vivo. Since no studies have been carried out on their metabolism or excretion characteristics, the aim of this study was to examine the pharmacological effects and excretion characteristics of oxonic acid, amide of oxonic acid and of 5-azauracil in rats.

METHODS

In vitro studies: uric acid and the inhibitors were used as aqueous stock solutions diluted with 20 M borate pH 9.3 for the assay; the reaction was started by adding 2 μl of enzyme (Uricase, Boehringer Mannheim). 2,6,8-trichloropurine was used as standard. The total volume was 1.0 ml, the temperature $37^{\circ}C$, the wavelength 293 nm. The initial velocity of the reaction was recorded with CARY 118.

In vivo studies: male Wistar rats (company's own breeding) with a body weight of 250 $\pm$ 15.5 g were used. Oxonic acid used as K^+-oxonate (potassium salt of 2,4-dioxo-1,2,3,4-tetrahydro-1,3,5-triazine-6-carboxylate), amide of oxonic acid (oxonamide = 2,4-dioxo-1,2,3,4-tetrahydro-1,3,5-triazine-6-carboxylic acid amide), and 5-azauracil(2,4-dioxo-1,2,3,4-tetrahydro-1,3,5-triazine) were injected i.p. at a dosage of 250 $mg \cdot kg^{-1}$ as a suspension in 0.25% methyl cellulose. The control rats received the vehicle alone. Blood samples were taken retro-orbitally at 2-, 4-, 6-, and 24-hour intervals following injection of the uricase inhibitors, and assayed for uric acid (Urica-quant-method).

Kinetics studies were carried out in male Wistar rats (mean body weight: 270 $\pm$ 14.5 g), anesthetized with Inactin® (100.0 $mg \cdot kg^{-1}$ i.p.). For test preparation, jugular vein and carotid artery were cannulated for infusion and blood sampling. Laparotomy was performed to cannulate both ureters for urine collection. Throughout the experiment, 10 % mannitol solution was infused at 2.5 $ml \cdot h^{-1}$ per 100 g body weight. 30 min. after beginning of the infusion, the test preparations were injected i.v. at a dosage of 250 $mg \cdot kg^{-1}$ body weight. In the first hour after the injection, the urine was collected every 5 min., and in the second hour every 15 min. Oxonate and oxonamide were dissolved in H_2O with addition of 1 N NaOH.

For analytical procedure, compounds were used as received (Pharma Synthese HOECHST AG, according to Piskala[8]). All solvents were analytical reagent grade (E. Merck, Darmstadt, West Germany). Silica gel 60 F 254 (for oxonamide) and cellulose F_{254} plates (for oxonate) were used. Plates were spotted by hand. Absorbance was determined by scanning TLC plates with a spectrodensitometer (Zeiss PM Q III, West Germany) using the reflectance mode. Peaks were electronically integrated. Urine (0.1 - 2.0 µl) was spotted directly on the plates. The plates were developed in water/ethanol 1:1 for oxonate (cellulose plates) and for oxonamide (silica gel plates). The plates were developed allowing the solvent to migrate about 10-12 cm. The plates were air-dried for 10 min. and scanned at 272 nm.

RESULTS

The inhibitory activity of the tested compounds in vitro on uricase reaction is shown in Table 1. The ratio of the

initial velocity of the inhibited and not inhibited reaction
was the lowest with oxonamide as compared with other compounds.
The inhibitor constant K_i (dissociation constant of the enzyme
inhibitor complex) of oxonamide, 2.95×10^{-7}, was distinctly
lower than that of oxonate, 6.2×10^{-7}.

Table 1. Quantitative differences of inhibitor properties

V = initial velocity of the inhibited reaction
(V at 5×10^{-5} M uric acid and 1×10^{-5} M inhibitor)

V_o = initial velocity of not inhibited reaction

K_i = inhibition constant

	OXONATE	OXONAMIDE	5-AZAURACIL	TRICHLOROPURINE
V/Vo	0.313	0.176	0.703	0.310
% Inhibition	68.7	82.4	29.7	69.0
K_i	6.2×10^{-7}	2.95×10^{-7}	3.33×10^{-6}	6.48×10^{-7}

The Lineweaver-Burk diagramm in the Fig. 1 shows that all
tested inhibitors act competitively in the assayed concentration range.

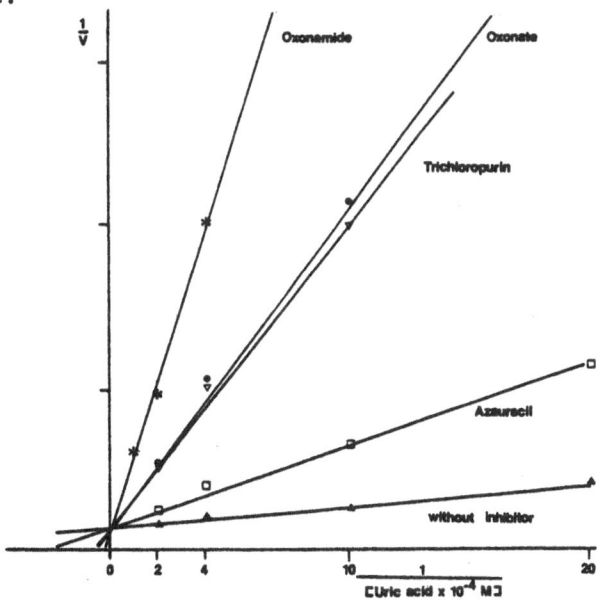

Fig. 1. Competitive inhibition of uricase by oxonamide, oxonate,
trichloropurine and 5-azauracil (10 μM)

Oxonamide was found to be superior to oxonate as a uricase
inhibitor in vivo (Fig. 2). However, 2 hours after the i.p. injection of tested compounds (250 mg·kg^{-1}), oxonate raised the
serum uric acid level to a mean value of 195.8 $\overset{+}{-}$ 29.5 μmol/l.

At that point, oxonamide was ineffective. Four hours after the
injection, both compounds increased the serum uric acid level to
the range of 124.1 to 150.0 μmol/l.

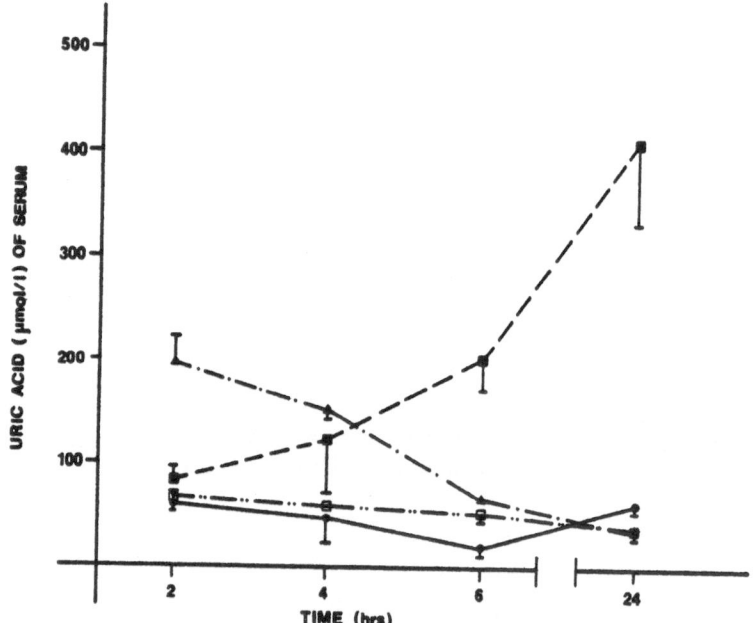

Fig. 2. Effect of 5-azauracil (□—·—·—□), oxonate (▲— · —▲)
       and oxonamide (■- - - -■) on serum uric acid levels in the
       rat as compared to control (●———●). Each point re-
       presents the mean value of 6 rats.

Thereafter the serum uric acid level enhanced steadily
and reached a peak level (410.3 $\overset{+}{-}$ 78.4 μmol/l) 24 hrs after
injection of oxonamide. Whereas after injection of oxonate the
serum uric acid returned to normal levels in approximately 6 hrs.,
5-azauracil did not affect the serum uric acid.

After intravenous injection of 250 mg·kg^{-1} oxonate or its
amide, measurable concentrations appeared in the urine. Both
drugs reached highest concentrations during the first five
minutes after administration (Fig. 4 and 5). Oxonate could be
detected and quantified only until 50-60 min. after administration (Fig. 5). Azauracil seemed to be highly metabolized and
could only be detected in small amounts which could not be quantified. Urinary excretion of oxonate was very rapid, peak ex-

cretion occurred during the first five minutes after administration. However, most of the drug was excreted during the first
50 minutes after administration. The oxonamide was also excreted
very rapidly, 56 % of the dose being excreted during the first
two hours after drug administration (Fig. 3).

The half-lives of renal elimination of both drugs are given
in Table 2.

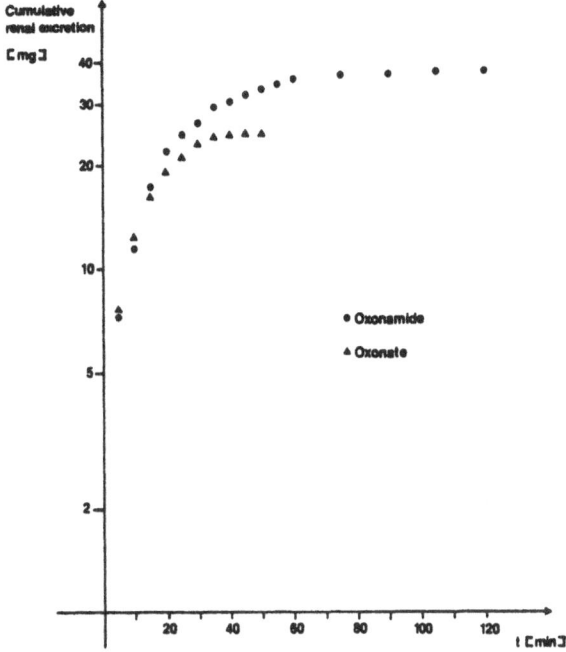

Fig. 3. Cumulative renal excretion of unchanged oxonate and
        oxonamide after intravenous administration of
        250 mg·kg⁻¹ body weight (n = 4)

Table 2. Half-lives of renal elimination of oxonate and
         oxonamide in rats

Oxonate		Oxonamide	
$t^{1/2}$(1st phase)	11.4 min.	$t^{1/2}$	25.4 min.
$t^{1/2}$(2nd phase)	4.64 min.		

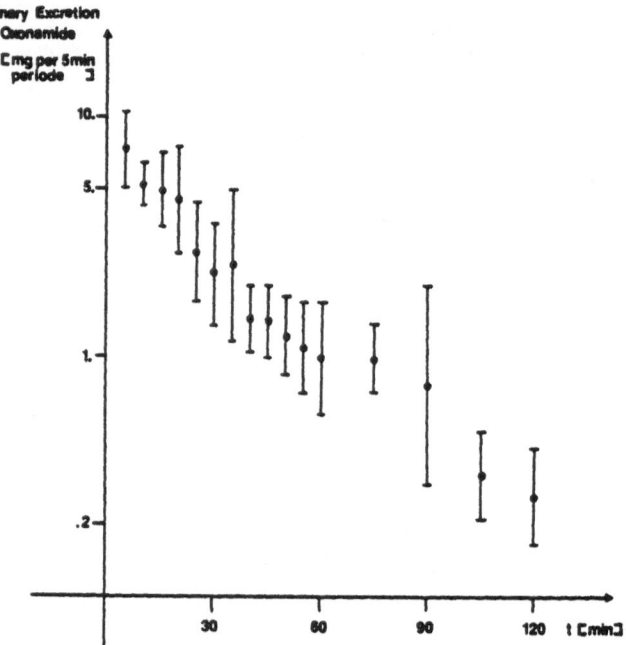

Fig. 4. Renal excretion of unchanged oxonamide after intravenous
injection of 250 mg·kg^{-1} body weight (n = 4)

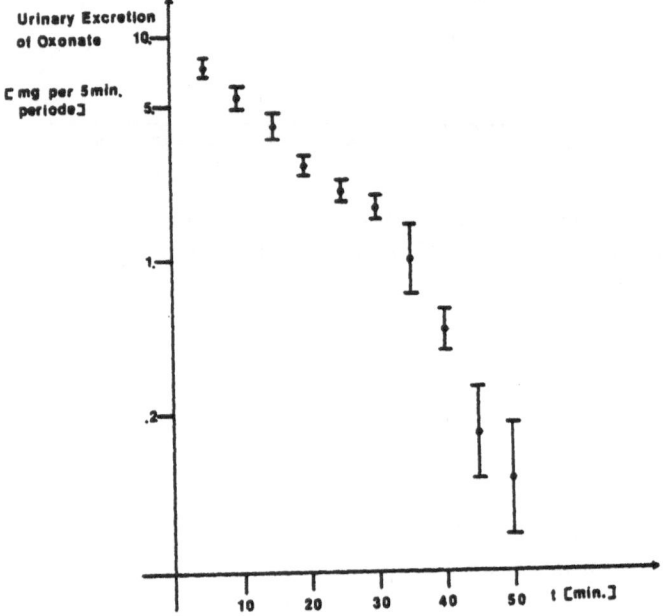

Fig. 5. Renal excretion of unchanged oxonate after intravenous
injection of 250 mg·kg^{-1} body weight (n = 4)

DISCUSSION

In vitro studies demonstrated that oxonamide is a potent inhibitor of uricase, the inhibitor constant K_i of oxonamide, 2.95×10^{-7}, being distinctly lower than that of oxonate, 6.2×10^{-7}, 5-azauracil, 3.33×10^{-6}, and trichloropurine, 6.48×10^{-7} (Tab. 1). Previously Fridovich[4], Johnson[1], Stavric[2] and Stavric[5] showed already these effects; additionally they found oxonate and oxonamide did not inhibit xanthine oxidase. The Lineweaver-Burk diagramm in Fig. 1 shows that all tested compounds act as competitive inhibitors of uricase. Iwata[7] tested various hypoxanthine and xanthine analogs. 2,8-diazahypoxanthine was the most inhibitory. It caused 90 and 64 per cent inhibition of uricase at concentrations of 1×10^{-5} M and 1×10^{-6} M, respectively. An in vivo study showed that 2,8-diazahypoxanthine given to rats $(250 \text{ mg} \cdot \text{kg}^{-1}$ i.p.) caused maximum increase of urate 1 hr. after treatment (238 μmol/l); the changes lasted for at least 5 hrs. In this study, oxonamide was found to be superior to oxonate as a uricase inhibitor in vivo (Fig. 2). In contrast to Johnson[6], in whose study the peak level of urate after oxonamide $(250 \text{ mg} \cdot \text{kg}^{-1}$ i.p.) was at 6 hrs., urate in serum increased steadily and reached its peak level at 24 hrs. in our study. However, in both studies the effect of oxonamide lasted for 24 hrs. As shown by Iwata[7] in in vivo studies, 2,8-diazahypoxanthine was as effective as oxonate.

Thereafter, we postulated that the long-lasting effect of oxonamide was due to its prolonged excretion. The kineticsstudies indeed showed that oxonamide excretion by the kidneys was different from that of oxonate. Oxonamide was eliminated by the kidneys with a half-life of 25.4 min. which was double of that of oxonate during 0-30 min. after administration. From 30-50 min., oxonate excretion was even four times higher than that of oxonamide (Table 2). This significant difference leads to the conclusion that the different duration of action of these two compounds on uricase inhibition may be explained by their different elimination kinetics. From the excretion data of oxonate, it is further supposed that oxonate excretion was saturated during 0-30 minutes after administration, a fact which was not observed for oxonamide. The very rapid excretion of both compounds during the first 30 minutes may be explained by tubular secretion which supplies high amounts of the compound for renal elimination. The more rapid renal excretion of oxonate is probably due to a higher affinity of this carboxylic acid to the anionic transport system in the proximal tubule. As only 50-60 % of the unchanged compounds were excreted via kidneys, metabolic conversion or extrarenal excretion seems to occur. No measurable amounts of oxonic acid were found in urine after administration of the amide, which shows that oxonamide exerts its action mainly in its unchanged form.

ACKNOWLEDGEMENTS

    The authors are grateful to Miss M. Meister,
Mr. R. Schrader and Mr. A. Schilling for their excellent technical assistance.

REFERENCES

1. W. J. Johnson, B. Stavric, and A. Chartrand, Uricase inhi-
    bition in the rat by s-triazines: an animal model for hyper-
    uricemia and hyperuricosuria, Proc. Soc. Exptl. Biol. Med.
    131:8 (1969)

2. B. Stavric, W. J. Johnson, and H. C. Grice, Uric acid nephro-
    pathy: an experimental model, Proc. Soc. Exptl. Biol. Med.,
    130:512 (1969)

3. G. M. Fanelli, Jr., and K.H. Beyer, Jr., Uric acid in non-
    human primates with special reference to its renal transport,
    Ann. Rev. Pharmacol., 14 : 355 (1974)

4. I. Fridovich, The competitive inhibition of uricase by
    oxonate and by related derivatives of s-triazines, J. Biol.
    Chem., 240:2491 (1965)

5. B. Stavric, and E. A. Nera, Use of the Uricase-Inhibited Rat
    as an Animal Model in Toxicology, Clin. Toxicology 13(1):
    47(1978)

6. W. J. Johnson, and André Chartrand, Allantoxonamide: A potent
    new Uricase Inhibitor in vivo, Life Sciences, 23:2239 (1978)

7. H. Iwata, I. Yamamoto, I. Gohda, E. K. Morita, M. Nakamuro
    and K. Sumi, Potent competitive Uricase Inhibitors – 2,8-
    Diazahypoxanthine and related Compounds, Bio. Pharmacology,
    22:2237 Pergamon Press (1973)

8. A. Piskala and J. Gut, Synthes of oxonic acid (2,4-dioxo-
    1,2,3,4-tetrahydro-1,3,5-triazine-6-carboxylic acid) and
    the acetylation of 2,4-dioxohexahydro-1,3,5-triazines, Coll.
    Czech. Chem. Commun. 27:1562 (1962)

TIENILIC ACID IN THE TREATMENT OF GOUT AND HYPERTENSION

T. Gibson, C. Potter, H.A. Simmonds, V. Rogers,
R.I. Gleadle

Guy's Arthritis Research Unit, Department of Medicine,
Guy's Hospital, London, SE1 9RT and Smith Kline and
French, Welwyn Garden City, Herts.

Hypertension is not infrequently associated with gout.[1] In
this context, diuretics are often precluded because they invariably
accentuate hyperuricaemia. The advent of tienilic acid, a diuretic
with hypotensive and uricosuric properties may simplify the treatment of hypertension associated with hyperuricaemia,[2] and may also
obviate the need for a concurrent hypouricaemic drug. The present
study assessed the effects of tienilic acid on gouty patients.

SUBJECTS AND METHODS

Eleven patients with a mean age of 56 (range 39 - 60) and
recurrent gouty arthritis entered the study. Hypertension (standing
diastolic B.P. > 95mm Hg) was present in 5 subjects. All discontinued
their usual treatment one week before the initial evaluation. Colchicine 0.5mg twice daily was substituted and continued for the duration
of the investigation. No amendments were made to their usual diets.
At each assessment body weight and B.P. after 5 minutes recumbency
and 2 minutes standing were recorded. Blood samples were obtained
for routine haematology, liver function, electrolytes, creatinine,
urea, uric acid,[3] fasting glucose, triglyceride[4] and cholesterol.[5]
Liver function tests, electrolytes, creatinine and urea were measured
by standard autoanalyser techniques. Urine was collected for 24 h.
under toluene and paraffin for estimation of pH, electrolytes, uric acid, creatinine, protein, ammonium and titratable
acid.[6] Clinical and laboratory assessments were made one week and
immediately before beginning treatment with tienilic acid 125mg
twice daily. These were repeated after 2 and 4 weeks of treatment.

Two consecutive daily urine collections were obtained at 2 and 4
weeks and their mean results utilised in the analysis. Values
were compared statistically by Student's t test.

RESULTS

There was a slight but statistically insignificant fall of
body weight during treatment (Table 2). Blood pressure values of
the group as a whole did not alter. Of those with hypertension,
one was withdrawn because of a gouty attack. In the remaining 4
patients, supine and standing diastolic values declined progressively
over 4 weeks (Table 1). Measurements of blood are detailed in
Table 2. Significant falls of serum uric acid and potassium were
recorded. There was a slight and insignificant rise of blood urea.
Urine estimations are outlined in Table 3. Uric acid and potassium
excretion increased and there was a significant decline of calcium
excretion. Urine pH and hydrogen ion excretion were not altered.
Urate clearance increased dramatically but there was no significant
alteration of creatinine clearance (Fig. 1).

Attacks of gout occurred in 4 patients, necessitating with-
drawal of treatment in one. Transient loin pain and dysuria was
noted by one subject at the time of his gout. This patient completed
the study taking only 125mg tienilic acid daily. Mild dyspepsia
was noted by another patient. No other adverse effects were noted.
Haemoglobin, packed cell volume, white cell and platelet counts
remained normal. No fresh abnormalities of liver function were
observed and there was no significant proteinuria during the period
of treatment.

TABLE 1   Mean blood pressure $\pm$ S.D. recordings
                 of 4 patients with hypertension

TIME		-1	0	2	4
Supine (mm/Hg)	Systolic	160 $\pm$ 18	150 $\pm$ 9	137 $\pm$ 10	151 $\pm$ 10
	Diastolic	101 $\pm$ 6	101 $\pm$ 6	95 $\pm$ 9	92 $\pm$ 12
Standing (mm/Hg)	Systolic	150 $\pm$ 14	138 $\pm$ 6	132 $\pm$ 4	142 $\pm$ 5
	Diastolic	102 $\pm$ 6	102 $\pm$ 5	98 $\pm$ 5	94 $\pm$ 11

TABLE 2 Mean $\pm$ S.D. of body weight and measurements of blood values (mmol/l unless otherwise indicated)

TIME (WEEKS)	-1	0	2	4
Body wt. (kg)	86 $\pm$ 12.9	86.2 $\pm$ 12.8	85.3 $\pm$ 13.3	84.2 $\pm$ 15
Haemoglobin (g/dl)	15.3 $\pm$ 0.9	15.0 $\pm$ 0.7	14.8 $\pm$ 1.1	15.2 $\pm$ 0.6
Uric Acid	0.42 $\pm$ 0.08	0.44 $\pm$ 0.09	0.2 $\pm$ 0.06*	0.23 $\pm$ 0.1*
Urea	5.9 $\pm$ 2.1	5.6 $\pm$ 1.3	6.4 $\pm$ 1.8	6.8 $\pm$ 1.8
Creatinine (μmol/l)	96 $\pm$ 19	97 $\pm$ 27	101 $\pm$ 20	100 $\pm$ 26
Sodium	140 $\pm$ 3.3	141 $\pm$ 3.1	141.2 $\pm$ 2.9	141 $\pm$ 2.8
Potassium	4.2 $\pm$ 0.3	4.1 $\pm$ 0.4	3.6 $\pm$ 0.19*	3.4 $\pm$ 0.45*
Bicarbonate	28.8 $\pm$ 3.6	27 $\pm$ 4.4	26.8 $\pm$ 5.0	29.2 $\pm$ 4.3*
Calcium	2.3 $\pm$ 0.14	2.36 $\pm$ 0.08	2.4 $\pm$ 0.09	2.4 $\pm$ 0.09
Phosphate	1.00 $\pm$ 0.18	1.04 $\pm$ 0.31	0.93 $\pm$ 0.12	0.94 $\pm$ 0.07
Triglyceride	2.8 $\pm$ 1.0	2.8 $\pm$ 1.3	2.95 $\pm$ 1.2	3.2 $\pm$ 1.4
Cholesterol	6.8 $\pm$ 0.8	6.9 $\pm$ 1.1	6.7 $\pm$ 0.63	7.0 $\pm$ 0.96
Glucose	4.9 $\pm$ 0.7	4.88 $\pm$ 0.58	4.96 $\pm$ 0.73	5.3 $\pm$ 1.0

* p < 0.001 compared with week 0

DISCUSSION

Several studies have already illustrated the hypotensive and urico-suric properties of tienilic acid.[7,8] The current investigation has confirmed the striking effect on uric acid clearance and has demonstrated that this is sufficient to reduce the hyperuricaemic levels of gouty patients by 50%. Interpretation of the effects on blood pressure of the hypertensive gouty patients was limited by the small number of observations and the open nature of the study. Nevertheless, a downward trend of standing and supine diastolic

TABLE 3   Mean $\pm$ S.D. of urine measurements

TIME (WEEKS)	-1	0	2	4
Uric Acid (mmol/24h)	2.76$\pm$0.83	3.01$\pm$0.93	3.45$\pm$1.5	4.08$\pm$1.52
Sodium (mmol/1)	154$\pm$62	163$\pm$60	159$\pm$65	167$\pm$67
Potassium (mmol/24h)	52.7$\pm$21	55.2$\pm$21.6	56.6$\pm$21.5	66.3$\pm$29
Calcium (mmol/24h)	4.23$\pm$1.56	4.97$\pm$2.05	3.32$\pm$1.6*	3.7$\pm$2.0
Phosphate (mmol/24h)	24.9$\pm$7.7	28.2$\pm$9.5	25.2$\pm$11.4	28$\pm$10.7
pH	5.2$\pm$0.43	5.3$\pm$0.38	5.2$\pm$0.38	5.3$\pm$0.75
Ammonium (mmol/24h)	40$\pm$9.9	40.8$\pm$8.9	40$\pm$8.5	42.8$\pm$11.1
Titratable Acid (mmol/24h)	21.4$\pm$9.1	21.5$\pm$9.0	22.3$\pm$9.1	26.2$\pm$7.9
Urate Clearance (ml/min)	4.6$\pm$2.04	4.88$\pm$1.46	13.2$\pm$6.9**	13.2$\pm$4.4**
Creatine Clearance (ml/min)	83$\pm$26	93$\pm$15	88$\pm$19	88$\pm$24

*p < 0.005; p < 0.001 compared with week 0

blood pressures was noted.

Precipitation of acute gout is a well recognised complication of hypouricaemic therapy and the high incidence in this study, despite colchicine prophylaxis, probably reflects the rapid and profound effect of tienilic acid on blood uric acid levels. Loin pain and dysuria in one patient provided presumptive evidence of uric acid crystal deposition in the renal tract. This is a potential risk which requires safeguards, particularly amongst patients with gout who have increased urate body pools and may normally exhibit increased excretion of uric acid.

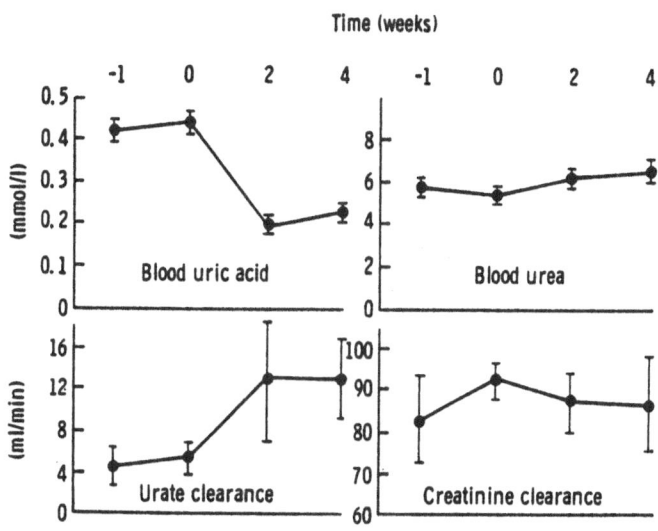

Fig. 1   Effect of tienilic   acid on blood uric acid, urea and
          clearance of urate and creatinine

REFERENCES

1. Grahame, R. and Scott, J.T. Clinical survey of 354 patients
    with gout. Ann. Rheum. Dis. 29 : 461 (1970)
2. Roberts, C.J., Marshall, A.J., Heaton, S., Barritt, D.W.
    Comparison of natriuretic, uricosuric and anti-hypertensive
    properties of tienilic acid, bendrofluazide and spironolactone.
    Brit. Med. J. 1 : 224 (1979)
3. Simmonds, H.A. A method of estimation of uric acid in urine and
    other body fluids. Clin. Chim. Acta 15 : 375 (1967)
4. Wahlefeld, A.W. Triglycerides. Determination after enzymatic
    hydrolysis in Methods of enzymatic analysis. Bergmeyer, H.U.
·   Academic Press, New York (1974)
5. Searcy, R.L. and Bergquist, L.M. A new color reaction for the
    quantitation of serum cholesterol. Clin. Chim. Acta 5 : 192
    (1960)

6.   Chan, J.C.M. The rapid determination of urinary titratable
      acid and ammonium and evaluation of freezing as a method of
      preservation. Clin. Biochem. 5 : 94 (1972)
7.   Nemati, M., Kyle, M.C., Freis, E.D. Clinical study of ticry-
      nafen. A new diuretic, antihypertensive and uricosuric agent.
      J. Am. Med. Assoc. 237 : 652 (1977)
8.   Lemieux, G., Beauchemin, M., Gougoux, A., Vinay, P. Treatment
      of arterial hypertension with tienilic acid - a new diuretic
      with uricosuric properties. Can. Med. Assn. J. 118 : 1074
      (1978)

BENZBROMARONE AS A LONG-TERM URICOSURIC AGENT

Rodney Bluestone, M.B., M.R.C.P.
James Klinenberg, M.D. and Ian K. Lee, M.B., Ch.B.

Chief, Rheumatology Section, Wadsworth VA Hospital,
Professor of Medicine, UCLA Medical Center
Chairman, Department of Medicine, Cedars-Sinai
Medical Center, Professor of Medicine UCLA Medical
Center
Director of Clinical Research & International
Coordinator, Mead Johnson Pharmaceutical Division

Benzbromarone is a potent uricosuric drug with a rapid onset and relatively long duration of action. It belongs to a new class of antihyperuricemic compounds and has the following structural formula and name:

3- (3,5-dibromo-4-hydroxybenzoyl) - 2 - ethylbenzofuran

It was discovered by the Labaz Pharmaceutical Company in Europe over a decade ago and studies beginning in 1967 showed it to possess potent antihyperuricemic properties. It has been commerically available in various European countries since 1971 under the names of Uricovac or Desuric and is about to be made available in Japan under the name of Urinorm.

The potent antihyperuricemic property of benzbromarone was confirmed in studies in the United States beginning in 1973 and monitored by the Mead Johnson Pharmaceutical Company. In order to increase the drug's absorption from the intestinal tract these latter studies which included a full Phase I, II and III program utilized the micronized form of benzbromarone.

In these Phase I & II studies benzbromarone was shown to be
safe and highly effective in reducing serum uric levels, and to
possess advantages over currently available uricosuric agents, e.g.
it had minimal interaction with aspirin and counteracted the hyperuricemic effect of chlorothiazide.

The Phase III program was designed as a single multicenter
study, the objectives of which were to determine the safety and
efficacy of micronized benzbromarone tablets in long-term usage in
gouty and/or hyperuricemic patients. Safety was determined by (a)
regular, frequent patient/investigator consultations and the com-
pletion of a progress report; and (b) regular clinical and biochemical laboratory determinations. Efficacy was determined by
the ability of micronized benzbromarone (a) to reduce abnormal
serum uric acid levels to below 6 mg% or over 30% from pre-drug
levels (Initial Efficacy), and (b) to show no systematic upward
linear trend for the remainder of the study (Long-Term Efficacy).
Twenty-one investigators from North and South America were included in this multicenter study and all followed a basically
identical protocol.

Gouty and/or hyperuricemic patients eligible for admission to
the study were over 21 years of age, but if female, not of child-
bearing potential. Gouty patients were defined as those who sat-
isfied two of the following requirements:

a. Serum uric acid during the washout period of above 7.0 mg/
    100 ml in males or post-menopausal females, or above 6.0
    mg/100 ml in pre-menopausal females on two separate
    occasions.
b. Presence of tophi.
c. Demonstration of urate crystals in synovial fluid or of
    urate deposition in tissues by chemical or microscopic
    examination.
d. A history of attacks of painful joint swelling with an
    abrupt onset of severe pain and complete remission with
    treatment within ten days (the treatment may be colchicine
    or any nonsteroidal anti-inflammatory agent).

Hyperuricemic patients were those who had pre-drug serum uric
acid values of at least 9 mg/100 ml for males and 8.5 mg/100 ml for
females on two separate occasions. Patients were excluded only if
they showed evidence of any significant degree of any acute disease.

A total of 408 patients (333M, 75F) aged 18-85 (mean 53 years)
with documented gouty episodes of sustained hyperuricemia received
the drug for varying periods of time up to 24 (mean 10.5) months.
A single daily oral dose of 40 mg of drug was initially given to
all patients and the dose then raised to 80 mg in 148 patients.

All patients were seen monthly, questioned about their health and
examined. Regular laboratory testing included the serum urate
concentration, complete blood counts, blood chemistry (SMA 12) and
urinalysis. Electrocardiographic exams were obtained every 6 months
and an ophthalmologic exam performed every 6 or 12 months.

RESULTS

(A) Initial Efficacy

A total of 379 patients were evaluated for short-term efficacy.
When entered, they displayed a mean baseline serum urate value of
8.75 mg/100 mg and a mean "first" value (defined as the first reading after 20 days of treatment) of 5.65 mg/100 mg (a decrease of
3.1 mg/100 ml; p< .0001).

When investigators were considered individually, 19 had patients who displayed a statistically significant reduction of serum
urate (p< 0.0002). The one-way ANOVA (treatment=investigator)
yielded a significant F (p< .0001). There were individual investigators that differed from each other (Duncan multiple t); but
except for one test center, results were fairly comparable with the
range of -1.96 to -4.64 in the various investigators' serum urate
decreases. Differences that exist were mainly due to the large
number of investigators participating and the small variation.
Investigators reporting the lowest mean difference in reduction
also reported the lowest baselines and vice versa.

(B) Long-Term Efficacy

The dose of benzbromarone could be increased from 40 to 80 mg/
day after the "first" serum urate reading on or about the 16th day
of treatment. "Mean" serum urate values were then obtained on 381
patients beyond the "first" reading for up to 24 months; 233 were
maintained on 40 mg/day and 148 on 80 mg/day of the drug. Pooled
together the mean serum urate baseline was 8.75 mg/100 ml and the
mean of the "mean" measurements were 5.75 mg/100 ml (a drop of
3.0; p< .0001). Of the 233 patients taking 40 mg/day, a mean reduction of 2.82 mg was noted (p< .0001); of the 148 patients taking
80 mg/day, a mean reduction of 3.26 mg was noted (p< .0001). There
was a significant difference (p< .0001) between baselines and
between changes from baseline (p< .02) when comparing the 2 dosage
groups. When baseline was considered a covariate, slopes were not
significantly different and mean adjusted differences between the
2 dosage groups were no longer significant (p< .25).

Statistical analysis of long-term serum urate values beyond
day 20 showed that only 3 out of a total of 326 evaluable patients
showed any evidence of a linear upward trend throughout the study;
all 3 were taking the 40 mg dose throughout.

During the entire study a total of 219 acute gouty attacks
were reported. For a similar period prior to the commencement of
the study 346 acute gouty attacks had been noted. Tophi were present in 29 patients prior to study, and disappeared or greatly decreased in size in 22 of these by the end of the trial.

A total of 10 patients discontinued their drug prematurely
because of possible side effects noted as diarrhea (5), renal
calculi or colic (4), or persistent acute gout (1). No other
serious drug-related adverse effects were noted.

From these data we conclude that micronized benzbromarone
orally administered at a dose of 40 or 80 mg daily is safe for
extended use, and effective in reducing serum urate concentrations
in a sustained fashion.

1. Zollner N, Dofel W, Grobner W: The effect of benzbromarone
    on urinary uric acid excretion in normal adults. Klin Wochen-
    schr 48, 426-432, 1970
2. Mertz DP, Sulzberger I, Klopfer M: Diabetes mellitus, hyper-
    lipidemia, fatty liver, hypertension with primary gout and the
    effect of benzbromarone. Munch Med Wochenschr 112, 241-247,
    1970
3. Kotzaurek R, Hueber EF: Clinical experience with 2-ethyl-3
    (4-hydroxy-3,5-dibrombenzol-benzofuran (benzbromarone) in the
    treatment of gout and hyperuricemia. Wien Klin Wochenschr
    118, 1014-1018 (No 23), 1968
4. VanBogaert P: Clinical trial of L2214 in the treatment of gout
    or rheumatological syndromes associated with hyperuricemia.
    Belg Rheum Med Phys 24, 295-315, 1969
5. Masbernard A, Guilbaud J, Droniou J: Clinical experience in
    long-term treatment of primary gout with benzofuranne iodine
    and brome-derivatives. Soc Med Chir Hop 3, 393-410, 1971
6. Broekhuysen J. Pacco M, Sion R, Demeulenaere L, Van Hee M:
    Metabolism of benzbromarone in man. Eur J Clin Pharmacol
    4/2, 125-130, 1972
7. Oshima Y, Yoshimura T, Akaoka U, Mishida S: Clinical experi-
    ence with benzbromarone in gout. Ryumachi 9/1, 49, 1969
8. Bresnik W, Muller MM: Results obtained with long-term benz-
    bromarone therapy. Wien Klin Wochenschr 86/10, 283-287, 1974.
9. Delbarre F, Auscher C, Saporta L, De Gery A, Danchot J: Treat-
    ment of gout with a dibrominated derivative of benzofuran. Rev
    Rheum Mal Osteo-Artic 40/4, 273-275, 1973
10. Sorensen LB, Levinson DJ: Clinical evaluation of benzbromarone.
    A new uricosuric drug. Arthritis Rheum 19/2, 183-190, 1976
11. Sinclair DS, Fox IH: The pharmacology of hypouricemic effect
    of benzbromarone. J Rheum 2/4, 437-445, 1975
12. Yu Ta-F: Pharmacokinetic and clinical studies of a new urico-
    suric agent - benzbromarone. J Rheum 3/3, 305-312, 1976.

THE ACTION OF BENZBROMARONE IN RELATION TO AGE, SEX AND ACCOMPANYING DISEASES

H. Ferber [x], U. Bader [x], F. Matzkies [xx]

[x] Clinical Research, Heumann-Pharma, D-8500 Nürnberg
[xx] Metabolic Unit (Head Prof. Dr. Dr. h. c. G. Berg)
University Erlangen/Nürnberg, D-8520 Erlangen

INTRODUCTION

As has repeatedly been reported, hyperuricaemia is very frequently accompanied by other diseases as hypertension, adiposity, nephropathy, hyperlipoproteinaemia and diabetes mellitus (1). No investigation of the dependence of the reduction of uric acid on the accompanying diseases, age and sex has been carried out. The aim of this study was to show the effect of benzbromarone in reducing uric acid levels, in relation to existing accompanying diseases, age and sex.

PATIENTS AND METHOD

In a field study, a total of 2220 patients treated with 50 mg benzbromarone (Narcaricin® mite) and/or 100 mg benzbromarone (Narcaricin®) were investigated. 104 patients had to be eliminated from the study on account of incomplete collected data and a further 132 patients with existing renal insufficiency (serum creatinine $>1,5$ mg/100 ml) also had to be removed from the study. A total of 1984 patients comprising 1378 males and 606 females were evaluated. Further details can be seen in Table 1.

330 patients were changed over to 100 mg benzbromarone (Narcaricin®). The mean initial level was $7,41 \pm 1,49$ mg/100 ml. The final check after 2 months revealed a mean level of $6,18 \pm 1,36$ mg/ 100 ml. Figure 3 shows the results in percent for the individual uric acid levels of the patients changed to 100 mg benzbromarone.

After reaching a steady state, the initial level in the case of the patients further treated with 50 mg benzbromarone was, in

Table 1

	Hyperuricaemia n = 1984
Patient, male	1378
Patient, female	606
Age	54 $\pm$ 13,10
Body-weight, kg	78,54 $\pm$ 12,12
Body-stature, cm	170,55 $\pm$ 8,50

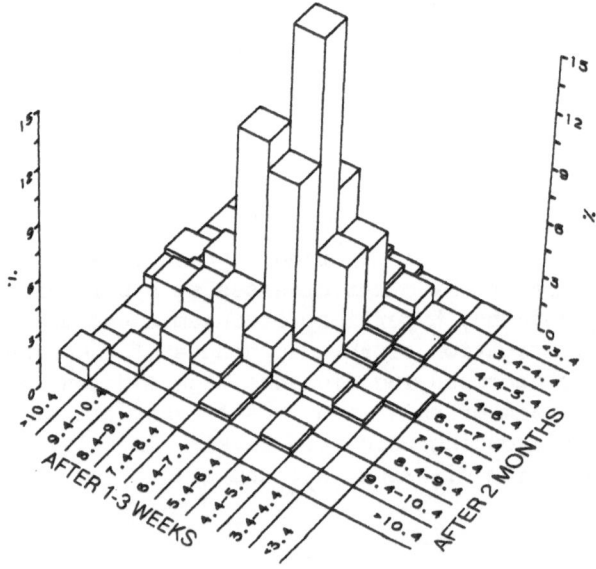

Figure 1.

1654 cases, a mean 5,80 $\pm$ 1,39 mg/100 ml. The mean uric acid level
established at the final measurement after 2 months, was 5,14 $\pm$
1,17 mg/100 ml. The percentage distribution of the uric acid levels
reached in the various classes is shown in Figure 4 for 1654
patients.

Of the group of patients treated throughout the study with
50 mg benzbromarone, the percentage of patients with normalized
uric acid levels after 8 weeks of therapy is 87,7 %. In the group
of patients treated in the first 1 to 3 weeks with 50 mg benzbromarone and then changed over to 100 mg benzbromarone, the uric
acid levels were reduced to the normal range in 55,8 % of the cases.

To investigate possible dependencies of the serum uric acid

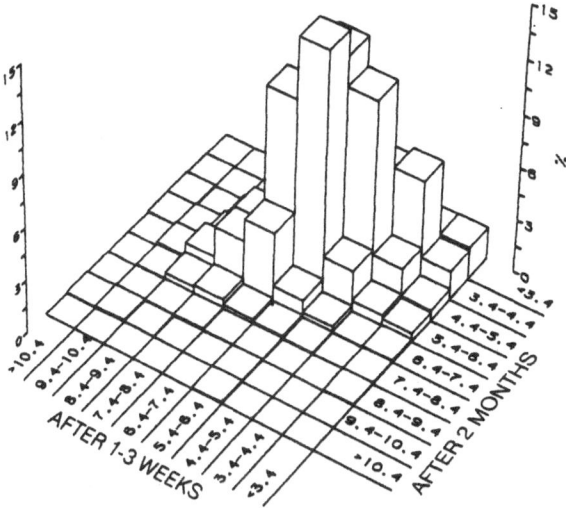

Figure 2.

levels prior to the start of therapy and their pattern of behaviour
during therapy with 50 mg benzbromarone and 100 mg benzbromarone
administered once daily, on the variables age, sex, hypertension,
adiposity, diabetes and hyperlipoproteinaemia, 2- and 6-factor
variance and co-variance analysis were carried out for repeated
measurements (7).The level of significance was determined as p=0,05.

    The age of the patients had no influence on the level of uric
acid concentration determined prior to initiation of treatment (Fig.5)

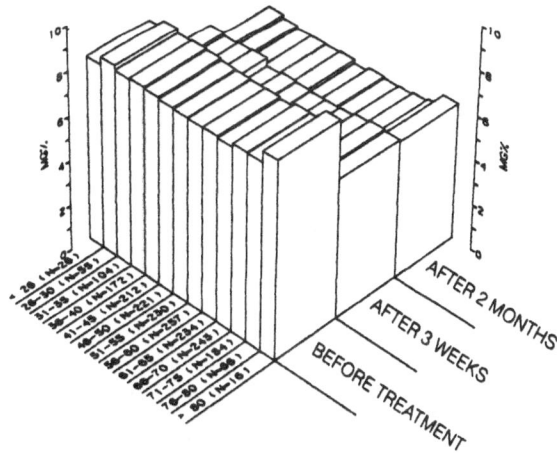

Figure 3.

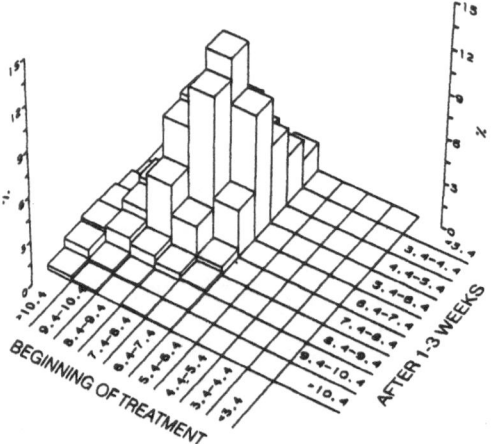

Figure 4.

The proportion of male patients, at 1378 out of 1984, exceeds
that of female patients markedly (n=606). The factors sex, hypertension and hyperlipoproteinaemia have a significant influence on
the level of the uric acid concentrations prior to treatment. When
the differences in initial levels are eliminated, the co-variance
analysis reveals a pattern of behavour of the uric acid levels
that is related only to sex (see Fig. 6).

Only patients with a serum uric acid level of 6,4 mg % and
more were incorporated in the study. These patients were
stabilized in 1 to 3 weeks with one 50 mg benzbromarone tablet
taken daily. Those patients who after this time still persented
with a serum uric acid level of 6.4 mg % or more, were then
treated with 100 mg benzbromarone administered once daily. Those
patients who, after 1 to 3 weeks, manifested serum uric acid levels
below 6,4 mg %, continued to receive a once-daily dose of 50 mg
benzbromarone (Narcaricin ® mite). A final examination was carried
out after a treatment period of two months. The occurrence of side
effects was established both after 1 to 3 weeks and after 2 months.
Accompanying conditions were adiposity in 1175 patients (59,22 %),
hypertension in 953 patients (48,03 %) diabetes mellitus in 341
patients (17,18 %) and hyperlipoproteinaemia in 847 patients (42,69%).

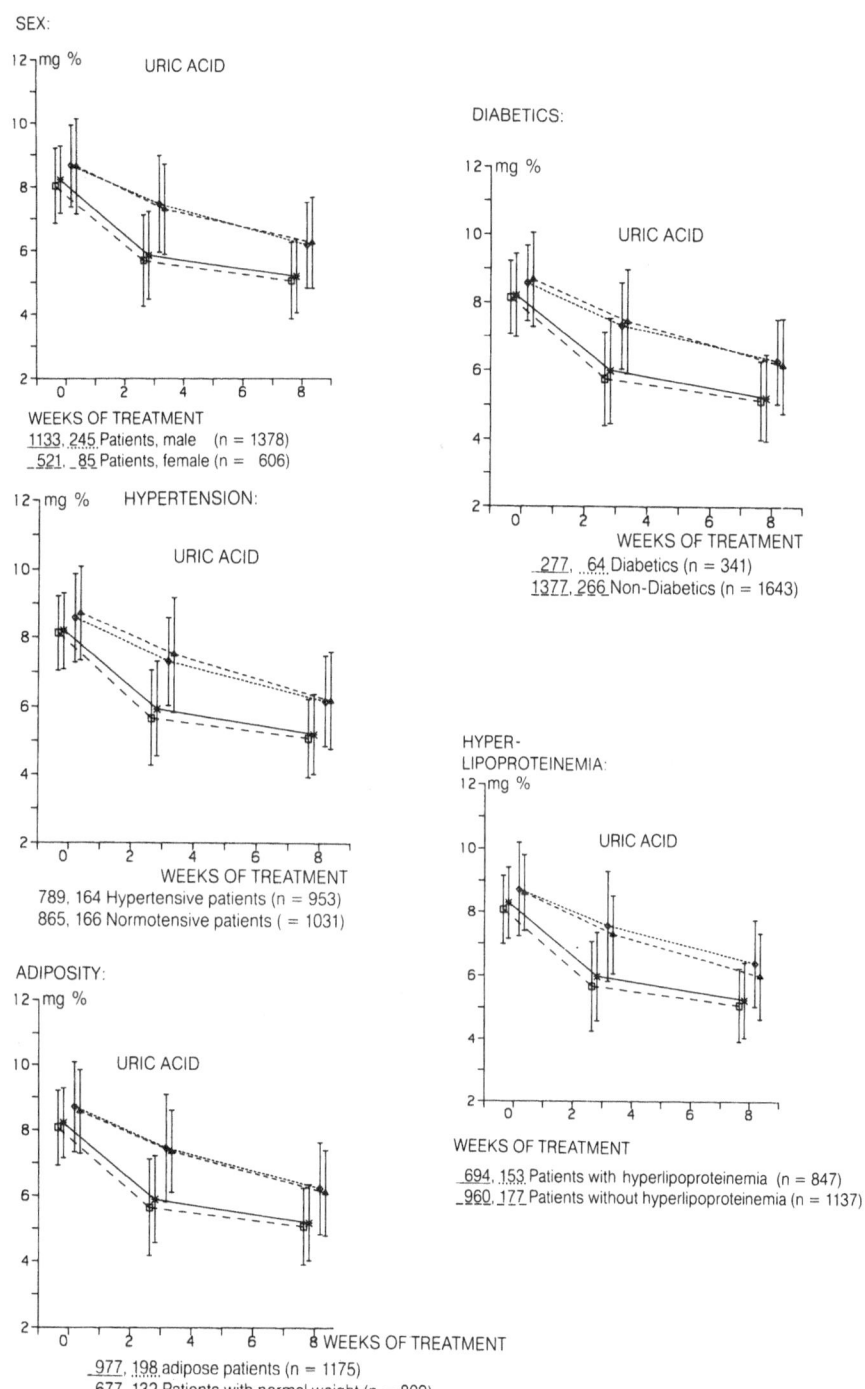

Figure 5.

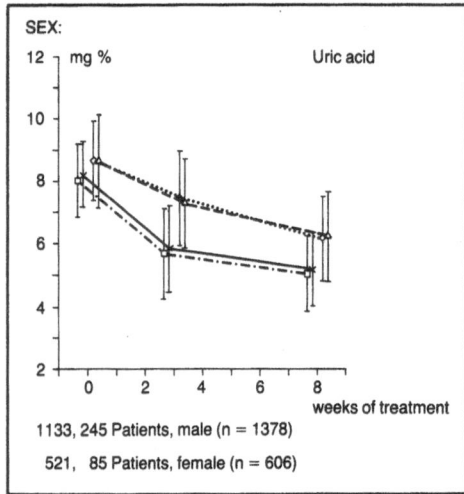

1133, 245 Patients, male (n = 1378)

521, 85 Patients, female (n = 606)

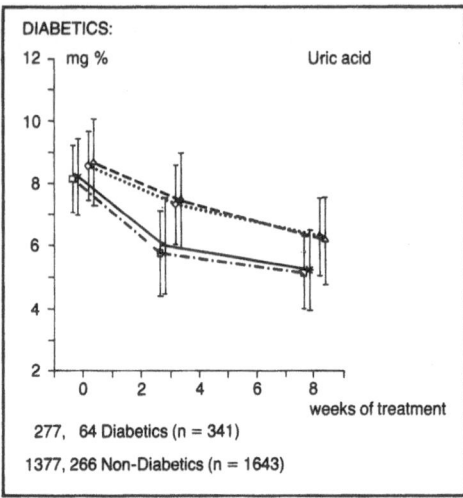

277, 64 Diabetics (n = 341)

1377, 266 Non-Diabetics (n = 1643)

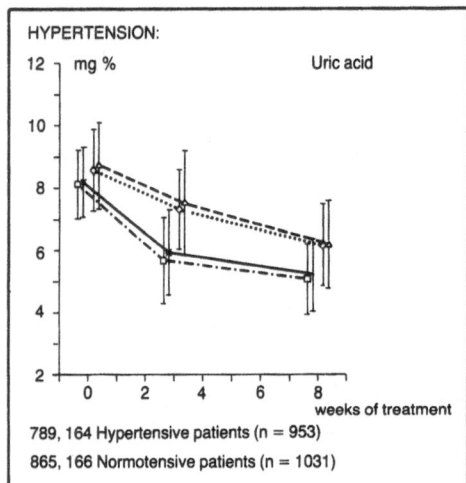

789, 164 Hypertensive patients (n = 953)

865, 166 Normotensive patients (n = 1031)

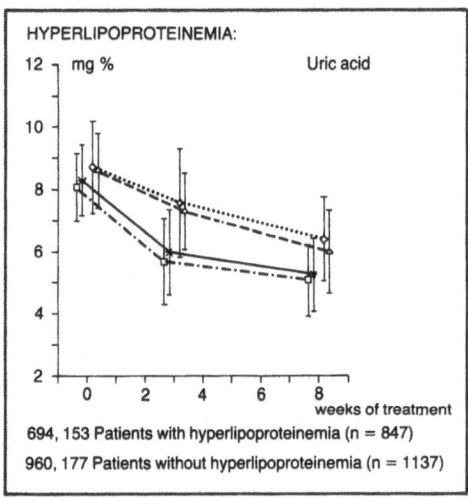

694, 153 Patients with hyperlipoproteinemia (n = 847)

960, 177 Patients without hyperlipoproteinemia (n = 1137)

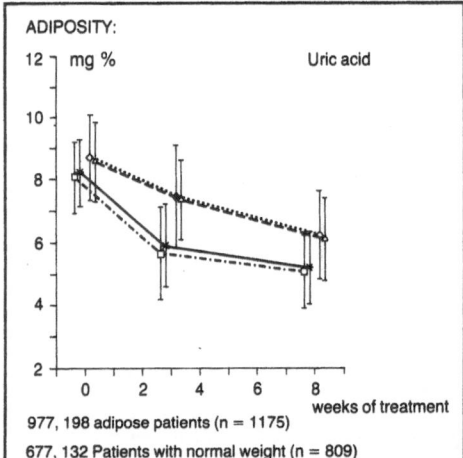

977, 198 adipose patients (n = 1175)

677, 132 Patients with normal weight (n = 809)

Fig. 6 Uric acid levels prior to, during and after 8 weeks of treatment with Narcaricin mite and/or Narcaricin. Breakdown by sex and risk factors, arithmetic mean levels and standard deviations——, —·—·— = treatment with 50 mg benzbromarone (Narcaricin® mite), throughout the study n = 1654. ....,———— = first 1-3 weeks of therapy with 50 mg benzbromarone, subsequent treatment with 100 mg benzbromarone to the 8th week, n = 330.

RESULTS AND DISCUSSION

Following the fortuitous discovery of the uricosuric action of benzofuran derivatives (2) this class of substances has now becomce firmly established in the treatment of hyperuricaemia (3,4,5,6). Hitherto, the pattern of behaviour of uric acid reduction in relation to the risk factors, age and sex, had not been investigated. Our study had two objectives; the first was to establish whether, on an out-patient basis, the administration of 50 mg benzbromarone led to an effective reduction of uric acid in the first 1 to 3 weeks; the second was to establish, in a large group of patients, the relation of the reduction to age, sex, hypertension, diabetes, adiposity and hyperlipoproteinaemia. The mean initial level of uric acid was 8,24 ± 1,16 mg/100 ml. After reaching the steady state (1-3 weeks), check measurement revealed a mean level of 6,06 ± 1,53 mg/100 ml. Fig. 2 shows the percentage assessment of the patients in the individual classes of uric acid level.

The original postulation that hypertension might have an influence on the effect of uric acid reduction by benzbromarone, was not confirmed.

SUMMARY

The uric-acid lowering effect was investigated in a group of 2220 patients. 1984 of these were employed for statistical evaluation purposes. On average, the uric acid level was reduced from 8,24 ± 1,16 mg/100 ml to 5,32 ± 1,265 mg/100 ml. In 82 % of all the cases, the uric acid level at the end of the treatment period was below 6,4 mg %; both in patients treated throughout with 50 mg benzbromarone (Narcaricin[R]mite) and in those changed over to 100 mg benzbromarone (Narcaricin[R]) after 1 to 3 weeks. The lowering of the uric acid levels was in no way related to hypertension, adiposity, hyperlipoproteinaemia, diabetes mellitus or age.

Acknowledgment: We are grateful to Mrs. S. Häckler for preparing the material and to Dr. W. Haase for the statistical evaluation at the Institut für Numerische Statistik, Köln.

Table 2

Side effects	No. of patients
Urticaria	6
Diarrhoea	22
Renal colics with concrement	2
Gastro-intestinal disturbances	8

References:
1. G. Babucke, D. P. Mertz, Häufigkeit der primären Hyperurikämie unter ambulanten Patienten, Münch.Med.Wschr. 116:875 (1974).
2. F. Delbarre, C. Auscher, B. Amor, Action uricosrique et anti-goutteuse de certains du benzofuranne, Presse Med.73:255 (1965).
3. N. Zöllner, G. Stern, W. Gröbner, Über die Senkung des Harnsäure-spiegels im Plasma durch Benzbromaronum. Klin.Wschr.46:1318 (1968).
   -, N. Griebsch, J. K. Fink, Über die Wirkung von Benzbromaron auf den Serumharnsäurespiegel und die Harnsäureausscheidung des Gicht-kranken. Dtsch. med. Wschr. 95:2405 (1970).
4. W. Bresnik, Ein einfaches Verfahren zum Nachweis von Harnsäureab-lagerungen bei Gichtkranken. Therapiewoche 25:4620 (1975).
5. H. Bröll, H. Sochor, G. Tausch, R. Eberl, Über die Langzeittherapie mit Benzbromaron bei Arthritis urica, Wien.med.Wschr.125:546 (1975)
6. F. Matzkies, Wirkungen und Nebenwirkungen von Benzbromaron bei der Initialbehandlung von Hyperurikämie und Gicht, Fortschr. Med. 96: 1619 (1978).
   -, G. Berg, R. Minzlaff, Hyperurikämiebehandlung mit Tagesdosen von 50 und 25 mg Benzbromaron, Fortschr. Med.95:1748 (1977).
7. B. J. Winer, Statistical principles in experimental design 2nd ed. (McGraw-Hill, pp. 907) New York.

COVALITIN® A NEW DRUG FOR THE TREATMENT OF URIC LITHIASIS

Tina Covaliu

Post and Telecommunication Administration Hospital
Medical Clinic II
Bucharest, Romania

FUNDAMENTALS OF THE COVALITIN TREATMENT

It is a known fact that the fundamental role of the kidney is to maintain within constant limits the volume and composition of the blood plasma. This function is performed by specific mechanisms acting on the components who's concentration is to be controlled.

Urine formation cannot be considered a process of simple filtration, diffusion or osmosis, since it involves the active participation of the renal tissue.

The study of this process directed our attention to the mineral components of the blood and urine particularly cations and anions. Thus, besides water which represents the main component, we considered the most important cations from a quantitative point of view, i. e. , Na^+, K^+, Ca^{2+}, Mg^{2+}, also Fe^{2+}, Cu^{2+}, Zn^{2+} and Pb^{2+} which occurred in smaller amounts as well as the anions Cl^-, HCO_3^-, HPO_4^{2-}, $H_2PO_4^-$ and SO_4^{2-}. According to the available data these elements and mineral components totalize 310 mEq/liter of which 155 mEq/liter cations and 155 mEq/liter anions.

Under normal conditions the organism reaches an ionic equilibrium which promotes and also reflects its good metabolical activity.

The main function of the kidney, i. e. to maintain within normal

295

limits the value of the blood components, is performed by specialized mechanisms regulating the concentration and elimination of these components.

Interdependence of ionic concentration, in case of normal blood and urine ion content, is expressed by the ratio :

$$\frac{[Na^+] + [K^+] + [OH^-]}{[Ca^{2+}] + [Mg^{2+}] + [H^+]} = \text{Constant}$$

The study of various types of urolithiasis revealed modifications in the ionic concentration.

According to Le Chatelier's principle "the modification affecting one condition of a system in equilibrium results in the adjustment of the equilibrium such as to diminish the initial modification". Considering the above mentioned principle concentration evaluations of all ions were carried out, which revealed that, the modification in the concentration of one ion brings about a change in all other concentrations, conducting to an ionic disequilibrium. As a function of factors generating it, this disequilibrium - which is characteristic of all types of urolithiasis - results in the excess or deficiency of blood electrolytes which may be of either anionic or cationic nature.

The study of hydro-electric disequilibrium role in the formation of urinary calculi revealed that a certain type of disequilibrium is accompanied by a specific mineralogical structure of the stones. Since renal calculi exhibit a characteristic mineralogical structure, it may be concluded that a fine analysis of this structure could give valuable indications on the nature and type of hydro-electric disequilibrium.

According to our opinion the urinary cations, completely disregarded until now, play in the majority of stone formers a much more important role in the lithogenetic process as compared to anions (i.e. acid radicals). From this point of view urinary calculi may be classified in two groups. The first group comprises the calculi forming compounds with calcium cation as a main component and also ammonium magnesium phosphate which is never found alone but in combination with calcium phosphate. The second group consist of calculi made of organic compounds (cystine, xanthine, uric acid) which are initiated especially by the action of anions.

Basic components of the second group calculi are :
• uric acid, $C_5HN_4(OH)_3$

- ammonium urate, $C_5(NH_4) N_4(OH)_3$
- cystine, $[SCH_2 CH(NH_2)] COOH_2$
- xanthine, $C_5 H_2 N_4 (OH)_2$

Such a classification is supported also by the results of re-searches concerning the matrix of urinary calculi. All the calculi belonging to the first group display identically structured matrixes, while in the second group the matrix evidences variations either from a chemical point of view (in calculi of uric acid or urates), or consid-ering its structure (in the cystine calculi).

All these results conduct to the conclusion that lithiasis appear-ance and evolution is preceded in nearly all instances, by a concen-tration change affecting all the ions existing in the hydric compart-ments, as well as uric acid and cystine.

PHARMACOLOGICAL PROPERTIES AND MECHANISM OF ACTION

According to the above mentioned ionic theory of urolithiasis the changes affecting ionic concentration rates are reflected in initiation and subsequent evolution of calculi. This fact conducted to our concept that restoration of electrolytic equilibrium should act as an important protection factor against lithogenesis in stone forming patients, providing meantime a conclusive evidence for the validity of ionic theory.

With this end in view we directed our attention to the synthetic ion exchange resins which are widely used in different branches of the industry, in pharmacology and still in a very small extent in medicine.

Processing of ion exchange resins according to a romanian technology enabled after a great number of experiments the prepara-tion of Covalitin.

The therapy of urinary lithiasis with ion exchange resins is applied in Romania since 1962. The treatment - based both on the litholytic effect and protective action against lithogenesis - is admin-istered indirectly, i.e. orally.

COVALITIN® is supplied in three forms marked 1, 2, 3 according to the existing types of lithiasis.

Description and action of the COVALITIN® products

Product characteristics	COVALITIN 1	COVALITIN 2	COVALITIN 3
How supplied	Capsules of 0.50 mg in bottles containing 10 packages of 9 capsules each		
Description	Active ingredient : Magnesium sodium and potassium ions Chelated on a synthetic resin.	Ammonium, sodium and potassium ions	Calcium, sodium and potassium ions
Pharmaco-logical action	COVALITIN® acts on the mineral metabolism restoring the electrolytic equilibrium. ● Prophylactic effect in calcium oxalate lithiasis Promotes calculi passage.	● Prophylactic effect in uric acid lithiasis ● Acts on the concentration of uric acid	● Prophylactic effect in calcium and magnesium phosphate lithiasis
Indications	Microlithiases, reno-ureteral lithiases (with therapeutical indications) prevention of post-surgical recurrence. No adverse reactions were noted in case of associated affections. ● Calcium and magnesium oxalate lithiasis Combined therapy with the three COVALITIN® products may be applied.	● Uric acid and urate lithiasis	● Calcium and magnesium phosphate lithiasis
Contra-indications	Although well tolerated in the organism, COVALITIN® may induce electrolytic disequilibrium when administered without medical supervision. It is contra-indicated in pregnancy and severe renal insufficiency.		

COVALITIN® ACTION ON CALCULI AS OBSERVED UNDER MICROSCOPE

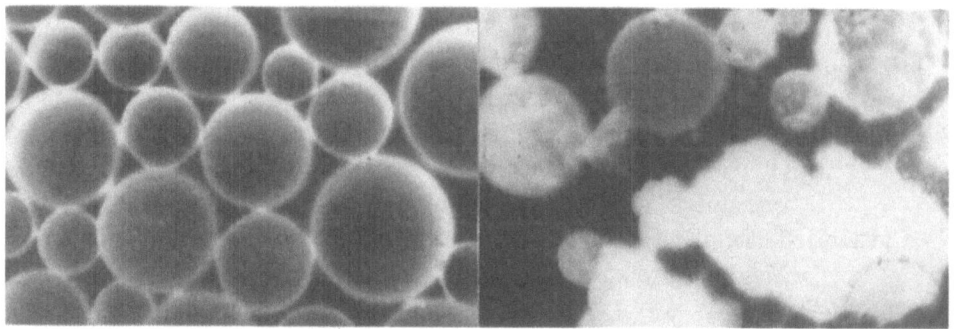

Fig. 1 The aspect of COVALITIN®   Fig. 2 COVALITIN® in contact
with calculus

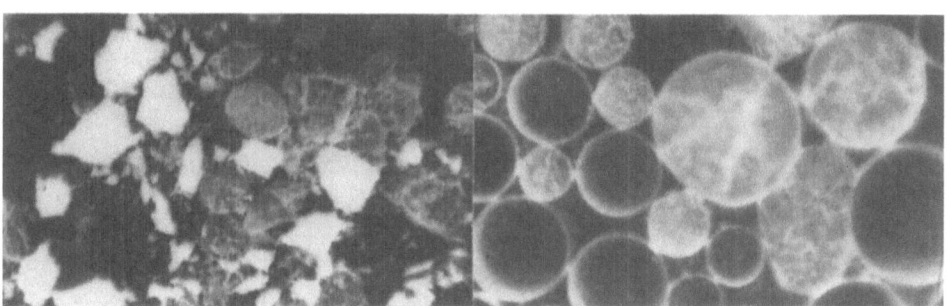

Fig. 3 Calculus fragmentation   Fig. 4 Calculus components taken
under the action of drug   over by the drug

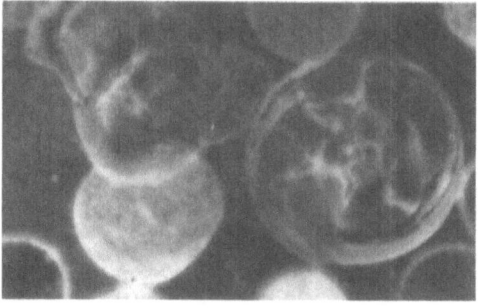

Fig. 5 Drug inactivation by calculus components.
The patient passes the calculi

REFERENCES

Burghele, Th, Covaliu, Tina, 1966, Zur Harnstein Prophylaxe mit Ionenaustauschern, Presented at : Urology Conference, Görlitz.

Covaliu, Tina, 1964, Early diagnosis of urinary lithiasis, Rev. Roum. Méd., 1:259.

Popescu Buzeu, M., Rügendorff, E.W., Covaliu, Tina, Mihăilă, V., 1965, The present state of knowledge of urinary calculus chemolysis, Arch. de l'Un. Méd. Balkan, 3:407.

Covaliu, Tina, 1965, Zur Kalziumoxalat Harnstein-prophylaxe mit Magnesium, Ztsch.f. Urol., 57:573.

Covaliu, Tina, 1967, Le rôle des désordres ionique dans la pathogenie de la lithiase urinaire, Journ. d'Urol. et Nephrol., 73:719

Burghele, Th. Dinculescu, T., Cociaşu, E., Covaliu, Tina, Cicotti, L., Mihăilă, V., 1965, Effect of Căciulata mineral waters in the treatment of urinary lithiasis, St. Cercet. de Balneolog. şi Fizioterap., 8:229.

Burghele, Th., Dinculescu, T., Covaliu, Tina, Cociaşu, E., Mămularu, Gh., 1967, The effect of mineral water from Căciulata in the conservative treatment of urinary lithiasis, St. Cercet. de Balneolog. şi Fizioterap., 8:522.

Covaliu, Tina, Marinescu, I., 1968, Electrolytic re-equilibration activity of ion exchange resins, a protective factor in urinary lithiasis, Rev. Sanit. Milit., 6:971.

Burghele, Th., Covaliu, Tina, 1967, Zur Harnsteinprophylaxe mit Ionenaustauschern, Der Urologie, 6:234.

Marinescu, L., Covaliu, Tina, 1968, Activity of ion exchange resins in urinary lithiasis, Rev. Sanit. Milit., 6:971

Covaliu, Tina, 1971, A medicinal product for the prophylaxis and treatment of lithiasis and procedure for its preparation, Patent, OSIM, Bucharest.

HYPOXANTHINE SALVAGE IN MAN: ITS IMPORTANCE IN URATE OVER-

PRODUCTION IN THE LESCH-NYHAN SYNDROME

N. Lawrence Edwards, David P. Recker, and Irving H. Fox

Human Purine Research Center
Departments of Internal Medicine and Biological Chemistry
University of Michigan, Ann Arbor, Michigan, 48109

The deficiency of hypoxanthine-guanine phosphoribosyl-transferase (HGPRT) is associated with massive overproduction and overexcretion of uric acid[1,2]. The mechanism for increased uric acid production in this enzyme deficiency has been investigated in vivo by the administration of (^{14}C)glycine[1] and in vitro by the conversion of (^{14}C)formate to formylglycineamide ribonucleotide (FGAR) in fibroblasts and lymphoblasts[3,4]. The deficiency of HGPRT results in a 20-fold increase in the rate of incorporation of (^{14}C)glycine into urinary uric acid and a 4-fold increase in FGAR formation from (^{14}C)formate (Figure 1). These studies appear to indicate an elevated rate of purine synthesis de novo in the enzyme deficient state.

The explanation of excessive uric acid production on the basis of increased de novo purine synthesis does not emphasize a potentially important contributor to uric acid overproduction in HGPRT deficiency. The absence of this enzyme leads to an inability to resynthesize IMP from hypoxanthine via the salvage pathway (Figure 1). As a result, any hypoxanthine formed in this disorder can only be oxidized to uric acid. To assess the relative contribution of a disorder of hypoxanthine salvage to uric acid overproduction in HGPRT deficiency, we have used (8-^{14}C)adenine intravenously to compare this pathway isolated from de novo synthesis in subjects with deficient or normal enzyme activity[5].

Tracer doses of (8-^{14}C)adenine were administered intravenously to 9 patients with normal enzyme activity, 3 patients with partial deficiency of HGPRT, and 6 patients with the Lesch-Nyhan syndrome to examine the degree of hypoxanthine reutilization in vivo

301

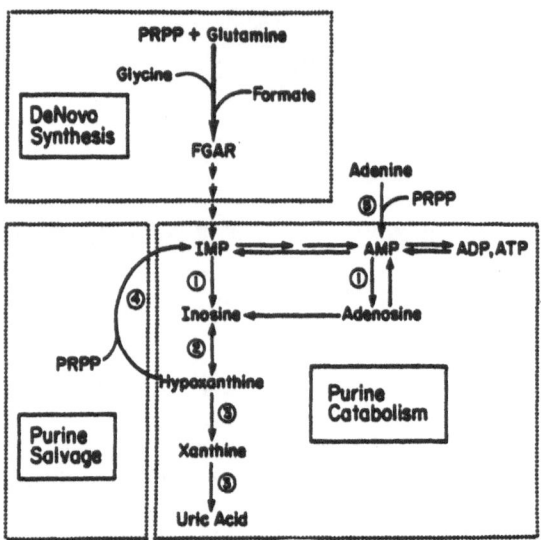

Fig. 1. Uric acid production is regulated by interdependent
         metabolic processess. (A) De novo biosynthesis is a
         pathway for the synthesis of the purine ring from non-
         purine precursors. (B) Purine nucleotide catabolism
         proceeds through 5'-nucleotidas (1), purine nucleoside
         phosphorylase (2), and xanthine oxidase (3). (C)
         Purine salvage is catalyzed by HGPRT (4) and adenine
         phosphoribosyltransferase (5) (From Edwards, et al[5]).

(Figure 2). The mean cumulative excretion of radioactivity 7 days
after the adenine administration is 5.6 ± 2.4, 12.9 ± 0.9 and
22.3 ± 4.7 percent of the infused radioactivity for control
subjects, partial HGPRT deficient subjects, and the Lesch-Nyhan
patients, respectively.

      The increased urinary excretion of radioactivity following
([14C])adenine infusion in subjects deficient in HGPRT may reflect
diminished reutilization of hypoxanthine to IMP (Figure 1).
Alternatively, this may reflect an increase in the rate of adenine
nucleotide degradation in the enzyme-deficient subjects. This
latter hypothesis was examined by comparing the purine nucleotide
degradative pathway in the three patient groups under the conditions
of acute substrate load caused by the rapid infusion of fructose
(0.5 gm/kg). The infusion of D-fructose results in a cascade of
purine nucleotide degradation as a result of the depletion of intracellular ATP and inorganic phosphate during the hepatic phosphorylation of fructose[6].

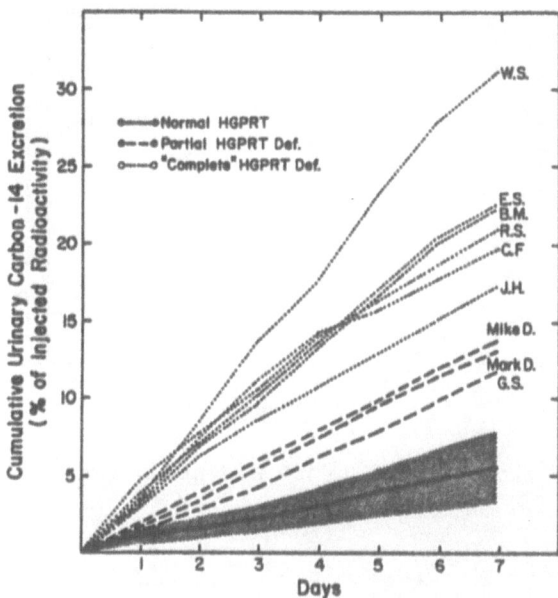

Fig. 2.   Cumulative carbon-14 excretion after intravenous (8-^{14}C)-
adenine administration (From Edwards, et al[5]).

The three patient groups demonstrated similar rapid elevations
of plasma urate following fructose administration (Figure 3). Baseline plasma urate concentrations in 7 control subjects increased
by 1.6 ± 0.8 mg/dl, by 2.2 ± 1.0 in 6 patients with partial HGPRT
deficiency and by 2.1 ± 0.3 mg/dl in 6 patients with complete
enzyme deficiency.

Urinary total purine excretion increased after D-fructose in
all patients studied, although the pattern of purine compounds
excreted is altered in the enzyme-deficiency subjects (Figure 4).
The mean increases in total urinary purine excretion (sum of uric
acid, hypoxanthine, xanthine, and inosine) for the 3 hours after
intravenous fructose is 7.5, 18.6, and 17.3 nmole/g creatinine in
the 7 control subjects, 6 patients with a partial enzyme deficiency,
and 6 patients with the Lesch-Nyhan syndrome, respectively. The
increased purine excretory response to a fructose infusion in
the enzyme deficient subjects suggests a full potential to increase
purine nucleotide degradation and thus no pertubation of the adenine
nucleotide catabolic process.

The data described suggest that the increased rate of radio-
isotope excretion after (8-^{14}C)adenine administration in hypoxanthine-guanine phosphoribosyltransferase deficiency represents
decreased reutilization of hypoxanthine. From these observations,

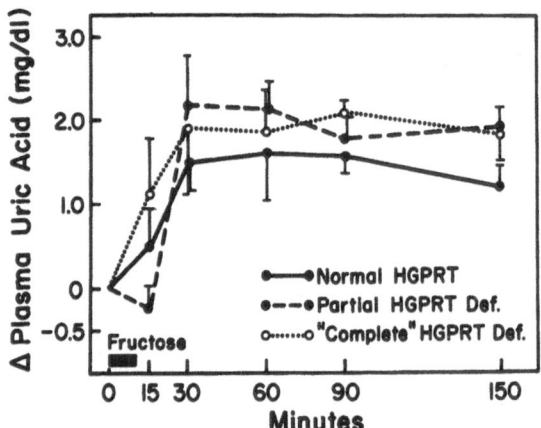

Fig. 3.   Post-fructose plasma urate elevation (from Edwards, et al[5]).

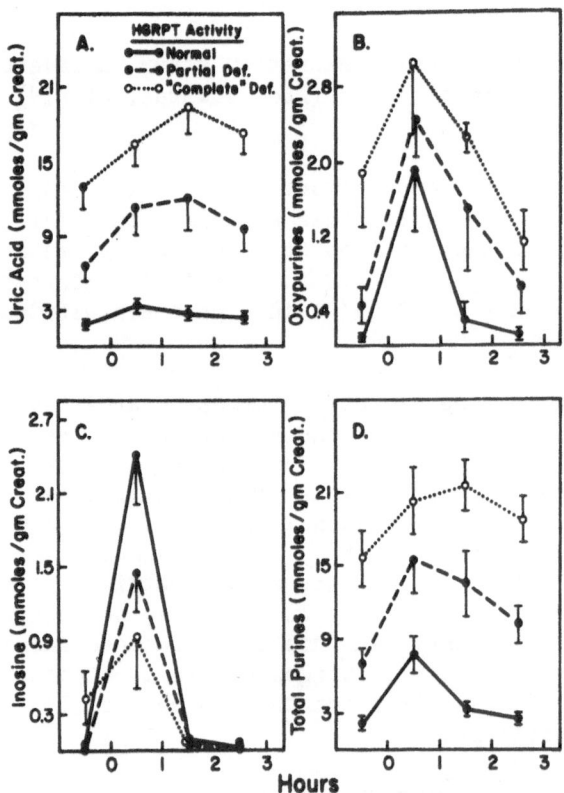

Fig. 4.   Urinary purine excretion after fructose infusion. Values
          expressed are the mean ± S.E. for each group (from Edwards,
          et al[5]).

estimations concerning the role of hypoxanthine salvage are possible. The control subjects with normal erythrocyte hypoxanthine-guanine phosphoribosyltransferase levels excrete 5.6 ± 2.4 percent of the infused radioactivity per week. In contrast, patients with the Lesch-Nyhan syndrome and an inability to reutilize hypoxanthine excrete 22.3 ± 4.7 percent of the infused radioactivity per week. This 16.7 percent difference in radioactivity excretion between the normal and Lesch-Nyhan patients represents that portion of purine breakdown products that are normally reutilized via the salvage pathway. Such numbers lead to the estimation that approximately 75 percent ((16.7/22.3) x 100) of hypoxanthine is normally reutilized each day.

CONCLUSION

1. A daily urinary excretion of 0.8 percent of the administered radioactivity results from the turnover of the labeled adenine nucleotide pool.
2. A 4-fold increase of urinary radioactivity excretion occurs in patients with the Lesch-Nyhan syndrome and support the role of impaired hypoxanthine salvage in the purine overexcretion associated with HGPRT deficiency.
3. Our data do not support the possibility that the increased radioactivity excretion in the HGPRT deficient subjects results from an elevated rate of adenine nucleotide degradation.

ACKNOWLEDGMENTS

    The authors wish to thank Jumana Judeh for her excellent typing of the manuscript. This work is supported by USPHS grants AM19674 and 5M01RR42 and a grant from The American Heart Foundation and The Michigan Heart Association. N.L.E. is the recipient of a Clinical Associate Physician Award for the General Clinical Research Center Branch of the National Institutes of Health.

REFERENCES

1. M. Lesch and W. Nyhan, A familial disorder of uric acid metabolism and central nervous system function, Am. J. Med. 36:561 (1964).
2. W. N. Kelley, M. L. Greene, F. M. Rosenbloom, J. F. Henderson, and J. E. Seegmiller, Hypoxanthine-guanine phosphoribosyltransferase deficiency in gout, Ann. Int. Med. 70:155 (1969).

3. F. M. Rosenbloom, J. F. Henderson, I. C. Caldwell, W. N. Kelley,
    and J. E. Seegmiller, Biochemical basis of accelerated purine
    biosynthesis de novo in human fibroblasts lacking hypo-
    xanthine-guanine phosphoribosyltransferase. J. Biol. Chem.
    243:1166 (1968).
4. A. W. Wood, M. A. Becker, and J. E. Seegmiller, Purine
    nucleotide synthesis in lymphoblasts cultured from normal
    subjects and a patient with Lesch-Nyhan syndrome. Biochem.
    Genet. 9:261 (1973).
5. N. L. Edwards, D. Recker, and I. H. Fox, Overproduction of uric
    acid in hypoxantine-guanine phosphoribosyltransferase
    deficiency: Contribution by impaired purine salvage. J.
    Clin. Invest. 63:922 (1979).
6. I. H. Fox and W. N. Kelley, Studies on the mechanism of fructose
    induced hyperuricemia in man. Metab. Clin. Exp. 21:713
    (1972).

ASPECT OF PURINE METABOLIC ABERRATION ASSOCIATED WITH URIC ACID OVERPRODUCTION AND GOUT

L.C. Yip, Ts'ai-Fan Yü and M.Earl Balis

Sloan-Kettering Institute for Cancer Research
1275 York Avenue, New York, New York   10021

Mount Sinai School of Medicine
Fifth Avenue & 100th Street, New York, New York 10029

It has been reported from these laboratories that of 425 patients with gouty arthritis, uric acid stones or both, hypoxanthine guanine phosphoribosyltransferase (HPRTase) was deficient in only 1.6% of the cases, including five members of one family (1). This result, together with others, indicates that HPRTase deficiency per se is a rather uncommon cause of gout. We have also shown that correlation of symptomatology with the amount of residual HPRTase activity as in erythrocytes is not always good. In some instances, HPRTase activity has been extremely low, as in Lesch-Nyhan (L-N) disease, and yet the patients have been normal neurologically (2).

In some subjects with no detectable erythrocyte HPRTase we have found 10-15% of normal activity in leucocyte lysates (3). The intact and disrupted leucocytes of the hemizygotes, in the absence of added phosphoribosyl pyrophosphate (PRPP), have the same rate of inosinate synthesis as normal cells. Under these circumstances enzyme concentration is not rate-limiting whereas the concentration of the co-substrate, PRPP is. The result from the intact cell assay is more representative of the cells in vivo function than the lysate, which may explain the important modification of clinical symptomatology the relatively mild hyperuricosuria, and the presence of mosaicism in circulating blood cells of the heterozygotes (3).

With the above in mind, and assuming that overproduction of PRPP could stimulate the first step in purine biosynthesis (4) (i.e. PRPP + glutamine → PRA) leading to excess purine and urate, we assayed endogenous PRPP levels of gouty subjects with apparently normal erythrocyte enzyme levels. We have also estimated functional PRPP in vivo by measuring the conversion of purines into nucleotides in intact cells and to relate these observations to a possible cause

of urate overproduction in patients with apparently normal levels
of erythrocyte purine metabolizing enzymes.

We have previously shown that the intracellular level of PRPP is
not necessarily proportional to the concentration of its synthetic
enzyme, PRPP synthetase (5,6). Variants of PRPP synthetase in gouty
subjects have been reported (7) as have several patients with abnormally high levels of erythrocyte PRPP synthetase activity (8).

The 38 patients under study were divided into four groups according to.intact leucocyte Hx incorporation. Group 1 (0.02-1.10
nmoles/mgprotein/hr) with a mean of 0.05, one forth of the mean of the
the normal subjects 0.23. Group 2 (0.10-0.16) have a mean of 0.14
about one half normal. Group 3 (0.13 - 0.27) have a mean of 0.19.
Group 4 (0.26 - 0.40) have a mean of 0.35 which is above the normal
value. The same groupings, but with more distinctly separated value
were observed if the classification of patients was determined by the
ratio of intact versus lysate HPRTase activity (Table I). The incorporation of ^{14}C Adenine into leucocytes is greater than that of Hx.
However, patients can be similarly classified by the ratio of intact/
lysate incorporation of Ad.

Table II shows that patient 1, who is in the 10% normal range
of erythrocyte HPRTase activity, has a normal intact leucocyte HPRTase
activity, but the ratio of intact to lysate enzyme activity is abnormal. Patients 2, 3 and 4 in Table II have 1% of normal erythrocyte
HPRTase. Their ratio of intact/lysate activity is more than 10 times
normal. An even higher ratio is seen with patients 5, who has 0.1%
normal HPRTase, is hyperuricemic and has severe gout.

TABLE I
Leucocyte Purine Phosphoribosyltransferase Activities of Patients
with Gout and Hyperuricemia (nmoles nucleotide/mg protein/hr).

Patients		Plasma Urate	UA-N	Intact		Lysate*		Intact/Lysate	
Group	#	mg/100mg	Tn (%)	HPRT	APRT	HPRT	APRT	HPRT	APRT (x10²)
1	11	10.2 ±1.9	2.0 ±0.5	0.05 ±0.02	0.51 ±0.27	4.24 ±1.26	5.55 ±2.02	1.30 ±0.37	10.00 ±3.32
2	12	10.2 ±2.0	1.9 ±0.6	0.14 ±0.02	0.73 ±0.22	5.17 ±0.77	6.54 ±2.41	2.57 ±0.31	12.85 ±6.13
3	9	9.6 ±2.4	1.7 ±0.5	0.19 ±0.02	1.37 ±0.66	4.05 ±1.27	4.27 ±1.20	4.80 ±0.50	22.57 ±4.30
4	6	10.4 ±1.7	2.9 ±0.7	0.35 ±0.05	1.61 ±0.30	4.77 ±1.13	4.63 ±1.16	7.47 ±1.50	35.00 ±2.33
Normal	12	6.5 ±1.0	1.5 ±0.3	0.23 ±0.05	0.66 ±0.18	4.20 ±1.64	4.45 ±0.97	4.80 ±0.70	15.90 ±6.00

* Lysate activity = nmoles nucleotide/mg protein/min.

TABLE II
Leucocyte HPRTase Activity of Partially Deficient HPRTase Patients.

Patients	Intact nmoles/ mg protein/hr	Lysate nmoles/ mg protein/min	Intact/Lysate (10^2)
(1) Ra	0.17	0.80	22
(2) Bo	0.03	0.06	59
(3) D'a	0.02	0.05	50
(4) Fa	0.02	0.04	60
(5) Il	0.01	0.01	131

TABLE III
Endogeneous Erythrocyte PRPP of Normal and Gouty Subjects.

Subjects	PRPP pmole/mg protein
Normal (6)	145+19
Patients from group 1 (7)	69+38
Patients from group 2 (7)	190+58
Patients from group 3 (6)	232+13
Patients from group 4 (5)	324+132
Patients with HPRTase partial deficiency (5)	702+341

Erythrocyte PRPP concentrations of patients randomly chosen from each group were assayed. Differences in mean value among groups (Table III) are in agreement with the intact lysate ratios. Group I patients have endogenous levels significantly lower than normal.

The stimulation of PRPP synthesis by glucose, frutose and mannose and the inhibition by iodoacetate and dinitrophenol indicate the importance of glycolytic intermediates in regulating ribose-5-phosphate availability for PRPP synthesis (9). We have reported that one of these, 2,3-DPG, is both an inhibitor of PRPP synthetase and an effector for HPRTase and suggested a regulatory role in purine metabolism, at least in erythrocytes (10). The erythrocyte concentrations of 2,3-DPG show an inverse relationship to endogenous PRPP levels (Table IV). This suggests that abnormal 2,3 DPG in some patients might be a contributing factor in abnormal PRPP function.

TABLE IV
2,3-DPG Content of Blood from Normal and Hyperuricemic Patients.

Subject	2,3-DPG umoles/ml
Normal (6)	1.74 + 0.35
Patients from group 1 (3)*	2.57 + 0.7
Patients from group 2 (1)	2.75
Patients from group 3 (2)	1.65 + 0.07
Patients from group 4 (2)	1.31 + 0.5
Patients of partial HPRTase deficiency (5)	1.44 + 0.68
L-N patients (2)	1.55 + 0

* See Table I for classifications.

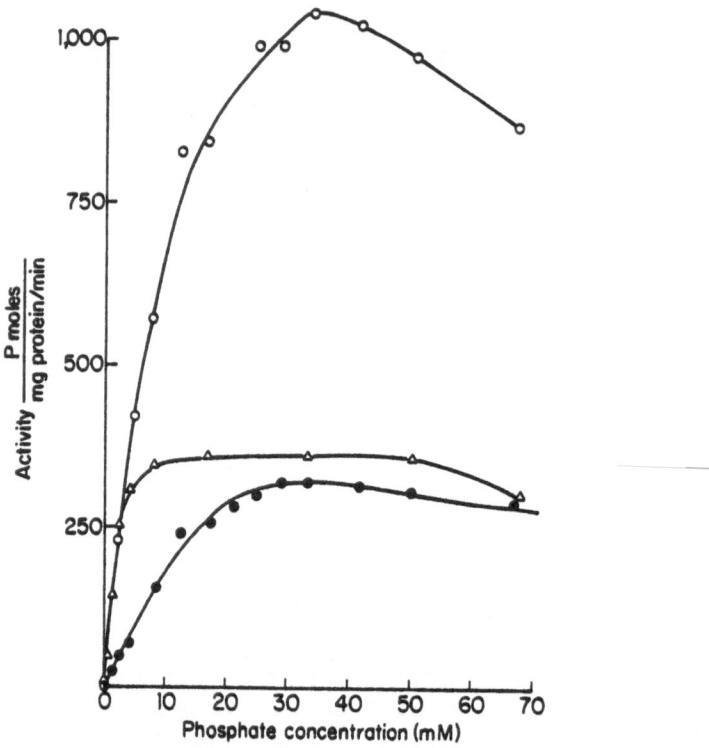

Fig 1: Phosphate requirements of PRPP synthetase from normal and
gouty subjects. Fresh erythrocytes lysate of normal subject (●),
patient SS (Δ) and patient SG (o) were assayed for PRPP synthetase
activity at various phosphate concentrations.

        In 18 control male subjects, the mean value for erythrocyte PRPP
synthetase was 309 pmoles of AMP/mg protein/ min. In 16 control female subjects, the mean value was 284. With four exceptions, 50 of
the patients screened gave results within the normal range but with
a slightly higher mean of 344. Patient SG has the highest erythrocyte
PRPP synthetase activity, 888 pmoles/mg protein/min. He also had the
highest intact leucocyte purine nucleotide synthesis rate, 0.4 nmoles/
mg protein/ hour, and a high ratio of 0.1 of intact vs lysate leucocyte HPRTase activity. The endogenous PRPP level of SG, 262 pmoles/
mg protein, was also nearly twice the normal mean value. Further experiments have shown that the PRPP synthetase of SG was different from
normal in phosphate requirement (Fig 1) and heat stability (Fig 2).
The PRPP synthetase of SS, who had a normal intact leucocyte purine
incorporation rate of 0.24 nmoles/mg protein/hour, was on the high
side but was much more active at physiological phosphate concentration
(Fig 1) than normal.

        The result presented here shows that enzyme assay in unphysiologic environments and under optimal reaction condition i.e. saturated substrate concentration, do not necessarily correlate with the actual enzyme behavior in vivo. We have screened 38 gouty subjects with no detectable erythrocyte purine metabolizing enzyme defect and have observed that intact leucocytes from more than half the patients have low capacity to utilize extracellular purine through the HPRTase-catalyzed reaction. Thus, they behave metabolically like HPRTase-deficient cells. Similar results were obtained by the assay of Ad incorporation into intact leucocytes. In the usual determination of erythrocyte enzyme activity, the lysates undergo extensive dilution and are thus subjected to denaturation or changes in effector or inhibitior concentration. In the intact leucocyte assay such difficulties are eliminated and the results are truer reflections of the in vivo cellular function of the enzymes. The use of the ratio of intact cell to lysate enzyme activity as an index to classify patients retains the accuracy of intact cell assay, takes any exogeneous variation on enzyme activity into consideration, and at the same time, eliminates any error in the determination of cell mass.

PRPP is a critical metabolite in purine and pyrimidine nucleotide biosynthesis. We have observed this in our studies with patients classified as Group 1 and Group 2. The extent of over-production seems to correlate with the abnormality of their intact cell enzyme activities and their endogenous PRPP levels (Table I and III). Group III patients have normal intact/lysate ratios. Most of them

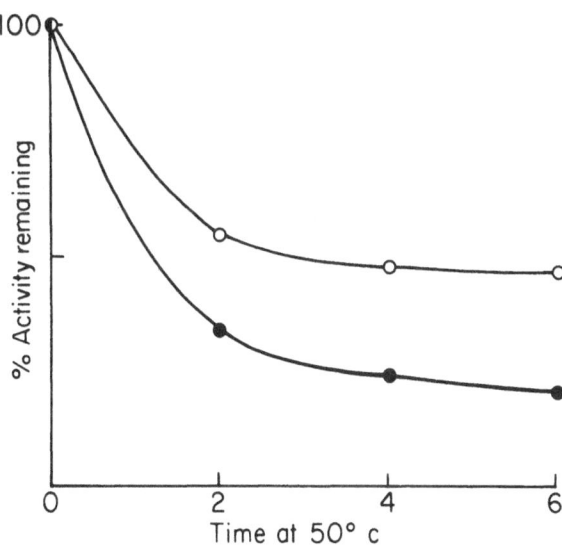

Fig 2: Heat stability of PRPP synthetase from normal and gouty subjects. Fresh erythrocyte lysates of normal (●) and patients SG (o) were assayed for PRPP synthetase activity after incubation at 50°C.

have normal urinary urate values and no family history of gout. The
above observations support the hypothesis that both low and high
function PRPP levels can lead to excessive urate formation.

Decreased PRPP concentration had not been considered a cause of
purine overproduction. However, since the Km of PRPP for HPRTase from
erythrocytes is higher than other PRPP utilizing enzymes and since
the latter is extremely sensitive to end product inhibition (11),
under conditions of partially limited PRPP, Hx would be expected to
spill, but no reduction of de novo purine synthesis would occur. Thus
these intact leucocyte studies suggest a new potential biochemical
cause of hyperuricemia.

REFERENCES:
1.   T.F.Yü, M.E. Balis, T.A. Krenitsky, J. Dancis, D.N. Silvers,
      G.B. Elion and A.B. Gutman. Rarity of x-linked partial hypo-
      xanthine-guanine phosphoribosyltransferase deficiency in a
      large gouty population, Ann Intern Med 76:255, 1972.
2.   M.E. Balis, L.C. Yip, T.F. Yu, A.B. Gutman, R. Cox and J. Dancis
      Unstable HPRTase in subjects with abnormal urinary oxypurine
      excretion. in Purine Metabolism in Man,(A. De Vries and O.
      Sperling, eds) Vol 41A, New York 1973.
3.   J. Dancis, L.C. Yip, R.P. Cox, S. Piomelli and M.E. Balis
      Disparate enzyme activity in erythrocytes and leucocytes,
      J Clin Invest 52:2068, 1973.
4.   O.W. Jones, D.M. Ashton and J.D. Wyngaarden. Accelerated turn-
      over of PRPP a purine nucleotide precursor in certain gouty
      subjects. J Clin Invest 41:1805, 1962.
5.   L.C. Yip, J. Dancis, B. Mathieson and M.E. Balis. Age-induced
      changes in AMP: Pyrophosphate phosphoribosyltransferase and
      IMP: Pyrophosphate phosphoribosyltransferase from normal and
      Lesch-Nyhan erythrocytes. Biochemistry 13:2558.
6.   L.C. Yip, S. Roome and M.E. Balis. In vitro and in vivo age-
      related modification of human erythrocyte phosphoribosyl
      pyrophosphate synthetase. Biochemistry 17:3286, 1978.
7.   W.N. Kelley and J.B. Wyngaarden. Enzymology of Gout in: Adv in
      Enzymology, A Meister, ed. 14:1, 1974.
8.   T.F. Yu, M.E. Balis and L.C. Yip. Over production of uric acid
      in primary gout. Arthritis and Rheumatism 18:695, 1975.
9.   J.F. Henderson and K.Y. Khoo. Synthesis of PRPP from glucose
      in ascites tumor cells in vitro. J Biol Chem 240:2349, 1965.
10.   L.C. Yip and M.E. Balis. Inhibitory effects of 2,3-DPG on enzymes
      of purine nucleotide metabolism. Biochem Biophys Res Commun
      71:14, 1976.
11.   J.B. Wyngaarden. Glutamine PRPP amidotransferase. Current Topics
      in Cell Regulation 5:135, 1972.

PROPERTIES OF A MUTANT HYPOXANTHINE-PHOSPHORIBO-SYLTRANSFERASE IN A PATIENT WITH GOUT

Wolfgang Gröbner and Wolf Gutensohn

Medizinische Poliklinik und Institut für Anthropologie
und Humangenetik der Universität
D 8000 München 2, Fed. Rep. Germany

INTRODUCTION

In addition to complete absence of hypoxanthine-phosphori-bosyltransferase (HPRT) activity leading to the Lesch-Nyhan-syndrome partial deficiencies of this enzyme have been descri-bed in a number of gouty patients without neurological symptoms. Although a detailed biochemical analysis of the mutant HPRT has been reported only for a relatively small number of cases the picture of a considerable genetic heterogeneity in this group of patients with gout has emerged[1, 2, 3]. Regarding this situation the detailed study of a single case seems justified and will be presented.

MATERIALS AND METHODS

The biochemical methodology used throughout the study consisted of more or less standard procedures and was descri-bed in detail elsewhere[4, 5].

The patient (male, 46 years) was suffering from severe gout since 1950 with tophi on joints and ears and repeated re-nal colics since 1957. Since 1967 he is under allopurinol thera-py. A detailed case history has been published[6].

RESULTS AND DISCUSSION

The following enzyme activities were determined in the patient's hemolysate: HPRT 4.1 nmoles/h/mg protein (Controls 79.49 $\pm$ 5.89 nmoles/h/mg; n = 13); adenine-phosphoribosyltransferase (APRT) 14.1 nmoles/h/mg (Controls: 15.97 $\pm$ 3.06 nmoles/h/mg; n = 13); orotidine-5-phosphatedecarboxylase (ODC) 1.56 nmoles/h/mg (Controls: 0.128 $\pm$ 0.068 nmoles/h/mg; n = 37). Thus residual HPRT activity was about 5% of normal; APRT was not increased, whereas the considerable increase in ODC is due to the allopurinol treatment. The presence of an inhibitor of HPRT in the patient's hemolysate was excluded by mixing-experiments with normal lysate.

The sedimentation constant of HPRT in hemolysates was determined by sucrose gradient ultracentrifugation with hemoglobin as internal marker. The mutant and normal enzymes gave identical values of $S_{20,w}$ = 3.9 $\pm$ 0.12 (n = 11). A similar case has been reported by Fox et al[3].

Normal and mutant enzyme were also compared by isoelectric focusing in a preparative sucrose gradient. A normal profile showed major peaks at pI 5.62 and 5.79 and shoulders at pI 5.43; 5.49 and 5.98, whereas the patient's HPRT split into two components with pI 5.75 and a more prominent peak at pI 4.55. A shift towards a more acidic pI has also been described for some other mutant enzymes[7].

Stability of the enzymes was checked at low and high temperatures. On storage at 4°C the patient's HPRT was more stable with 100% of the original activity retained after 6 days and 71% after 15 days. The respective values for normal enzyme were 68% after 6 days and 43% after 15 days. Heat inactivation at 80°C revealed a remarkable difference between the patient's hemolysate and his fibroblasts. In his hemolysate HPRT was distinctly more stable than HPRT from control lysates, whereas in his fibroblast lysate the enzyme proved to be more labile than that of 4 independent fibroblast lines. This difference remaines unexplained.

An antiserum raised in rabbits against highly purified human HPRT was used for an immunotitration experiment. Normal and patient hemolysates with equivalent concentrations of protein and hemoglobin resp. were compared. It was clearly shown that equivalent amounts of both lysates concumed equivalent volumes of antiserum. From this it is concluded that the patient's erythrocytes contain the mutated enzyme in amounts comparable to those of normal HPRT in normal erythrocytes.

In kinetic experiments normal and mutant HPRT in hemolysates were compared. The patient's enzyme showed an increase in K_M for the purine substrates (for hypoxanthine to 77 μM versus 8 μM for the normal enzyme, for guanine to 37 μM compared with 11 μM for normal enzyme) but a decrease for phosphoribosyl-pyrophosphate (PRPP; 66 μM compared with 170 μM for normal HPRT).

Reversible inhibition was checked with purine nucleotides and nucleosides in 0.5 mM concentration. The patient's enzyme proved to be more resistant to inhibition by the nucleotides but more sensitive to inhibition by nucleosides when compared with control HPRT. Inhibition of mutant HPRT was for IMP 19% versus 43% for controls; for GMP 31% vs 61%; for inosine 39% vs. 30%; and for guanosine 53% vs. 22%.

Irreversible inhibition of HPRT was studied by chemical modification using the following reagents: p-Chloromercuribenzoate and 2,2′-dinitro-5,5′-dithiodibenzoate directed against sulfhydryl groups; 2,4,6-trinitrobenzenesulfonic acid directed against primary amino groups; and phenylglyoxal, diacetyl and periodate-oxidized GMP directed against arginyl-residues. In all experiments under various preincubation conditions, with hemolysates as well as with partially purified enzyme preparations the patient's HPRT-activity turned out to be more resistant towards chemical modification than that of the normal enzyme. On the other hand chemical modification seems to be a specific process both for the normal and the mutant enzyme, since protection against modification by the substrate PRPP/Mg^{++} is equally effective for both enzymes. The specificity of this protective effect of PRPP/Mg^{++} for normal HPRT has been discussed in more detail in another contribution to this volume[8]

In summary we conclude from these observations that the HPRT of this patient is the product of a structural mutation which has altered the surface charge and the stability but not the molecular size of the enzyme. The enzyme is present in normal amounts in his erythrocytes with its catalytic activity lowered to 5% of normal. The active center shows changed affinities towards substrates and products and concomittantly has become less susceptible to active-site-directed chemical modification.

ACKNOWLEDGMENT

This work was supported by the Deutsche Forschungsgemeinschaft.

REFERENCES

1. W. N. Kelley, M. L. Greene, F. M. Rosenbloom, J. F.
    Henderson, and J. E. Seegmiller, Hypoxanthineguanine-
    phosphoribosyltransferase deficiency in gout, Ann. Int.
    Med. 70:155 (1969).
2. B. Bakay, W. L. Nyhan, N. Fawcett, and M. D. Kogut,
    Isoenzymes of hypoxanthineguaninephosphoribosyltrans-
    ferase in a family with partial deficiency of the enzyme,
    Biochem. Genet. 7:73 (1972).
3. I. H. Fox, I. L. Dwosh, P. J. Marchant, S. Lacroix, M. R.
    Moore, S. Omura, and V. Wyhofsky, Hypoxanthine-gua-
    nine-phosphoribosyltransferase. Characterization of a
    mutant in a patient with gout, J. Clin. Invest. 56:1239 -
    1249 (1975).
4. W. Gröbner, and N. Zöllner, Eigenschaften der Hypoxanthin-
    guaninphosphoribosyltransferase (HGPRTase) bei einem
    Gichtpatienten mit verminderter Aktivität dieses Enzyms,
    Klin. Wochenschr. 57:63 - 68 (1979).
5. W. Gutensohn, and H. Jahn, Partial deficiency of hypoxan-
    thine-phosphoribosyltransferase: evidence for a structu-
    ral mutation in a patient with gout, Eur. J. Clin. Invest.
    9:43 - 47 (1979).
6. N. Zöllner, F. D. Goebel, G. Öhlschlägel, and W. Gröbner,
    Juvenile Gicht mit verminderter Aktivität der Hypoxan-
    thin-guanin-phosphoribosyltransferase und Phäochromo-
    cytom. Teilweise Persistenz von Tophi trotz 12 jähriger
    harnsäuresenkender Therapie, Dtsch. Med. Wochenschr.
    103:1044   1049 (1978).
7. I. H. Fox, and S. Lacroix, Electrophoretic variation in the
    partial deficiency of hypoxanthine-guanine-phosphoribo-
    syltransferase, J. Lab. Clin. Med. 90:25 - 29 (1977).
8. W. Gutensohn, and H. Jahn, Chemical modification of hypo-
    xanthine-phosphoribosyltransferase and its protection
    by substrates and products, This Volume (1979).

VARIATION IN HUMAN HPRT AND ITS RELATIONSHIP TO NEUROLOGIC AND BEHAVIORAL MANIFESTATIONS

B. Bakay, E. Nissinen, L. Sweetman, U. Francke, and
W. L. Nyhan

Department of Pediatrics
University of California, San Diego
La Jolla, California 92093

Patients with the Lesch-Nyhan syndrome[1] have a distinctive phenotype, an essential quantitative feature of which is a virtually complete deficiency of the activity of hypoxanthine, guanine phosphoribosyl transferase (HPRT; E.C.:2.4.2.8)[2,3]. The clinical phenotype has a prominent central nervous system component in which there are two distinct sets of symptoms. On the one hand the patients have neurological abnormalities, choreoathetosis, spasticity, increased deep tendon reflexes and occasional seizure disorders. Most have developmental retardation. On the other hand they have bizarre compulsive, aggressive behavior, the hallmark feature of which is self-mutilative biting. They also have hyperuricemia, a consequence of a marked increase in the rate of synthesis of purine de novo[1] and as a consequence they have all of the clinical manifestations of the patient with gout. In contrast there are patients with partial deficiencies of HPRT. These patients may have up to 50% of normal HPRT activity but usually the activity is 5 percent or less[4,5]. They have hyperuricemia and they may have renal stones[6], but they do not have abnormalities of the central nervous system.

It has not yet been possible to make more than this broad general distinction of two clinical phenotypes. It has not been possible to correlate the amount of residual HPRT found in the variant HPRT with the clinical expression[7]. However the spectrum of patients studied has so far been small. We have been impressed with the behavior of the patient with the Lesch-Nyhan syndrome, and we have expressed the opinion that self-mutilation is an integral part of the syndrome. Others have said that the behavior may or may not be expressed. Certainly if one studies infants who are under one year of age many patients with the Lesch-Nyhan syndrome do not mutilate. So there is an ontogeny to this part of the phenotype, and we have

seen patients who began to mutilate at 3 or 5 years of age, but in
our experience sooner or later they all have mutilated. Thus it was
of interest to seek out the basis for the alternative point of view.
This led us to the patient of Catel and Schmidt[8],[9] who has been cited
as the exception that disproves the rule, a patient with the Lesch-
Nyhan syndrome in whom the intelligence and behavior were normal.

We have recently, through the courtesy of Dr. H. Manzke, had the
opportunity to see and study this patient. H.Chr.B. was 22-years-
old at that time. Slow motor development had been first noted at 6
months, and he was first studied in Kiel, Germany at 18 months. He
was hypotonic then and could not sit or stand. Uric acid crystals
were found in the diapers, and the concentration of uric acid in the
serum was as high as 15.5 mg/dl. He had athetoid movements and later
developed dysarthric speech.

When we saw him he had finished high school with good grades and
graduated from a junior college with a major in Science. He had normal height. He could get about by himself in a wheel chair. He
could walk only with assistance. He had dysarthric speech, but he was
understandable in English and German. He could control choreoathetosis when still, but always developed involuntary movements on in-
tention. He could write legibly. Muscle tone was increased, as were
the deep tendon reflexes. There was no evidence of self-mutilation,
nor had he ever had any desire to indulge in this type of behavior.
His personality was pleasant, and there was no evidence of the compulsive, aggressive behavior seen regularly in patients with the
Lesch-Nyhan disease.

The concentration of uric acid in his serum ranged from 4-6mg/dl
while he was receiving 300 mg of allopurinol per day. Assay of HPRT
activity in erythrocyte hemolysates yielded no evidence of activity.
We assayed samples obtained directly in San Diego in our standard
assay[10] at 37°C and 60°C in 0.01 $\underline{mM}$ hypoxanthine and 1 $\underline{mM}$ PRPP.
Similar results were obtained with guanine as substrate.

The metabolism of purines was studied in H.Chr.B., normal individuals and patients with the Lesch-Nyhan syndrome by incubation of
fibroblasts suspended in buffer containing 2 µC/ml of 8-^{14}C-hypo-
xanthine (specific activity 42.4 mC/mM) or 8-^{14}C guanine (specific
activity 43.1 mC/mM) for 2 hrs at 37°C. HClO$_4$ soluble extracts were
analyzed chromatographically using the method for purines and their
metabolites we have recently described[11].

The profiles obtained are shown in Figure 1. Fibroblasts derived from normal individuals converted labeled hypoxanthine predominantly to the adenine nucleotides, AMP, ADP, and ATP. These adenine nucleotide pools constituted the greatest amount of purine

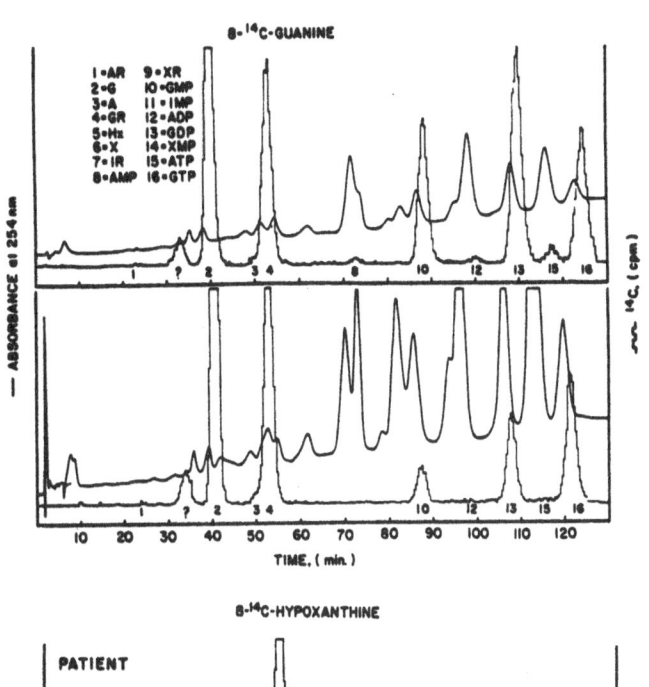

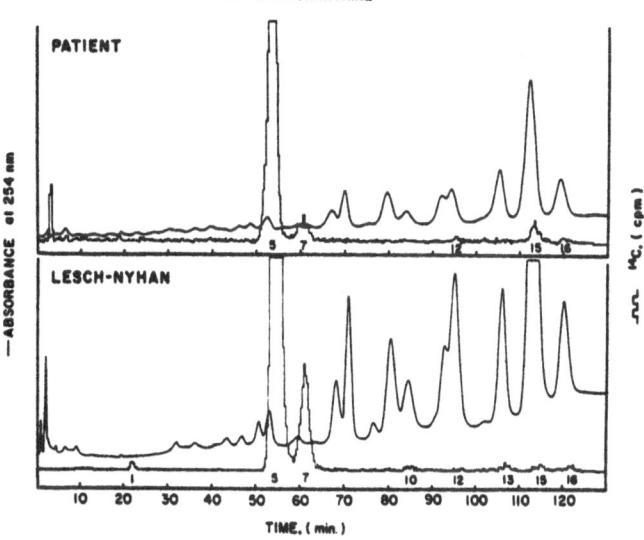

<u>Figure 1.</u> Profiles of purines and their metabolites from
        fibroblasts of the patient as compared with that of
        a normal individual and a patient with the classic
        Lesch-Nyhan syndrome. The cells were incubated with
        guanine-8-^{14}C or hypoxanthine-8-[14]C. The profile of
        the patient with the Lesch-Nyhan syndrome obtained
        in the presence of guanine-8-^{14}C was very similar to
        that obtained in the presence of hypoxanthine-8-[14]C.
        The normal individual incorporated large amounts of
        labeled hypoxanthine into nucleotides of adenine and
        guanine. The profiles obtained with labeled guanine are: on
        top, a normal individual, and below, the patient.

TABLE 1

METABOLISM OF 8-^{14}C-HYPOXANTHINE

Purine Metabolite	NORMAL N=6 n=8				L.N. N=3 n=7				H.Chr.B. N=1 n=2			
	nM uv	pM^{14}C	%	nCi/nM	nM uv	pM^{14}C	%	nCi/nM	nM uv	pM^{14}C	%	nCi/nM
AR	0.1	3.9	0.2	1.79	0.1	3.9	3.7	0.66	0	7.1	4.6	0
AMP	7.2	127.6	7.7	1.44	4.5	3.8	3.6	0.02	1.3	8.0	5.2	0.73
ADP	14.1	238.9	14.5	0.70	12.4	7.2	6.8	0.03	7.3	22.3	14.4	0.05
ATP	43.1	911.1	55.1	0.80	49.1	2.2	2.1	0.02	60.4	41.1	26.6	0.03
GMP	5.7	21.7	1.3	0.19	4.7	4.8	4.6	0.01	3.7	3.2	2.1	0.06
GDP	11.2	44.7	2.7	0.19	12.0	2.7	2.6	<0.01	11.8	5.8	3.3	0.01
GTP	8.9	122.9	7.4	0.52	8.3	4.4	4.2	0.02	11.7	15.0	9.7	0.05

n = number of determinations

N = number of patients

metabolites in all three cell types. Guanine nucleotides constituted the next largest. In normal cells as much as 78% of the isotope
of hypoxanthine accumulated in adenine nucleotides and 11% in guanine nucleotides. There was 8% in inosine. By contrast in L.N.
cells the major portion of the labeled hypoxanthine utilized was
converted to inosine. The specific activities of the adenine and
guanine nucleotides were 0.02 Ci/mM or less.

The amount of hypoxanthine utilized by H.Chr.B. cells was considerably smaller than that of normal cells. Nevertheless it was
clearly different from L.N. cells. Some 46% of the isotope of
hypoxanthine was converted to adenine nucleotides and 15% to guanine nucleotides and 34% to inosine.

The metabolism of $8-^{14}C$-guanine is shown in Table 2. Normal
cells converted 85% of the isotope of guanine to guanine nucleotides and 3% to adenine nucleotides. Only 10% was found in guanosine. L.N. cells converted 76% of the guanine metabolized to
guanosine. Almost none was found in guanine nucleotides.

In contrast, the cells of H.Chr.B., utilized 78% of the $8-^{14}C$-
guanine for the synthesis of guanine nucleotides. Some 17% was
found in guanosine. Thus, the metabolism of intact cells derived
from H.Chr.B. displayed a pattern of metabolism of hypoxanthine and
guanine under these conditions, which appear to be more physiological than the enzyme assay in a lysate, that was consistent with an
HPRT that was considerably more leaky than the enzyme found in L.N.
cells.

The specific activities of the purine metabolites may point the
way to the pathways employed.Following incubation with $8-^{14}C$-guanine,
the highest specific activity in normal cells was that of GTP followed by that of guanosine, and then inosine. In H.Chr.B. cells the
pattern was GTP, guanosine, inosine. These data indicate the relative importance of HPRT and nucleoside phosphorylase in the various
cell types. In this respect H.Chr.B. cells were more like normal
cells. Following inoculation with hypoxanthine the highest specific
activity in each cell type was inosine. In normal cells and H.Chr.B.
cells the next highest specific activity was in AMP while in L.N.
cells the next highest specific activity was that of adenosine.
These data suggest that the overall pattern of purine metabolism in
H.Chr.B. was very different from that of individuals with the Lesch-
Nyhan disease.

Another approach to the assessment of the functional activity
of HPRT in the various cell types was to examine the rates of growth
in selective media (Figure 2). The cells of H.Chr.B. and other cell
lines displayed similar rates of growth in MEM. Normal cells grew
as well in HAT medium as in MEM, but L.N. cells displayed virtually
complete inhibition of growth in HAT. The growth rate of H.Chr.B.

TABLE 2

METABOLISM OF 8-^{14}C-GUANINE

Purine Metabolite	NORMAL N=7 n=8				L.N. N=2 n=4				H.Chr.B. N=1 n=3			
	nM uv	pM^{14}C	%	nCi/nM	nM uv	pM^{14}C	%	nCi/nM	nM uv	pM^{14}C	%	nCi/nM
GR	2.6	190.7	10.4	3.43	0.7	164.7	78.6	10.65	1.1	83.0	17.0	2.84
GMP	6.7	289.5	15.9	1.62	5.7	1.4	0.7	<0.01	4.9	10.3	2.1	0.07
GDP	12.8	399.4	21.9	1.51	14.9	2.4	1.1	<0.01	17.4	45.6	9.4	0.12
GTP	8.0	856.6	46.9	4.13	8.1	9.7	4.6	0.05	19.1	325.8	66.8	1.49
AMP	10.2	7.7	0.4	0.04	5.7	4.7	2.2	0.01	2.5	3.9	0.8	0.35
ADP	14.4	10.1	0.6	0.03	13.9	3.9	1.8	0.01	7.2	3.2	0.7	0.05
ATP	34.0	34.0	1.9	0.05	44.1	15.6	7.4	0.02	40.3	7.9	1.6	0.01

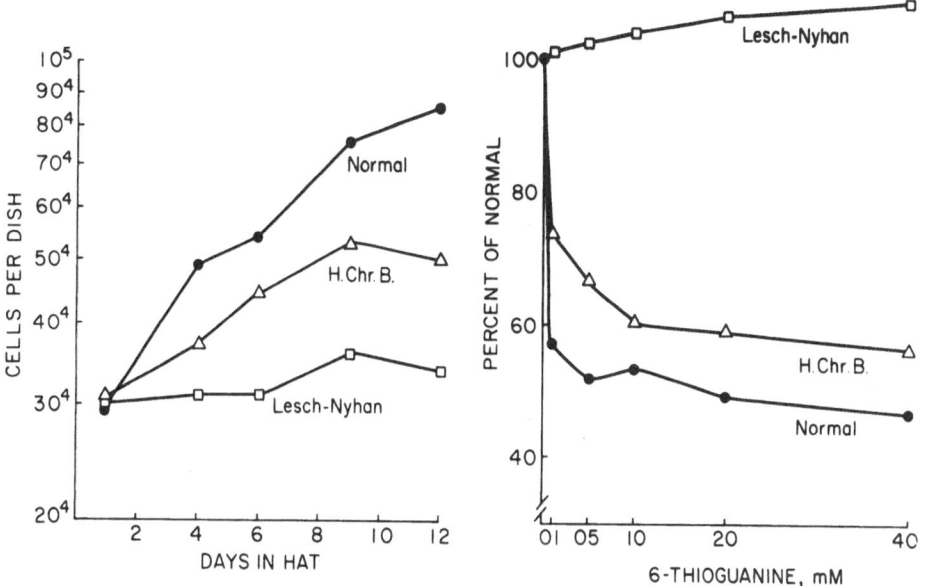

Figure 2. Growth of fibroblasts in selective media. HAT
symbolizes media containing hypoxanthine, aminopterin
and thymidine in which cells of patients with the
classic Lesch-Nyhan syndrome do not grow. Curves
similar to that obtained with 6-thioguanine, which
is nontoxic for patients with the classic Lesch-Nyhan
syndrome, were obtained with 8-azahypoxanthine and
8-azaguanine. Incubation was for 72 hours. The
points represent the means of 4 plates.

cells in HAT was intermediate between the other two cell types.

In the presence of 6-thioguanine the growth of normal cells was suppressed virtually completely. In fact normal cells began to die in this media. This purine analog had essentially no effect on the growth of L.N. cells. Its effect on the growth of H.Chr.B. cells was intermediate, but the pattern was more like that of normal cells.

These observations provide further evidence for heterogeneity of the HPRT locus. They indicate to us clearly that this important patient did not have the Lesch-Nyhan HPRT variant. Rather he had a variant enzyme with a certain amount of activity under relatively physiological conditions in vitro and presumably appreciable activity in vivo.

They suggest to us the importance of a close appraisal of the clinical phenotype. We believe it is likely that if a patient of sufficient age does not mutilate that it is likely he does not have the classic Lesch-Nyhan disease. We are interested in the assessment of such patients for the possible characterization of new variants of HPRT.

ACKNOWLEDGEMENTS

Aided by U.S. Public Health Service Research Grant GM-17702, from The National Institute of General Medical Services, UCSD General Clinical Research Center Grant RR-827, from the Division of Research Resources, National Institutes of Health and NF-13 from the National Foundation March of Dimes.

REFERENCES

1. Lesch, M., and Nyhan, W.L.: A familial disorder of uric acid metabolism and central nervous system function. Amer. J. Med. 36: 561 (1964).
2. Seegmiller, J.E., Rosenbloom, F.M., and Kelley, W.N.: An enzyme defect associated with a sex-linked human neurological disorder and excessive purine synthesis. Science 155: 1682 (1967).
3. Sweetman, L., and Nyhan, W.L.: Further studies of the enzyme composition of mutant cells in X-linked uric aciduria. Arch. Intern. Med. 130: 214 (1972).
4. Kelley, W.N., Greene, M.L., Rosenbloom, M., Henderson, J.R. and Seegmiller, J.E.: Hypoxanthine-guanine phosphoribosyltransferase deficiency in gout. Ann. Intern. Med. 70: 155 (1969).
5. Sweetman, L., Borden, M., Lesh, P., Bakay, B., and Becker, M.A.: Diminished affinity for purine substrate as a basis for gout with mild deficiency of hypoxanthine guanine phosphoribosyl transferase. Adv. Exp. Med. Biol. 76A: 361 (1977).

6. Kogut, M.D., Donnell, G.N., Nyhan, W.L., and Sweetman, L.: Dis-
    order of purine metabolism due to a partial deficiency of
    hypoxanthine guanine phosphoribosyl transferase. Amer. J.
    Med. <u>48</u>: 148 (1970).
7. Emerson, B.T., and Thompson, L.: The spectrum of hypoxanthine
    guanine phosphoribosyl transferase deficiency. Quart. J. Med.
    <u>42</u>: 423 (1973).
8. Catel, V.W., and Schmidt, J.: Über familiäre gichtische Dia-
    these in Verbindung mit zerebralen und renalen Symptomen bei
    einen Kleinkind. Dtsch. Med. Wschr. <u>84</u>: 2145 (1959).
9. Manzke, H., Harms, D., and Dorner, K.: Zur Problematik der Be-
    handlung der kongenitalen Hyperurikämie. Mschr. Kinderheilk.
    <u>119</u>: 424 (1971).
10. Bakey, B., Telfer, M.A., and Nyhan, W.L.: Assay of hypoxan-
    thine-guanine and adenine phosphoribosyl transferases. A
    simple screening test for the Lesch-Nyhan syndrome and related
    disorders of purine metabolism. Biochem. Med. <u>3</u>: 230 (1969).
11. Bakay, B., Nissinen, E., and Sweetmen, L.: Analysis of radio-
    active and nonradioactive purine bases, nucleosides and nu-
    cleotides by high-speed chromatography on a single column.
    Analyt. Biochem. <u>86</u>: 65 (1978).

HIGH HPRT ACTIVITY IN FIBROBLASTS FROM PATIENTS WITH LESCH-NYHAN SYNDROME DUE TO BACTERIAL "L-FORM" CONTAMINATION

I. Willers, S. Singh, K. R. Held, H. W. Goedde

Institut für Humangenetik der Universität Hamburg

Butenfeld 32, 2000 Hamburg 54, F. F. Germany

INTRODUCTION

The studies reported here demonstrate how an unnoticed infection with bacterial "L-forms" (L for Lister Institute) may lead to erroneous diagnosis or interpretation in metabolic genetic disorders. Bacterial "L-forms" are bacterias which have lost their cell wall and have become resistent due to prolonged treatment with antibiotics. The morphology of the colonies compared to the mycoplasmas is very similar and can only be distinguished by an experienced microbiologist (Blenk, 1977). During our studies with HPRT mutant fibroblasts, we observed a sudden increase of HPRT activity, although mycoplasmas were absent. Increase of HPRT activity due to infection with mycoplasmas is known (Stanbridge et al., 1975). Special microbiological tests by Dr. Blenk revealed the presence of an infection with bacterial "L-forms" which could be traced back to a particular charge of fetal calf serum used by us. This observation prompted us further to characterize this contamination by using immunoprecipitation techniques and by studying its effects on incorporation of hypoxanthine and cell growth in selection media.

MATERIAL AND METHODS

The fibroblasts were derived from human forearm biopsies, and cultured in F - 10 media containing 15 % fetal calf serum. 8 - ^{14}C hypoxanthine was purchased from Amersham Radiochemical Center, England. Phosphoribosyl pyrophosphate was obtained from Boehringer, Mannheim: 8 - azaguanine and azaserine from Serva, Heidelberg, West-Germany.

The following methods were used:

(a) Immunoprecipitation tests (Held et al., 1975). The antibody
    used was prepared from highly purified human erythrocyte HPRT
    (Münsch et al., 1977) and was a gift of Dr. Münsch.
(b) Microtest of ^{14}C hypoxanthine incorporation in cell cultures
    grown in microtiter plates (Willers et al., 1975).
(c) Selection experiments in 8 - azaguanine and HAT containing
    media (DeMars and Held, 1972).
(d) HPRT activity was assayed according to DeMars and Held (1972).
(e) The protein concentration was estimated according to
    Lowry et al., (1951).

RESULTS

(a) HPRT activity in infected fibroblasts of patients with
    HPRT mutations in comparison to the cells prior to infection

    Tab. 1 shows the differences in HPRT activity of the same
cell lines infected with bacterial "L-forms" and prior to infection. An increase of HPRT activity up to 25 fold can be
observed in the infected cells.

(b) Immunoprecipitation with HPRT antiserum

    The results are summerized in tab. 2. In contrast to the
findings in the non infected fibroblasts of normal controls, the
infected mutant fibroblasts showed no cross reacting material
indicating a different HPRT enzyme than that of human provenance.

Table 1. HPRT enzyme activities in fibroblasts of patients
          with HPRT Mutation and bacterial "L-forms"

HPRT activity (n moles/mg/h)

cell line	non infected	infected with "L-forms"	"L-forms"isolated from infected cultures
99	51	419	1277
155	26	654	1400
157	21	320	n. t.

HIGH HPRT ACTIVITY IN FIBROBLASTS
329

Table 2. Enzyme activities after immunoprecipitation with
          antiserum against purified human erythrocyte HPRT

cell line	HPRT activity (n moles/mg/h) μl antiserum				
	0	1.2	2.5	5	10
non infected controls					
151	309	200	142	88	41
136	175	147	129	31	17
125	210	147	129	111	60
infected HPRT mutants					
157	350	359	372	354	343
99	454	454	456	466	488
155	717	778	870	745	745

(c) Incorporation of hypoxanthine

    The incorporation studies in fibroblasts were performed
with an average of at least 12 determination at different
concentrations of ^{14}C hypoxanthine (Fig. 1). The uptake of the
substrate remaines unchanged in infected and non infected HPRT
mutant fibroblasts.

(d) Selection experiments

    The growth curves of infected mutant fibroblasts in
medium containing different concentrations of 8-azaguanine and
their growth in HAT medium are presented in Fig. 2. No change
in growth profiles of infected cells under these conditions could
be observed as compared to the data of non-infected HPRT mutant
fibroblasts.

DISCUSSION

    These studies demonstrate that HPRT mutant fibroblasts in-
fected with bacterial "L-forms" differ in some respects from the
uncontaminated cells. The absence of cross reacting material
against specific human HPRT antiserum in spite of high HPRT-
activity points to presence of an HPRT enzyme from non human
source. The HPRT activity in infected fibroblasts from Lesch-Nyhan

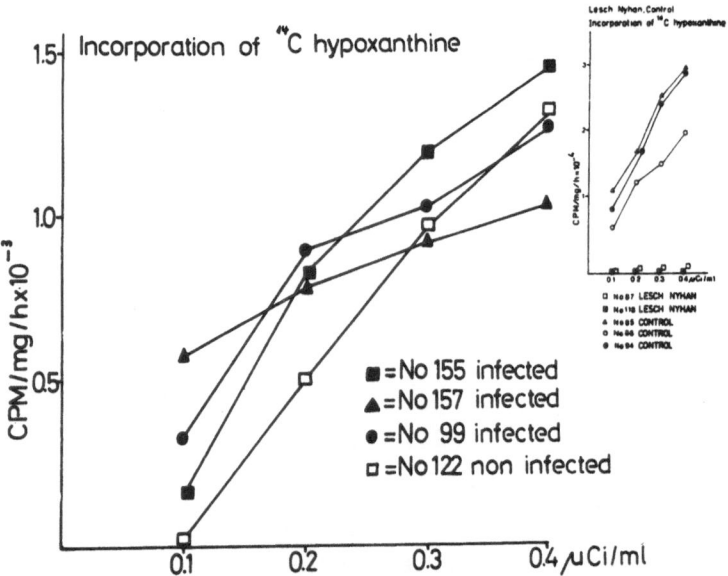

Fig. 1 Incorporation of hypoxanthine in infected and non infected fibroblasts with mutant HPRT. The inset shows the
incorporation in fibroblasts from normal controls in
comparison to Lesch Nyhan cells (scale: CPM/mg/10^{-4}).

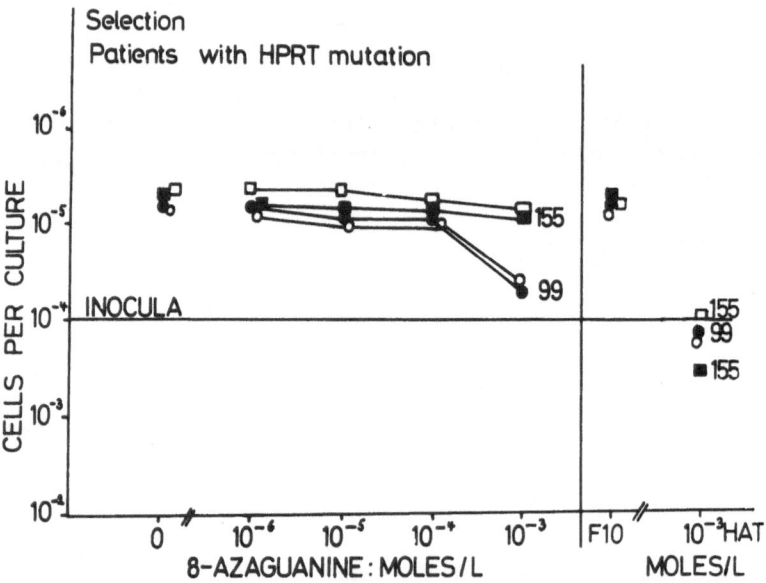

Fig. 2. Growth profiles of infected (■, ●) and non infected
(□, ○) fibroblasts in selective media.

patients is in the range or even higher than that of normal fibroblasts. Also the isolated bacterial "L-forms" of these cell cultures showed HPRT activities up to 3times higher than that of
normal fibroblasts. This provides additional support for the conclusion that the enzyme measurement in the contaminated cells is
of bacterial"L-form"origin. In spite of this contamination no
effect was noted on the hypoxanthine incorporation. thus the
HPRT mutant cells can still be diagnosed by means of incorporation
studies and autoradiographic techniques. The fact that no change
in resistance to 8-azaguanine and growth in HAT selection media
could be observed in infected HPRT mutant fibroblasts suggests
that these microorganisms are only bound to the outer host cell
membrane and are unable to convert hypoxanthine for the intact
HPRT mutant fibroblasts.

ACKNOWLEDGEMENTS

We are gratefull to Dr. H. Münsch for the gift of human HPRT
antiserum and Dr. H. Blenk for classifying and isolating the con-
taminants of the cell cultures. Mrs. B. Reßler and Mrs. G. Knaack
gave excellent technical assistance.

REFERENCES

Blenk, H., and Hofstetter, A., 1977, "Mykoplasmen", Hoechst AG,
     Frankfurt.
DeMars, R., and Held, K. R., 1972, The spontaneous azaguanine-
     resistant mutants of diploid human fibroblasts, Hum. Gen.,
     16: 87-109.
Held, K.R., Kahan, B., DeMars, R., 1975, Adenine phosphoribosyl-
     transferase immunoprecipitation reactions in human mouse
     and human hamster cell hybrids, Hum. Gen., 30: 23-34.
Muensch, H., and Yoshida, A., 1977, Purification and charac-
     terization of human hypoxanthine/guanine phosphoribosyl-
     transferase, Eur. J. Biochem. 76: 107.
Stanbridge, E. J., Tischfield, J. A., Schneider, E. L., 1975,
     Appearance of hypoxanthine guanine phosphoribosyltransferase
     as a consequence of mycoplasma contamination, Nature, 256:
     329-331.
Willers, I., Agarwal, D. P., Singh, S., Schloot, W., and Goedde,
     H. W., 1975, Hum. Gen., Rapid determination of hypoxanthine
     guanine-phosphoribosyltransferase in human fibroblasts
     and amiotic cells, Hum. Gen., 27: 323-328.

KINETICS OF A HGPRT MUTANT SHOWING SUBSTRATE INHIBITION

E. H. Harley, C. M. Adnams and L. M. Steyn

Department of Chemical Pathology, University of
Cape Town, Cape Town, Republic of South Africa

A twenty year old man, T.K., presented with recurrent episodes of renal colic, the first taking place when he was 13 years old. Mild neurological signs consisted of nystagmus, symmetrical hyperreflexia and moderate mental retardation. His serum uric acid was consistently higher than 0,71 mM/l (normal 0,12 - 0,5 mM/l) and he was found to have uric acid overproduction of typically 7,4 mM/day (normal 1,5 - 4,4 mM/day). In view of his age, the neurological signs and purine overproduction, red cell hypoxanthine-guanine phosphribosyl transferase (HGPRT) was assayed by the spectrophoto metric method of Johnson et al [1]. An activity of 23 nMol/hr/mg Hb was found compared with a normal value for this laboratory of 100 ± 9 nMol/hr/mg Hb.

Km values of HGPRT for hypoxanthine were determined in red cell lysates from the patient and normal controls by the isotopic assay of Emmerson et al [2]. The values obtained, 0,04 mM and 0,06 mM respectively, did not differ significantly from published results [3], Fig.1. Unexpectedly, the enzyme activities were similar to control values at low substrate concentrations and the low activity originally demonstrated was found to be a consequence of marked substrate inhibition at higher hypoxanthine concentrations. The Ki for hypoxanthine was 0,75 mM, Fig.1. PP-ribose-P gave normal kinetic data. The Km values were 0,16 mM and 0,25 mM for the mutant and control enzymes. Neither IMP, Fig.2, nor PPi were found to produce significant product inhibition.

In order to assess the relevance of these findings in cell free enzyme preparations to the metabolism of the intact cell, cultured skin fibroblasts from the patient were labelled with 1 µC/ml (17 µM/l)

KINETICS OF MUTANT HGPRT

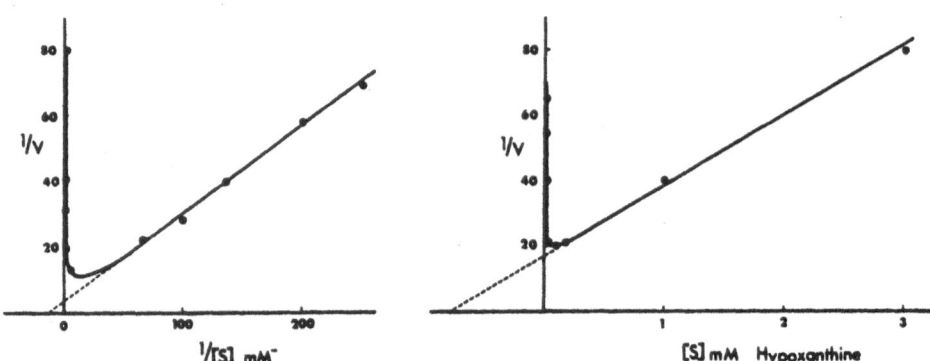

Fig. 1.   Determination of kinetic data for HGPRT in TK erythrocyte
           lysates.   Activities were determined by the isotopic assay [2]
           at different concentrations of hypoxanthine (0,004 - 3 mM)
           and constant concentration of PP-ribose-P (3 mM). Left : Plot
           of $1/v$ vs. $1/S$.   Right : Plot of $1/v$ vs S.   The double-
           reciprocal plot is shown here to illustrate the kinetics, but the
           Km value was determined from the statistically more accurate
           plot of S/v versus S.

(8-^{14}C) hypoxanthine and 0,5 μC/ml (Methyl-^{3}H) thymidine for 4,5
hours at 37°C. Ratios of ^{14}C to ^{3}H were determined in TCA
precipitable material as described by Rozen et al [4]. Control fibroblast
cultures were of similar passage number and cell density. The results,
Table 1, showed that the fibroblasts from the patient incorporated
hypoxanthine to only 15% of the extent found in the control cultures.
This value was found to vary between 10% to 40% in different
experiments. These results suggest that there is sufficient substrate in
the vicinity of the active site to effect inhibition of the enzyme in the
intact cell. It would be an unwarranted extrapolation to use the degree
of inhibition observed to give an accurate measure of intracellular
hypoxanthine concentration.

     A mutant HGPRT showing substrate inhibition has not been
demonstrated before. This has interesting implications for the treatment
of the patient since the activity of the enzyme varies with hypoxanthine
concentration.   Thus a therapeutic aim could be to devise measures
that avoid increased cellular and plasma hypoxanthine levels. It may
therefore be inappropriate to treat the patient with allopurinol which
increases plasma, and presumably cellular, hypoxanthine levels.   On

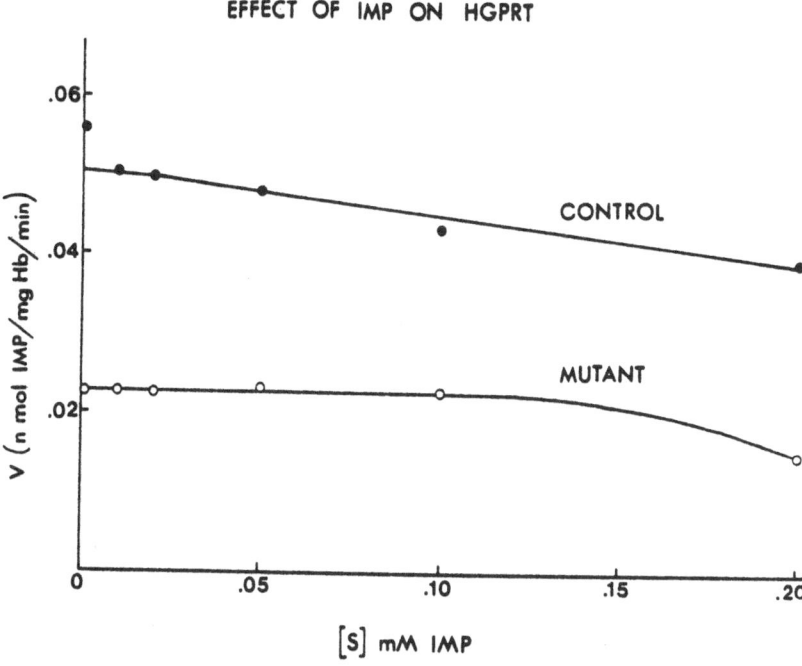

Fig.2   The effect of different concentrations of IMP (0,01 - 0,2 mM)
        on HGPRT activity in erythrocyte lysates from T.K. and a
        normal control

Table 1.   Incorporation of (methyl-^{3}H) thymidine and (8-^{14}C)
hypoxanthine into cultured skin fibroblasts from the patient and normal
controls.   The ratio of ^{14}C to ^{3}H in TCA precipitable material
was used as a measure of HGPRT activity.

		^{14}C	^{3}H	^{14}C : ^{3}H
		dpm	dpm	ratio
TK	Flask 1	380	8,410	0,045
	Flask 2	265	3,676	0,072
Control 1	Flask 1	2,800	6,039	0,463
	Flask 2	6,245	12,376	0,505
Control 2	Flask 1	2,455	2,453	1,001
	Flask 2	1,800	4,013	0,449

the other hand, dietary purine restriction may be more appropriate. Detailed balance   studies are in progress to assess these points.

The unusual features of this new mutant HGPRT may provide new insight into the following :

1.   The mechanisms operative at the active site of the enzyme.

2.   Purine metabolism and regulation of metabolism in the
       intact cell.

3.   The adoption of appropriate therapeutic measures in this
       particular patient.

This investigation was supported by grants from the S.A. Medical Research Council, the Atomic Energy Board, the Harry Crossley Foundation and the Cancer Research Trust.   L.M.S. is in receipt of a Guy Elliott Research Fellowship.

REFERENCES.

1.   L.A. Johnson, R. B. Gordon and B.T. Emmerson, Hypoxanthine-
       guanine phosphoribosyl transferase :   A simple spectrophotometric
       assay.   Clin. Chim. Acta.   80 : 203 (1977).
2.   B.T. Emmerson, C.J. Thompson and D.L. Wallace, Partial
       deficiency of hypoxanthine-guanine phosphoribosyl transferase :
       Intermediate enzyme deficiency in heterozygote red cells.
       Ann. Int. Med. 76 : 285 (1972).
3.   J.B. Wyngaarden and W.N. Kelley, Gout, in   The Metabolic basis
       of inherited disease, J.B. Stanbury, J.B. Wyngaarden and
       D.S. Fredrickson, eds, McGraw-Hill Book Company, New York,
       (1978).
4.   Rozen, S. Buhl, F. Mohyuddin, U. Caillibot and C.R.Scriver,
       Evaluation of metabolic pathway activity in cultured skin
       fibroblasts and blood leukocytes.   Clin. Chim. Acta 77 : 379
       (1977).

SPECTRUM OF 2,8-DIHYDROXYADENINE UROLITHIASIS IN COMPLETE APRT DEFICIENCY

H. A. Simmonds, T. M. Barratt, D. R. Webster, A. Sahota,
K. J. Van Acker, J. S. Cameron and M. Dillon
Clinical Science Laboratories, Guy's Hospital, London;
Hospital For Sick Children, London; University of
Antwerp, Belgium

The identification of 2,8-dihydroxyadenine (2,8-DHA), a uric acid analogue characterised by its extreme insolubility, as the principal component of so-called 'uric acid' stones in a young male child, was originally reported independently by workers in Paris and London[1,2]. The correct diagnosis resulted from the finding of nearly undetectable levels of the enzyme adenine phosphoribosyl-transferase (APRT : EC 2.4.2.7) in erythrocyte lysates in the first instance[1] and the detection of abnormal urinary levels of adenine (excreted together with 8, and 2,8-DHA in consequence of the enzyme defect) in the latter case[2]. Since that time a further three cases have been identified in London[2,3,4] (Dillon et al - in preparation). Another two are added to the list[1,5] in this symposium by the French workers, one of which has appeared in abstract[6] (Table 1).

Considerable data now exists from detailed studies in the first four homozygotes for the defect[1-5]. To date the defect has been found exclusively in children (both male and female) and appears to be inherited as an autosomal recessive trait, in confirmation of the earlier studies in heterozygotes[7] (Van Acker et al - this symposium). Clinical manifestations may be attributable solely to the extreme insolubility of 2,8-DHA. Children homozygous for the defect are otherwise clinically normal and show no evidence of immunodeficiency (Stevens et al - this symposium). An abnormal incidence of gout, or excessive hyperuricaemia, has not been associated with either homozygosity or heterozygosity for the defect[7] (Van Acker et al - this symposium).

Although the defect has now been characterised with certainty and in detail, several points have emerged. First, the number of new cases reported from two centres only suggest that the defect

Table 1.   Details at diagnosis of 2,8-dihydroxyadenuria

PATIENT	YEAR	SEX	AGE	ORIGINAL PRESENTATION	ONSET	INITIALLY DIAGNOSED BY	APRT ACTIVITY[+] (nmol/mgHb/h) (Normal 18-36)
LONDON							
B.Dh[2]	1975	M	2½	URIC ACID stones	Birth	ADENINE in urine	0.3
F.Dh[3]	1976	M	7	NONE (asymptomatic brother) (B.Dh)	Never ill		0.25
S.Rn[4]	1977	F	19m	STONES: L. ureter, R. pelvis	1 yr	2,8-DHA in stones	0.61
Sh.B[( )]	1979	F	4	ACUTE RENAL FAILURE (Comatose - anuric): BILATERAL OBSTRUCTIVE UROPATHY	2 yrs	2,8-DHA in stones	11.0**
PARIS							
P.Tho[1]	1974	M	3½	URIC ACID stones	2 yrs	APRT activity ABSENT in RBC's	0.002
This Sympo-	1979	M	2	STONES		See Hamet, Cartier, Vincent for details	
sium	1979	F		NONE (asymptomatic sister) (above)			
6	1979	?	?	STONES		Reported in abstract Reveilland et al.	

** Packed cell transfusion on admission   + RBC lysate

Table 2.   Renal abnormalities

CASE	MANIFESTATION	PLASMA CREATININE (a)	CREATININE CLEARANCE (1.73m^2)
1	Stones	44 μmol/l	56 ml/min[(b)]
2	Stones	71 μmol/l	105 ml/min
3	Asymptomatic	53 μmol/l	106 ml/min
4	Stones (L ureter, R pelvis)	120 μmol/l	90 ml/min*
5	Acute on chronic renal failure: bilateral ureteric stones.	815 μmol/l	8.5 ml/min*

(a) at diagnosis   * after surgery   (b) lost to follow-up

may be more frequent than the recent description of the defect
would indicate (Table 1). This fact, together with the original
misdiagnosis in some and the severe renal failure in the most recent
case (Table 1 & 2), should be highlighted. Two separate studies[7,8]
have put the incidence of heterozygosity for the defect as high as
0.5-1%. In other words, an incidence of homozygosity of the order
of ≃ 1 in 100,000, which would support the possibility of a greater
incidence and previous misdiagnosis. Two of the homozygotes have
been completely asymptomatic and only identified during the family
studies, so that possibly 25% or more homozygotes for the defect
will not be identified.

The problems of diagnosis are directly related to the frequent
misdiagnosis - 2,8-DHA can effectively masquerade as uric acid in
several different systems making its correct identification
extremely difficult by at least four methods used routinely[4].
Correct identification is possible but must depend on the access-
ibility of more sophisticated equipment or techniques for the
recognition of the real stone component[1,4,5]. However, it should
be underlined that one of the most frequent methods used in the
categorisation of homozygotes for the defect - the estimation of
APRT activity in erythrocyte lysates - may also produce fallacious
results (as in our most recent case) if blood transfusion has been
essential for the immediate correction of a co-existing clinical
condition (Table 1).

In this instance a new technique, isotachophoresis of the
urine, which will distinguish homozygotes for the defect by the
adenine excreted[4], will be helpful (Simmonds et al - this symposium).
The identification of adenine in the urine in this defect has been
found to be characteristic and diagnostic of the homozygous state;
heterozygotes, apart from a reduced APRT activity in erythrocyte
lysates (<50% control activity), are clinically and biochemically
indistinguishable from controls; their urine containing undetectable
levels of adenine, or any of its metabolites[3,4]. Likewise, their
intact RBC's (as distinct from lysates) at physiological levels of
adenine and inorganic phosphate (pi) have exhibited normal APRT
activity, when homozygote activity was negligible under such
conditions.

Urinary adenine may also be identified by HPLC but again
problems may be encountered in young children consuming chocolate
products or coca cola where an 'adenine' peak may in reality be a
theobromine degradation product. A caffeine-free diet will be
essential in such instances (unpublished observations).

The importance of early recognition and correct diagnosis in
this condition is underlined by the wide spectrum of clinical manifestation encountered. This has varied from none whatsoever to
presentation in coma with acute on chronic renal failure necessit-

ating dialysis (Table 1), due to bilateral ureteric obstruction by
2,8-DHA stones (Dillon et al). In the latter instance the diagnosis
was obscured since 2,8-dihydroxyadeninuria is not at present listed
as a possible cause of coma/acute renal failure in young children
and the usual alternatives (haemolytic uraemic syndrome, Reyes
syndrome, poisoning etc) had first to be eliminated.

In the two affected female homozygotes for the defect, surgery
was essential (two successive operations) for the removal of the
stones (Table 2). In all stone formers symptoms of abdominal pain,
dysuria and/or urinary tract infection have been present for more
than 2 years in some instances. Three of five subjects initially
showed impaired renal function and the most recent case appears to
have suffered permanent renal damage.

Renal biopsy at the time of surgery in this case (Pincott, J. -
personal communication) presented a picture similar to that noted
in an animal model during a long-term study of the pathogenesis of
the renal lesion induced by an acute crystal nephropathy[9]. It was
in fact a composite of the lesion noted during the acute phase
(intratubular crystal blockage) superimposed on the lesion noted at
differing periods during the following months (tubular degeneration,
relocation of the crystals in the interstitium, interstitial
nephritis with amorphous deposits surrounded by giant cells,
fibrosis and glomerulosclerosis) in the animal model. The 2,8-DHA
crystals had a similar birefringence to their analogues xanthine
(the animal model) or uric acid. These biopsy findings in the child
would be consistent with the reverse situation to the chronic on
acute renal failure noted in our animal model[9] and hence acute on
chronic renal failure.

Early recognition thus appears vital to avert the development
of the renal lesion; the renal damage is preventable since the
stone formation may be controlled by allopurinol; preferably without
alkali. (Alteration of urinary pH within the physiological range by
alkali will not improve 2,8-DHA solubility[1,5] and may even be
contraindicated[4].

Although an endogenous origin for at least some of the adenine
excreted in this defect has not been established with certainty,
the pathway concerned with polyamine synthesis seems a likely
candidate. Nevertheless, exogenous adenine taken in the diet is
almost certainly contributary and should be reduced to a miminum;
the macrobiotic diet of the latest child presenting in coma was
rich in lentils and other grain - all foods with possibly the
highest adenine purine content[10]. A low purine diet is thus also
advised[3].

SUMMARY

APRT deficiency may be totally benign or life threatening. The importance of early recognition/diagnosis is thus stressed. Urolithiasis (2,8-DHA stones: the precipitating factor in all cases) is treatable. With early recognition and treatment allopurinol without alkali and a diet low in purine homozygotes have remained clinically and biochemically normal to date. 'Uric acid' stones in children must always be suspect and subjected to sophisticated analysis. Diagnosis from red cell APRT activity may also have its pitfalls.

REFERENCES

1. P. Cartier and M. Hamet, Une nouvelle maladie métabolique: le déficit complet en adenine phosphoribosyltransférase avec lithiase de 2,8-dihydroxyadénine, C. R. Acad. Sci. (Paris), 297:883 (1974).
2. H. A. Simmonds, K. J. Van Acker, J. S. Cameron and W. Snedden, The identification of 2,8-dihydroxyadenine, a new compoment of urinary stones, Biochem. J., 157:485 (1976).
3. K. J. Van Acker, H. A. Simmonds, C. F. Potter and J. S. Cameron, Complete deficiency of adenine phosphoribosyltransferase, New Eng. J. Med., 297:127 (1977).
4. H. A. Simmonds, C. F. Potter, A. Sahota, J. S. Cameron, G. A. Rose, T. M. Barratt, D. I. Peters, D. G. Arkell and K. J. Van Acker, Adenine phosphoribosyltransferase deficiency presenting with supposed 'uric acid' stones: pitfalls of diagnosis, J. Roy. Soc. Med., 71:791 (1978).
5. H. Debray, P. Cartier, A. Temstet and J. Cendron, Child's urinary lithiasis revealing a complete deficit in adenine phosphoribosyltransferase, Pediat. Res., 10:762 (1976).
6. R. J. Reveillaud, M. Daudon, M. F. Protat, G. Ayrole and V. Vinther-Harder, Purine metabolism and urolithiasis, Kidney Int., 3:1 (1979) (abst).
7. L. A. Johnson, R. B. Gordon and B. T. Emmerson, Adenine phosphoribosyltransferase: a simple spectrophotometric assay and the incidence of mutation in the normal population, Bioch. Genet., 15:265 (1977).
8. I. H. Fox, S. Lacroix, G. Planet and M. Moore, Partial deficiency of adenine phosphoribosyltransferase in man, Medicine, 56:515 (1977).
9. D. A. Farebrother, P. Hatfield, H. A. Simmonds, J. S. Cameron, A. S. Jones and A. Cadenhead, Experimental crystal nephropathy (one year study in the pig), Clin. Nephrol., 4:243 (1975).
10. A. J. Clifford and D. L. Story, Levels of purines in foods and their metabolic effects in rats, J. Nutr., 106:435 (1976).

COMPLETE ADENINE PHOSPHORIBOSYLTRANSFERASE (APRT) DEFICIENCY IN

TWO SIBLINGS : REPORT OF A NEW CASE

P. CARTIER, M.HAMET, A. VINCENS, J.L. PERIGNON

Laboratoire de Biochimie, CHU Necker-Enfants Malades

156. rue de Vaugirard 75015 Paris, France

INTRODUCTION

Since the description of the first case of complete adenine phosphoribosyltransferase (APRT) deficiency with 2,8 dihydroxyadenine (2.8 DHA) urinary lithiasis (1). 3 other observations of complete APRT deficiency have been reported (2, 3). 3 of the 4 patients reported so far had 2.8 DHA urinary lithiasis ; 2 patients were brothers (2).

We report the 5th and 6th observations of the disease, in two siblings ; in the youngest child, the deficiency was detected during an investigation for urolithiasis, in the older sister it was demonstrated on the occasion of the family study.

CASE REPORT

L.B. is a Caucasian male child born in May 1975, the 4th child of unrelated parents (who, however, were born in vicinous villages in Algeria). Birth was normal, and the child was well until the age of 2, when an urine examination before a vaccination lead to the discovery of an urinary tract infection. Two months later, on recurrence of this infection, an IVP was performed and a diagnosis of left pyelo-ureteral stenosis was made. The surgical intervention (June 1977) discovered numerous calculous debris, and a greyish-yellow very friable calculus, diagnosed as calcium oxalate. After the operation, recurrent urinary tract infections occurred and, on several occasions, little brownish-yellow calculi were emitted. Finally, these calculi were submitted to chemical and physical analysis (infrared spectrophotometry, Dr Reveillaud) and shown to be made of 2.8 DHA, with traces of calcium phosphate. The determination of

RBC APRT activity (cf infra) confirmed the diagnosis of complete
APRT deficiency. At that time, the child had normal height (91 cm)
and weight (13.6 kg), and the physical and especially neurological
examination was normal. Blood urea (3.4 mmol/l), plasma creatinine
(53 μmol/l), serum urate (0.25 mmol/l) and urinary urate excretion
(2.5 to 4.5 mmol/1.73 m^2/24 hours) (Normal : 3.0 $\pm$ 0.9), were normal.

METHODS

Stone analysis was carried out by infrared and ultraviolet spectrophotometry and chromatography on a Dowex 50 cation-exchanger column ; urinary purines metabolites were separated with this same
chromatographic procedure. APRT, HxPRT, ADA, PNP, adenosine kinase
and AMP deaminase activities were determined by radiochemical methods
described elsewhere (4, 5, 6). "Adenosine phosphorylase" activity
was determined both by following adenosine synthesis from adenine and
ribose 4-phosphate, after Zimmerman (7), and adenosine phosphorolysis.
In the latter case, the reaction is coupled to phosphoribosyltransfer
by addition of PRPP and Mg^{2+} in the medium, and the final product
of the reaction is AMP. In the case of total APRT deficiency, purified APRT must be added to determine adenosine phosphorolysis activity. Optimum pH of "adenosine phosphorylase" was found to be pH 6
(data not shown). The solubility of 2,8 DHA as a function of pH was
studied with a radioactive ^{14}C product, obtained by chemical synthesis.

RESULTS

I. Activities of enzymes of purine metabolism in the child and his
family.

RBC APRT activities are reported on Fig. 1.

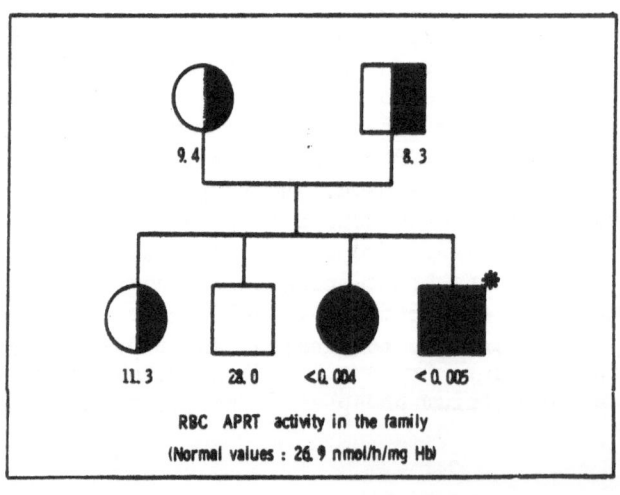

Fig. 1

RBC APRT activity in the family
(Normal values : 26.9 nmol/h/mg Hb)

This study led to the discovery of another child homozygous for the enzyme defect ; this girl, now aged 15, is apparently healthy, without any clinical or radiological evidence of nephrolithiasis. The genetic study is consistent with an autosomal recessive transmission of the defect, as in previous cases.

Other purine enzyme activities determined in RBC and/or lymphocytes were all within the normal range ; APRT and HxPRT activities in RBC and lymphocytes are reported in table I.

Table I : Phosphoribosyltransferase activities (RBC and lymphocytes)

	APRT RBC nmol/mg Hgb/h	APRT LYMPHOCYTES nmol/mg protein/mn	HxPRT RBC nmol/mg Hgb/hr	HxPRT LYMPHOCYTES nmol/mg protein/mn
NORMAL	26.9+4.9	3.2+1.1	101 + 21	1.8+0.5
LAURENT	<0.008	< 0.004	128	1.2
JOELLE	< 0.004	<0.002	115	1.8
CORINNE*	11.3	1.2	117	2.0
ERIC	28.0	2.3	113	1.6
FATHER*	8.3	1.0	121	1.7
MOTHER*	9.4	0.6	111	0.9
PATERNAL* GRANDMOTHER	.7.3	1.3	98	1.3

*: HETEROZYGOTE

Adenine and 2,8 DHA urinary excretion in APRT deficiency reflects a normal daily production of adenine of about 0.2 to 0.4 mmol/24 hours, whereas no metabolic pathway of production of adenine has ever been described in mammals. We focused on the adenosine phosphorylase activity described by Zimmerman et al., and we could measure this activity in the phosphorolysis direction by coupling the reaction with phosphoribosyltransfer. The table II shows adenosine phosphorylase activities determined in both directions in different members of the family of the propositus : normal enzyme activities were found in all of them. However, the phosphorolysis activity appears to be, at least "in vitro", dependent on APRT ac-

nmoles/min/ml GR pH 6.0	Adenosine cleavage (Ado ⟶ A)	Adenosine synthesis (A ⟶ Ado)
Normal	65.4 ± 12.8	104.1 ± 18.5
● Laurent	<5	98
(+ APRT)	61.5	-
● Joëlle	<5	118
◑ Corinne	37.9	-
Eric	58.8	-
◐ Father	21.1	102
◐ Mother	23.0	-

tivity, since a) it is not found in APRT deficient hemolysates (but can be restored upon addition of exogenous APRT), b) it is reduced in partially APRT deficient hemolysates ; the formation of free adenine was never observed.

II. The solubility of 2,8 DHA was studied at various pH values (fig. 2) ; this substance is highly insoluble in the normal range of urinary pH, which explains renal stone formation, and the inefficiency of urinary alkalinization as a preventive treatment of nephrolithiasis.

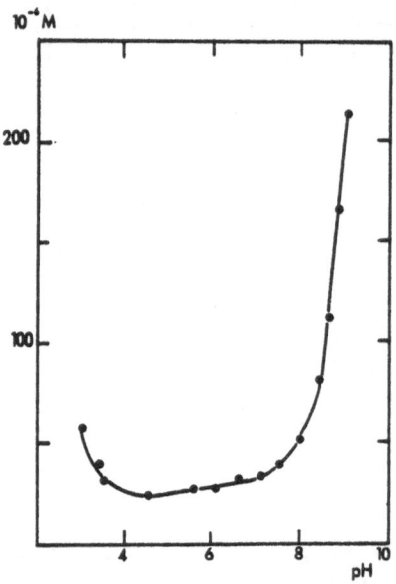

Fig. 2

III. Effect of allopurinol therapy

Adenine, 2,8-DHA and uric acid daily excretion in both patients are reported on table III.

Table III : Allopurinol and urinary oxypurines

	2,8 DHA µmol/24 h	A µmol/24h	2,8 DHA / A	2,8 DHA+A µmol/24h	UA mmol/24h
LAURENT Without therapy	278	147	L 89	425	0. 83-1. 50
Allopurinol:7 mg/kg/24h. Mean of 3 determinations	101	317	0. 33	418	1. 00
Allopurinol:14mg/kg/24h. Mean of 3 determinations	31	241	0. 13	271	0. 55
JOELLE Without therapy Mean of 3 determinations	445	300	L 49	744	2. 50
Allopurinol:200mg/24h. 1st DAY	118	458	0. 26	576	2. 22
2nd DAY	82	470	0. 17	552	L 90
3rd DAY	26	432	0. 06	458	L 17

DISCUSSION

These observations of APRT deficiency are very similar to the previous cases ; several points may be underlined :
- the children are clinically normal (apart from nephro-lithiasis), and the total purine production appears to be normal, both major differences from the Lesch-Nyhan syndrome (HxPRT deficiency)
- the mode of inheritance appears to be autosomal recessive
- Allopurinol therapy proves to be very efficient in reducing 2,8 DHA excretion ; the optimum dosage in L.B. (14 mg/kg/24 h) is similar to that finally used by Barratt et al. in their patient. However, in both our patients, we did observe a reduction in purine excretion, which contrasts with the findings of Barratt et al. (3) and Simmonds et al. (8). (We did not measure Hx and X elimination ; however, an unchanged total purine excretion would have implied a Hx + X excretion of 0.3 and 1.3 mmol/24 h in L.B. and his sister, respectively, an amount much higher than that previously reported (8).

Several questions remain unanswered :
        - the absence of 2,8 DHA nephrolithiasis in subjects with
complete APRT deficiency has already been mentionned (2), suggesting
that factors other than 2,8 DHA total daily excretion may interfere
in the constitution of renal stones.
        - "in vivo" production of adenine from adenosine could not
be proved by the present study ; however, this metabolic pathway
deserves to be considered, as well as the dietary origin for the
adenine suggested by Barratt et al. (3).
        - the site of production of 2,8 DHA has not yet been pre-
cisely determined.

REFERENCES

1   P. CARTIER and M. HAMET. Une nouvelle maladie métabolique : le
    déficit complet en adénine phosphoribosyltransférase avec lithia-
    se de 2,8 DHA, C. R. Acad. Sci. Sér. D. 279, 883 (1974).

2   K.J. Van ACKER, H.A. SIMMONDS, C.F. POTTER and J.S. CAMERON.
    Complete deficiency of adenine phosphoribosyltransferase. Report
    of a family. N. Engl. J. Med. 297, 127 (1977).

3   T.M. BARRATT, H.A. SIMMONDS, J.S. CAMERON, C.F. POTTER, G.A. ROSE,
    D.G. ARKELL and D.I. WILLIAMS. Complete deficiency of adenine
    phosphoribosyltransferase. A third case presenting as renal stones
    in a young child, Arch. Dis. Child. 54, 25 (1979).

4   P. CARTIER and M. HAMET.Les activités purine-phosphoribosyltrans-
    férasiques des globules rouges humains. Technique de dosage, Clin.
    Chim. Acta 20, 205 (1968).

5   P. CARTIER and M. HAMET. Dosage de l'activité adénosine désami-
    nasique dans les érythrocytes et les lymphocytes humains. Clin.
    Chim. Acta 71, 429 (1976).

6   J. CHALEON and P. CARTIER, manuscript in preparation.

7   T.P. ZIMMERMAN, N.B. GERSTEN, A.F. ROSS and R.P. MIECH. Adenine
    as substrate for purine nucleoside phosphorylase. Can. J. Biochem.
    49, 1050 (1971).

8   H.A. SIMMONDS, K.J. Van ACKER, J.S. CAMERON and A. McBURNEY. in
    "Purine metabolism in Man-II" pp. 304-311. MM. Müller, E. Kaiser
    and J.E. Seegmiller, eds, Plenum Press, N.Y. (1977).

INHERITANCE OF ADENINE PHOSPHORIBOSYLTRANSFERASE (APRT) DEFICIENCY

K.J. Van Acker, H.A. Simmonds, C.F. Potter and A. Sahota

Department of Paediatrics, University of Antwerp, Belgium and Purine Laboratory, Guy's Hospital, London,UK.

Recognition of the importance of the enzyme hypoxanthine-guanine phosphoribosyltransferase (HGPRT) in the control of purine metabolism lead to systematic investigations of the companion purine salvage enzyme, adenine phosphoribosyltransferase (APRT). This was followed by the discovery of individuals (considered to be heterozygous for the defect) with reduced APRT activity in erythrocyte lysates. Within a relatively short time, five such individuals and their families - a total of 82 individuals - 39 of whom had reduced erythrocyte APRT activity, were investigated (1-5). Disturbances of urate metabolism, accompanied in some instances by gout and/or urolithiasis, were found in 4 heterozygotes from three kindreds. HGPRT activity was normal in all. The possibility of an X linked inheritance could not be excluded from the first pedigrees (1). Subsequently the presence of identical disturbances in family members with normal erythrocyte APRT made any direct correlation between abnormal urate metabolism and APRT deficiency unlikely (Table); an autosomal recessive mode of inheritance was also proposed from detailed family investigations. The segregation of the defect varied widely.

The recent identification of individuals homozygous for

Table.   Abnormal urate metabolism, gout and urolithiasis
          in kindreds with heterozygotes for APRT-deficiency.

	Heterozygotes	Refer.	Normal APRT-Activity	Refer.
Total number	39		43	
Hyperuricemia/ Hyperuricosuria	4	(2,3,5)	8	(2,3,5)
Gout	3	(2,3,5)	2	(3,5)
Urolithiasis	1	(3)	3	(3)

APRT deficiency has enabled a better insight into the understanding of the role of APRT deficiency in urate metabolism as well as the mode of inheritance of the defect. We have recently reported studies in two such homozygotes and their immediate family (6). The investigations have now been extended to other family members. A total of 33 individuals, including the two homozygotes, have been studied over three generations. In addition, 14 individuals (3 generations) from another branch of this family have been investigated (Figure). Seventeen heterozygotes (10 female, 7 male) for the defect were found, all but one of them in the mothers kindred. The father was the only heterozygote in his family and his father could not be investigated: therefore the defect in this subject may be a spontaneous mutation. There was no evidence of consanguinity in this family. The results in the mothers kindred are consistent with the autosomal recessive mode of inheritance suggested from previous studies (5). Segregation varies widely.

With the exception of one hyperuricaemic teenage boy, uricaemia was normal in the homozygotes, heterozygotes and normal individuals from this family. Studies in the hyperuri-

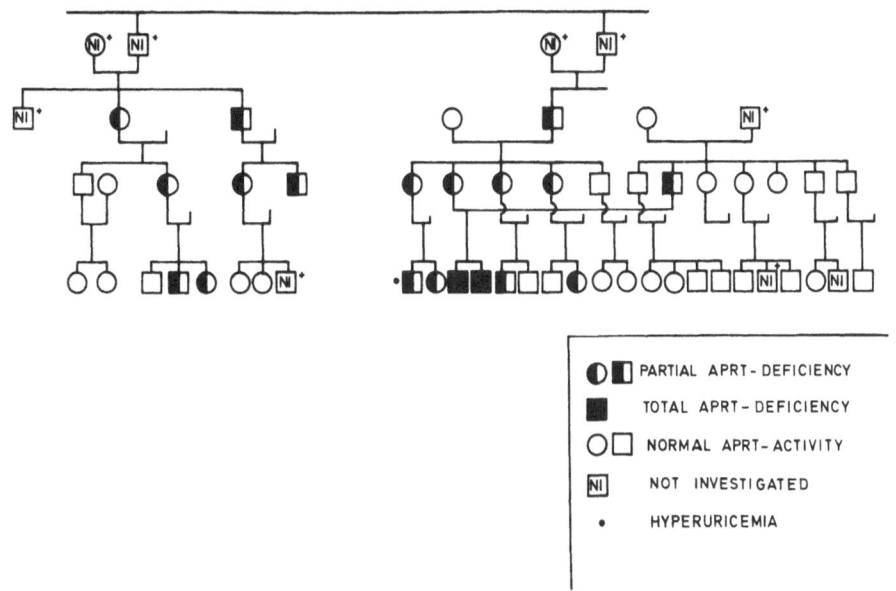

PARTIAL APRT-DEFICIENCY

TOTAL APRT-DEFICIENCY

NORMAL APRT-ACTIVITY

NOT INVESTIGATED

HYPERURICEMIA

Figure I. Pedigree of a family with APRT deficiency.

caemic boy are underway. Gout was absent in all family members. This also applies to urolithiasis with the exception, however, of the father of both homozygotes who presented with urolithiasis (consistence unknown). Urolithiasis (calcium oxalate) was also noted in the father of the homozygote of another family studied by us (9).

When these data are considered together with those from earlier studies in the two other families with homozygotes for APRT deficiency where only a restricted number of individuals could be investigated (7,8,9), the following conclusions may be drawn:

1. APRT deficiency is inherited as an autosomal recessive trait.
2. In contrast to HGPRT deficiency, the only clinical abnormality (not always manifest), 2,8 dihydroxyadenine stones, is found exclusively in the homozygous state: heterozygotes have no defect of purine metabolism.
3. Although hyperuricaemia and hyperuricosuria have been ob-

served in some heterozygotes (7,8,9), no cassal relationship between partial or complete APRT deficiency, abnormal urate metabolism and gout has been identified.

4. APRT activity thus appears to play no major part in the overall control of endogenous uric acid production in man.

5. The significance of stone formation in some heterozygotes (6-9) and its relationship to APRT deficiency requires further investigation.

REFERENCES

1. Kelley, W.N., Levy, R.I., Rosenbloom, F.M., Henderson, J.F. and Seegmiller, J.E., Adenine phosphoribosyltransferase deficiency: a previously undescribed genetic defect in man, J Clin Invest 47:2281-2289 (1968).

2. Fox, I.H., Meade, J.C. and Kelley, W.N., Adenine phosphoribosyltransferase deficiency in man. Report of a second family. Am J Med 55:614-620 (1973).

3. Delbarre, F., Auscher, C., Amor, B., de Gery, A., Cartier, P. and Hamet, M., Gout with adenine phosphoribosyl transferase deficiency. Biomedicine 21:82-85 (1974).

4. Cartier, P., Les déficits enzymatiques du métabolisme des purines. Symposium Interdisciplinaire sur l'hyperuricémie, Paris 1975.

5. Emmerson, B.T., Gordon, R.B. and Thompson, L., Adenine phosphoribosyltransferase deficiency: its inheritance and occurrence in a female with gout and renal disease. Aust N Z J Med 5:440-446 (1975).

6. Van Acker, K.J., Simmonds, H.A., Potter, C. and Cameron, J. S., Complete deficiency of adenine phosphoribosyltransferase. New Engl J Med 297:127-132 (1977).

7. Cartier, P. et Hamet, M., Une nouvelle maladie métabolique: le déficit en adénine-phosphoribosyltransférase avec lithiase de 2,8-dihydroxyadénine. C R Acad Sc (Paris) 279: 883-886 (1974).

8. Debray, H., Cartier, P., Temstet, A. and Cendron, J., Child's
    urinary lithiasis revealing a complete deficit in adenine
    phosphoribosyl transferase. Pediat Res 10:762-766 (1976).
9. Barratt, T.M., Simmonds, H.A., Cameron, J.S., Potter, F.C.,
    Rose, G.A., Arkell, D.G. and Williams, D.I., Complete de-
    ficiency of adenine phosphoribosyl transferase: a third
    case presenting as renal stones in a young child. Arch Dis
    Childh 54:25-31 (1979).

IMMUNOLOGICAL EVALUATION OF A FAMILY DEFICIENT IN

ADENINE PHOSPHORIBOSYL TRANSFERASE (APRT)

W.J.STEVENS,M.E.PEETERMANS,K.J.VAN ACKER

UNIVERSITEIT ANTWERPEN (UIA)

B-2610 WILRIJK, BELGIUM

The recent finding of immunodeficiency associated with
inherited defects of purine metabolism has led to an
intensive study of the latter in lymphocytes.Informa-
tion can be gained from two sources:analysis of lymphocyte function in patients with inborn errors of purine
metabolism and investigation of lymphocyte behaviour
after in vitro manipulation in these cells.

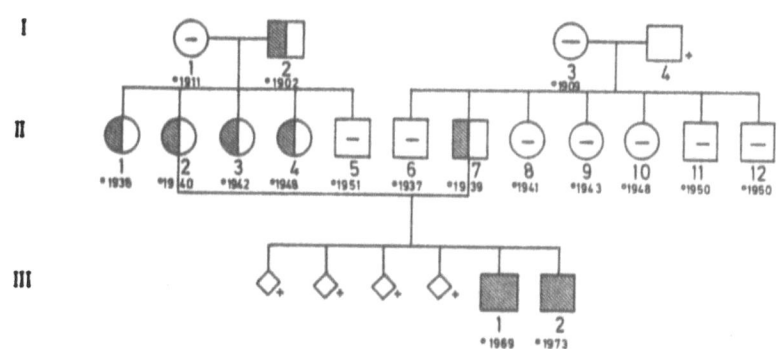

Fig.1.Pedigree of the family studied

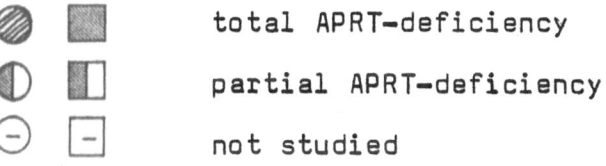

total APRT-deficiency

partial APRT-deficiency

not studied

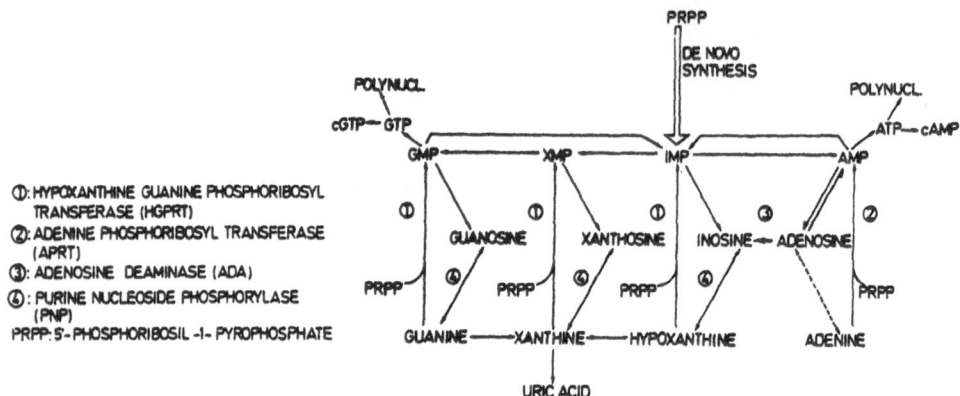

Fig.2 Schematic representation of purine metabolism

The present study was set up to obtain further information by investigating some immunological parameters in
a family with total or partial absence of APRT(fig.1),
a rare inborn error of metabolism (fig.2).
    T and B cell function was evaluated by counting
rosetteforming cells: E active rosettes(Eact), E total
(Etot) rosettes,EAC rosettes,by investigating lymphocyte
transformation to phytohaemagglutinin(PHA),concanava-
lin A(Con A) and pokeweed mitogen(PWM),by assaying serum
immunoglobulins and determining lymphocyte membrane
immunoglobulins by immunofluorescence.Delayed cutaneous
hypersensitivity was also assayed.

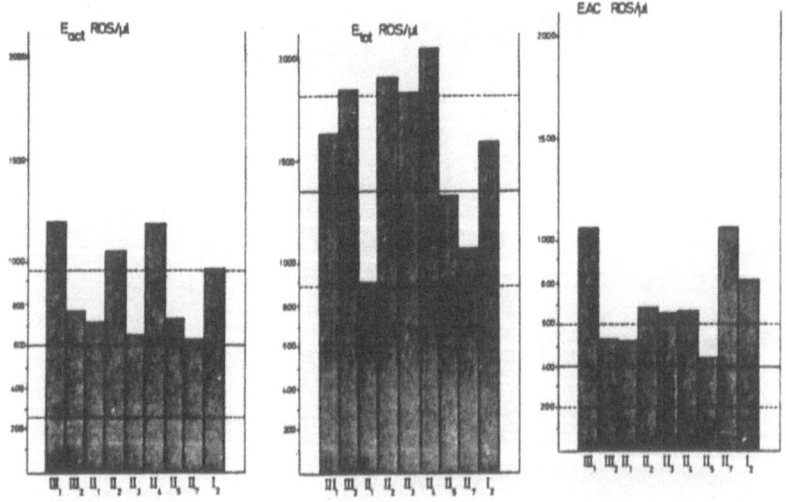

fig.3

Table 1. Peripheral blood lymphocytes, serum immunoglobulins and lymphocyte membrane immunoglobulins.

	Age (years)	Peripheral blood lymphocytes x10⁹/l	Serum immunoglobulins (mg/dl)				Lymphocyte membrane immunoglobulins (% of lymphocytes)								
			IgG	IgA	IgM	IgE	Ig	F(ab)	K	L	IgG	IgM	IgA	EgD	IgE
III_1	9	4.34	830	92	96	19	29	23	59	41	21	10	3	1	0
III_2	4	3.32	800	244	88	19	21	20	50	50	12	8	3	0	1
II_1	40	1.95	1170	178	85	-	22	-	-	-	21	-	-	-	-
II_2	38	3.54	780	226	134	8	15	13	57	43	9	8	1	0	1
II_3	37	2.44	1270	178	90	-	17	-	-	-	13	-	-	-	-
II_4	30	3.49	1000	226	108	-	16	-	-	-	17	-	-	-	-
II_7	39	2.28	-	-	-	-	29	-	-	-	14	-	-	-	-
I_2	76	2.11	900	240	62	-	23	-	-	-	17	-	-	-	-
II_5	27	2.21	600	160	56	-	21	-	-	-	26	-	-	-	-
Controls Mean		2.18	900*	145*	123*	<200	20	18	±60% of Ig	±40% of Ig	11	7.8	2.6	0-1	0-1
s.d.		0.628					6.1	5.9			5.0	3.8	2.0		
n		85					85	68			14	24	12		

*Serum pool of 100 regular blood donors.

-Not tested.

RESULTS

Table 1 represents the absolute number of peripheral
blood lymphocytes per litre,the serum immunoglobulin
levels and the results of lymphocyte membrane immunoglobulin assay.Lymphocyte counts were elevated in III_1,II_2,
and II_4.The immunoglobulin levels can be considered as
normal.Lymphocyte membrane immunoglobulins showed no major abnormalities.Agar gel electrophoresis as well as
immunoelectrophoresis were normal.Secretory piece of IgA
in saliva was assayed and found to be present in III_1,
III_2, and II_2.Isohaemagglutinins were present in the two
homozygotes.

Fig.3 represents the total number of rosetteforming cells.
No major decreases were found.

Results of lymphocyte stimulation are shown in fig.4 and
5.Although,on stimulation with PHA,lymphocyte transformation rates less than the mean - 1 SD of twenty-three
controls were observed,there were no values less than
the mean - 2 SD.

The two heterozygotes tested did not reach the mean
- 2 SD of thirty controls for Con A 6 µg/ml.One heterozygote (II_7) stayed below the mean - 2 SD for Con A 12
µg/ml.The same individual showed low stimulation in response to PWM.

Reaction to intradermal injection of PHA 1 µg/0.1 ml
was normal.III_1 had a positive skin reaction to strepto-
kinase-streptodornase.The reactions to tuberculin and
trichophytin were negative in the homozygotes.

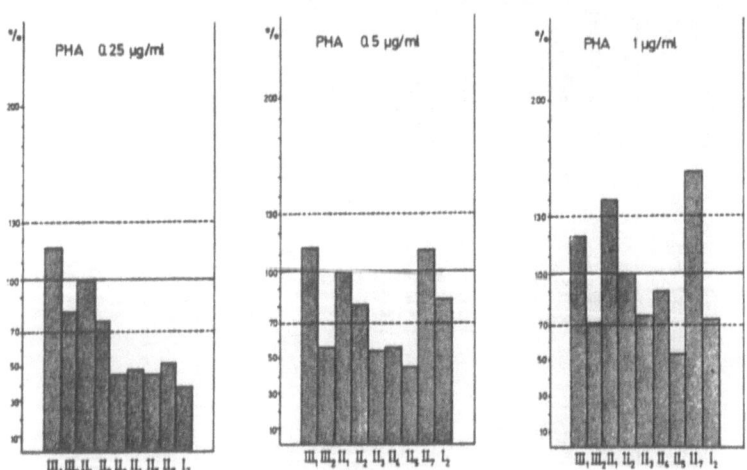

Fig.4 Lymphocyte transformation to PHA

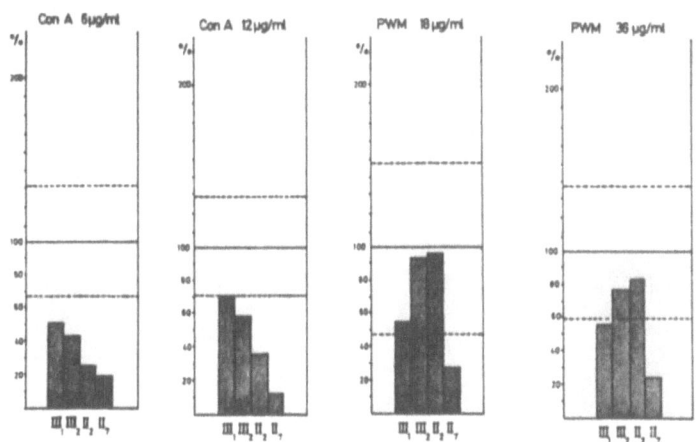

Fig.5 Lymphocyte transformation to Con A and PWM

CONCLUSION

The present study provided a unique opportunity to investigate the role of the APRT-mediated salvage pathway in the lymphocyte function.No major decrease of the
immunological function was found in the homozygotes.
Neither was recurrent infection a prominent feature in
these children.Considering their age,the negative skin
test results to tuberculin and trichophytin in the two
homozygotes were not abnormal.Although most of the results in the heterozygotes,as well as in the normal individual (II5),fell within normal limits,some abnormalities were observed.Thus the total number of EAC rosetteforming cells was increased in III1 and II7,though
the membrane immunoglobulins were normal.This might indicate that in these individuals part of the EAC rosettes were not formed by B-lymphocytes but by other cells,
wich have a receptor for C_3.We can not explain the low
response to Con A stimulation in the obligate heterozygotes II2 and II7.If this abnormality were linked to
the APRT-deficiency,one would also expect it to be present in the homozygotes: the values obtained in the
latter,although somewhat low,still were within 2 SD.
    We conclude that the APRT-mediated salvage pathway
is relatively unimportant for lymphocyte function.

This study will be reported extensively in Clinical
and Experimental Immunology,1979,36,364-370.

ACTIVITIES OF AMIDOPHOSPHORIBOSYLTRANSFERASE AND PURINE PHOSPHORIBOSYLTRANSFERASES IN DEVELOPING RAT BRAIN

Jennifer Allsop and R. W. E. Watts

Division of Inherited Metabolic Diseases
Medical Research Council Clinical Research Centre
Watford Road, Harrow, HA1 3UJ. England

The availability of purine nucleotides at their sites of physiological action depends upon the balance between purine synthesis de novo and the purine phosphoribosyltransferase catalysed purine salvage reactions. It was reported that purine synthesis de novo was considerably less active in brain tissue than in liver[1].

Neuroblasts and neuroglia proliferate most rapidly at separate specific time-points during development[2]. McKeran and Watts[3] suggested, on the basis of studies with a cellular model system, that hypoxanthine phosphoribosyltransferase (HPRT; EC 2.4.2.8) deficiency impairs brain cell proliferation, during these critical stages of development. Different enzyme activities develop at different stages of organogenesis. It is therefore of interest, in the case of brain damage by a specific inherted metabolic lesion in the brain itself, to relate the time at which the enzyme concerned normally becomes active with the periods in brain development, which might be susceptible to a deficiency of the enzyme. Ideally, one needs to study the rates of metabolic flow through the potentially affected metabolic pathway and any alternative possibly compensatory pathway in vivo. These are the purine salvage and purine de novo synthesis pathways in the present case. Amidophosphoribosyl-transferase (PRPP-amidotransferase; EC 2.4.2.14) catalyses the rate-limiting step on the purine synthesis de novo pathway, and we have compared the activity of this enzyme with that of HPRT and adenine phosphoribosyltransferase (APRT; EC 2.4.2.7) in different regions of the rat central nervous system at different stages of development.

METHODS

The rats (male Sprague-Dawley, age 1-8 weeks) were killed by asphyxiation with carbon dioxide. Samples of cerebral cortex, basal ganglia, cerebellum, the medullary-pontine region and spinal cord were dissected from the central nervous system and put on ice. They were washed free of blood with ice-cold 0.9% (w/v) NaCl, blotted dry, weighed and homogenised in 2 volumes of 0.9% (w/v) NaCl using a hand-held Duall all glass homogeniser with 12 passes of the pestle. The homogenates were centrifuged at 26,000xg$_{Av}$ for 1 h, and the super-natant fractions used for the enzyme assays. The whole brains and spinal cords of rats aged <24 h and 20 day foetuses were homogenised and analysed, but, for the 18 day foetal animals, tissue from two foetuses had to be combined.

The analytical methods are referenced elsewhere[4] except for PRPP[5]. The amounts of tissue used per assay were: HPRT, and APRT approximately 50 µg protein each; PRPP, approximately 250 µg protein; PRPP-amidotransferase, approximately 100 µg protein, this portion of the whole tissue homogenate was heated at 50° for 15 min before centrifugation in order to remove non phosphate requiring glutaminase activity. HPRT, APRT and PRPP were analysed in the presence of TTP (3.3 mM) to inhibit purine 5'-nucleotidase (EC3.1.3.5) activity[6].

RESULTS

Table 1 compares the activities of APRT, HPRT and PRPP-amido-transferase in the brain and some other tissues of 8 week old rats. Figs. 1 and 2 show the HPRT, APRT, PRPP-amidotransferase activities, and the HPRT/PRPP-amidotransferase activity ratios respectively, in 1-8 week old rats. Fig. 3 shows the data for the new born (< 24h old) and foetal rats. The PRPP concentrations were similar in all parts of the brain and decreased with increasing age. They were higher in the newly born (< 24h old) rats than in the 18 and 20 day foetuses.

DISCUSSION

The presence of PRPP-amidotransferase activity is taken, for the purposes of this discussion, as showing that purine synthesis de novo is possible in the tissues concerned (Table 1). In addition, if one assumes that enzyme activities measured in dilute tissue homogenates and under saturating substrate conditions, are related to the flow along the metabolic pathways they subserve, the results shown in Fig. 1 reflect the increasing importance of purine salvage as opposed to purine synthesis de novo in the normal rat's brain as it approaches maturity.

Table 1 Purine phosphoribosyltransferase (APRT; HPRT) and amidophosphoribosyltransferase (PRPP-amidotransferase) activities in the central nervous system and some other tissues of 8-week old rats.

Tissue	Enzyme Activity (n mol h^{-1} mg^{-1} protein) [Mean ± SEM (no. animals)]			Ratio $\dfrac{HPRT}{PRPP\text{-}amidotransferase}$
	APRT	HPRT	PRPP-amidotransferase	
Central Nervous System				
Cerebral Cortex	101 ± 6.67(14)	374 ± 26.3(14)	11.5 ± 5.69(16)	33
Basal Ganglia	134 ± 6.24(14)	417 ± 33.8(14)	12.7 ± 1.42(16)	33
Cerebellum	126 ± 9.06(14)	422 ± 20.1(14)	9.9 ± 5.0(16)	45
Medulla/Pons	116 ± 5.16(14)	362 ± 23.2(14)	9.1 ± 5.2(16)	40
Spinal cord	148 ± 9.70(12)	247 ± 18.2(14)	6.0 ± 0.8(16)	40
Liver	26 ± 3.0 (10)	373 ± 21.2(10)	21.1 ± 1.1 (9)	18
Spleen	203 ±32.9 (10)	306 ± 13.2(10)	6.3 ± 0.4 (9)	49
Kidney	42 ± 6.2 (10)	62 ± 4.9(10)	13.3 ± 0.7 (9)	5
Testis	82 ± 6.2 (10)	78 ± 2.5(10)	3.9 ± 1.3 (9)	20

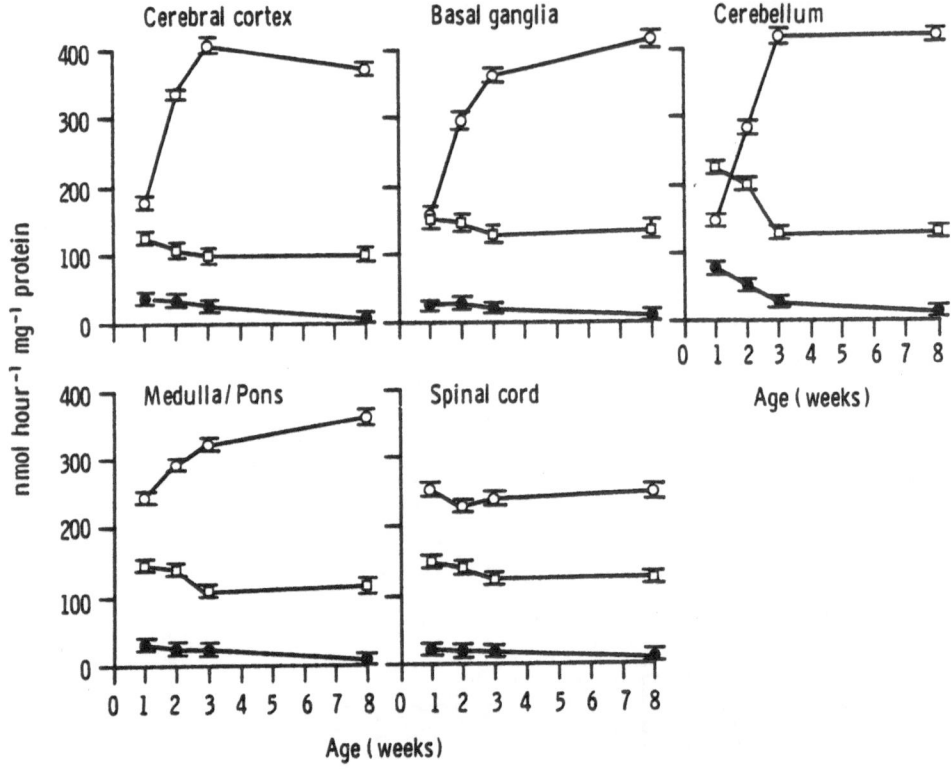

Fig 1. Purine phosphoribosyltransferase (HPRT and APRT) and amidophosphoribosyltransferase (PRPP-amidotransferase), activities of
   different parts of the central nervous system in rats age 1-8
   weeks. O-O, HPRT:□-□, APRT:●-●, PRPP-amidotransferase.
   Range = SEM

      Adams & Harkness[7] also found a post natal rise in the HPRT
activity of rat cerebral cortex and basal ganglia, but their
absolute values were lower than ours, and they did not study PRPP-
amidotransferase activities.

      The fresh weight of the brain increases most rapidly at about
6 weeks of age in man[2], and at the 10th day of postnatal life in the
rat[8]. The peak rates of neuroblast and glia proliferation are at
about the 18th week of gestation and 3 months of postnatal age in
the human[2]. These times correspond to about the 14th day of gestation and the 10th day of life in the rat. These stages of development precede the main rise in the HPRT/PRPP amidotransferase activity
ratio. We suggest that the purine nucleotide demands of the main
bursts of neuroblast and glia proliferation are met by purine synthesis de novo, and extrapolating our data to the Lesch-Nyhan syndrome patient, that this protects the brain against potential

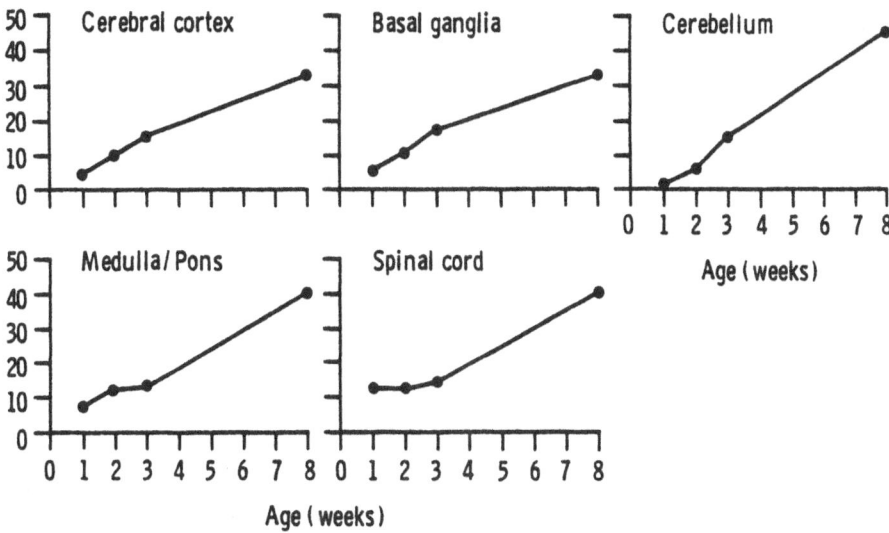

Fig 2. Hypoxanthine phosphoribosyltransferase: amidophosphoribosyl-
transferase activity ratios for different parts of the central
nervous system in rats age 1-8 weeks. Range = SEM.

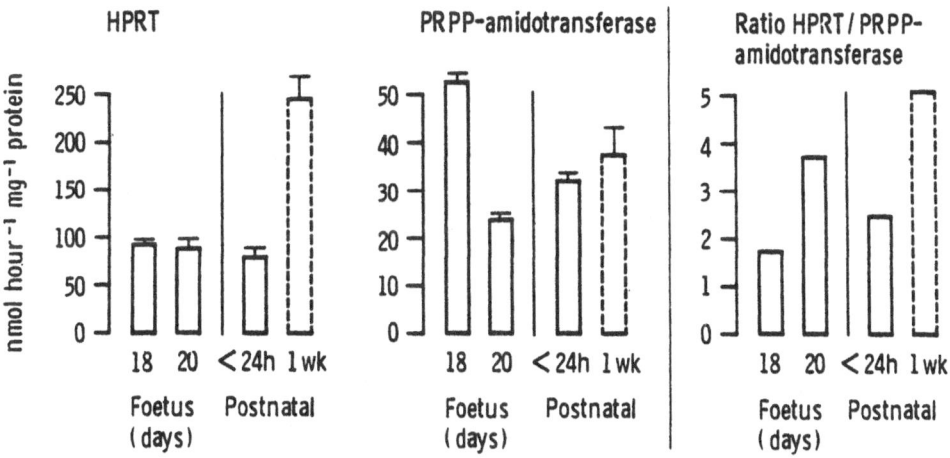

Fig 3. Purine phosphoribosyltransferase (H=HPRT: A=APRT), amidophosphoribosyltransferase (PRPP-amidotransferase), activities in homogenates of the whole central nervous system of foetal rats (18 and
20 days gestational age), neonatal (< 24 h old) rats and rats aged
1 week. Values for the 1 week old animals are means of the values
for separate regions of the central nervous system. Range = SEM.

structural damage by HPRT deficiency. This proposition is consistent
with the absence of structural changes in the brain in the Lesch-
Nyhan syndrome. However, the possibility that there is a specific
lesion of a small, as yet unidentified group of neurones with widely
projecting efferent pathways cannot be completely excluded, although
we think that this is unlikely.

We suggest that the present findings and the normal brain
morphology point to the manifestations of the Lesch-Nyhan syndrome
being due to lack of one or more purine nucleotides with direct or
indirect neuropharmacological functions. We propose that this
occurs when the brain PRPP-amidotransferase activity declines below
a critical level during post natal development, there being insufficient HPRT activity to maintain the supply of purine nucleotides by
the purine salvage pathway. The alternative hypothesis[3] is now
considered less likely.

REFERENCES

1. W.J. Howard, L.A. Kerson, and S.H.Appel, Synthesis de novo of
   purine in slices of rat brain and liver. J. Neurochem. 17:
   121 (1970).
2. J. Dobbing, Undernutrition and the developing brain. The
   relevance of animal models to the human problem. Amer. J. Dis.
   Child. 120: 411 (1970).
3. R.O. McKeran and R.W.E. Watts, Use of phytohaemagglutin stimulated
   lymphocytes to study effects of hypoxanthine-guanine phosphoribosyltransferase (HGPRT) deficiency on polynucleotide and
   protein synthesis in the Lesch-Nyhan syndrome. J. Med. Genet.
   13: 91 (1976).
4. A.D.B.Webster, M. North, J. Allsop, G.L.Asherson and R.W.E.Watts,
   Purine metabolism in lymphocytes from patients with primary
   hypogammaglobulinaemia. Clin. exp. Immunol. 31: 456 (1978).
5. I.H. Fox and W.N. Kelley, Human phosphoribosylpyrophosphate synthetase. Distribution, purification and properties. J. biol.
   Chem. 246: 5739 (1971).
6. W. Gutensohn and G. Guroff, Hypoxanthine-guanine phosphoribosyltransferase from rat brain, purification, kinetic properties,
   development and distribution. J. Neurochem. 19: 2139 (1972).
7. A. Adams and R.A. Harkness, Developmental changes in purine phosphoribosyltransferases in human and rat tissues. Biochem. J.
   160: 565 (1976).
8. A.N. Davison, J. Dobbing, Myelination as a vulnerable period in
   rat brain development. Br. med. Bull. 22: 40 (1966).

PURINE NUCLEOSIDE PHOSPHORYLASE DEFICIENCY; GENETIC STUDIES IN A DUTCH FAMILY

Gerard E.J.Staal[1], M.J.M.van der Vlist[1], R.Geerdink[2], J.M.Jansen-Schillhorn van Veen[2], B.J.M.Zegers[3] and J.W. Stoop[3]

[1] Medical Enzymology, State University Hospital, Utrecht, The Netherlands
[2] Unit of Clinical Genetics, State University Hospital, Utrecht, The Netherlands
[3] University Children's Hospital, "Het Wilhelmina Kinderziekenhuis", Utrecht, The Netherlands

INTRODUCTION

Purine-nucleoside phosphorylase (EC 2.4.2.1) catalyzes the phosphorolysis of (deoxy)inosine and (deoxy)guanosine. Patients deficient in purine nucleoside phosphorylase (PNP) show disturbances in thymus-dependent immunity and have normal or nearly normal humoral immunity [1-5]. PNP deficiency is inherited as an autosomally recessive trait. The gene for PNP is located on chromosome 14. PNP appears to be a trimer with a molecular weight of 93 800[6]. The subunits of the enzyme from human erythrocytes have the same molecular weight but different isoelectric points[6]. Giblett et al.[3] described a patient homozygous for PNP deficiency whose heterozygous parents exhibited an erythrocyte PNP with diminished activity and an altered electrophoretic mobility. Osborne et al.[7] described one family in which two different variant alleles for PNP deficiency were present. We were able too to study a case of PNP deficiency[5]. In the Dutch family with PNP deficiency no PNP activity could be detected in the erythrocytes and lymphocytes of the patient, whereas the PNP activity of the erythrocytes of the father and mother showed one-half normal enzyme activity. The parents are unrelated. It may be possible that each parent has a different mutation. The purpose of this study is to describe more in detail the possible genetic heterogeneity of PNP deficiency in the Dutch family. The results reported here are highly indicative that only one mutant allele is present in this family.

MATERIALS AND METHODS

Nucleoside phosphorylase activity was determined according to Kalckar[8]. A unit of activity is defined as the amount of enzyme required to catalyze the conversion of 1/umol inosine to hypoxanthine per min. The specific activity is expressed as units per gram Hb. All chemicals used were of analytical grade of purity.

RESULTS AND DISCUSSION

PNP Activity

The activity of purine nucleoside phosphorylase in the erythrocytes of controls is in the range of 17.0-30.0 Units/gHb[4]. Patient R.V. had no detectable PNP activity in her erythrocytes and lymphocytes. When guanosine or xanthosine was used as substrate (instead of inosine) no PNP activity could be detected too. When a normal hemolysate was mixed with the patient's hemolysate there was no inhibition of PNP activity, excluding the presence of an inhibitor in the patient's erythrocytes. The pedigree of the family is shown in fig. 1. The PNP activity of other family members are summarized in table I.

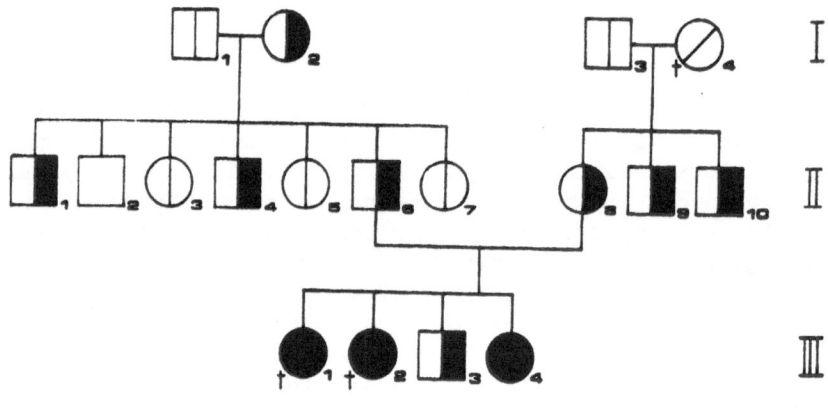

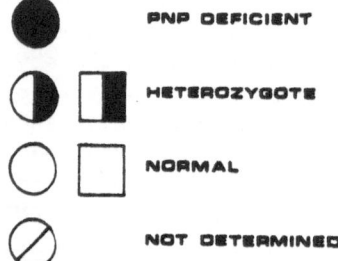

Fig. 1. The pedigree of the family with PNP deficiency.

Table I. Purine nucleoside phosphorylase activity in the
erythrocytes of family members of the patient
with PNP deficiency

			PNP activity (Units/gHb, normal range 17-30)
I	1	E.V.	22.2
	2	F.V.-H.	10.3
	3	R.J.	22.1
	4	J.	not determined
II	1	M.V.	10.4
	2	E.V.	22.2
	3	A.V.-H.	25.1
	4	G.V.	10.6
	5	W.B.-V.	20.6
	6	V.	14.7
	7	F.H.-V.	22.1
	8	V.-J.	13.0
	9	G.J.	11.2
	10	G.J.J.	11.1
III	1	H.V.	not determined
	2	A.V.	not determined
	3	H.V.	13.9
	4	R.V.	0

From table I it can be concluded that at least eight family members
are heterozygous for PNP deficiency; they have about one-half normal
enzyme activity. Purine analysis of stored frozen samples of serum
and urine of III,1 and III,2, both sisters of the patient (III,4),
showed that these children also had a PNP deficiency[5]. The pedigree
of the family demonstrates the mode of inheritance of PNP deficiency
which is autosomally recessive.

Starch-gel Electrophoresis

Starch-gel electrophoresis of PNP from hemolysates deriving from
the family and a control showed no enzyme activity in the patient and
a normal isoenzyme pattern for the relatives (see ref. 4).

pH Optimum

Fig. 2 shows the activity of PNP (from III,3, II,6 and II,8) at
different pH values. The pH optimum was found to be about 7.5. No
significant difference could be detected between the pH optima of the
normal and defective enzymes.

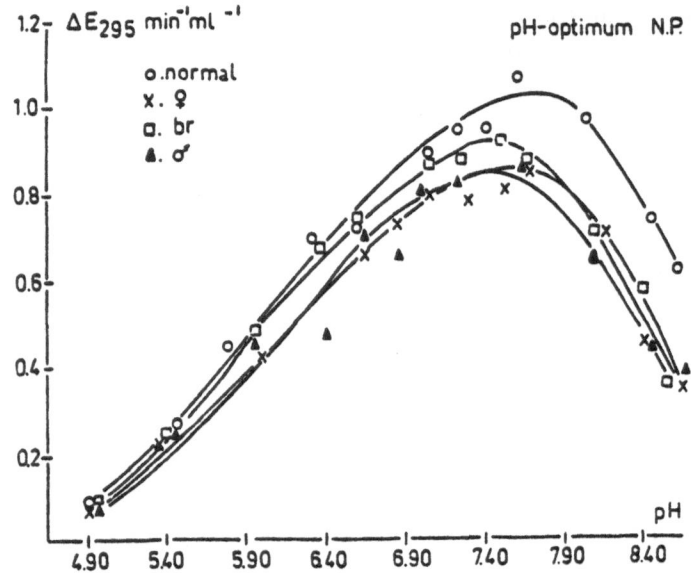

Fig. 2. pH optimum of PNP from normal control, father, mother and
brother of the propositus.

Heat Stability

Heat stability of PNP was studied by heating the enzyme from
hemolysate at $56^{o}C$ during 60 min. In this way the enzyme of the
father and the mother of our patient was investigated. Fig. 3 shows
the results. It can be concluded that PNP of the father as well as of
the mother is more labile compared with the normal enzyme. Furthermore the increased lability of PNP of both parents is quite the same.

Kinetics

The Lineweaver-Burk plots of the variation in the activity of
PNP at various inosine concentrations of normal erythrocytes and of
the red blood cells of I,2 and II,10 (see pedigree) are shown in
fig. 4. From these plots the K_m value of the normal enzyme was calculated as 30/uM, whereas the K_m values in case of the family members
are increased (K_m = 45/uM). Again the same abnormality is found in
both family members.

The PNP deficiencies reported here are probably of a common
genetic origin. The results indicate the presence of one mutant
allele. The normal data on starch-gel electrophoresis, the increased
heat lability, the increased K_m values for inosine and the normal pH
optimum suggest the presence of a single variant. In crude hemolysate
of our patient it could be demonstrated[6] that all forms of purine
nucleoside phosphorylase subunits are absent. In view of the

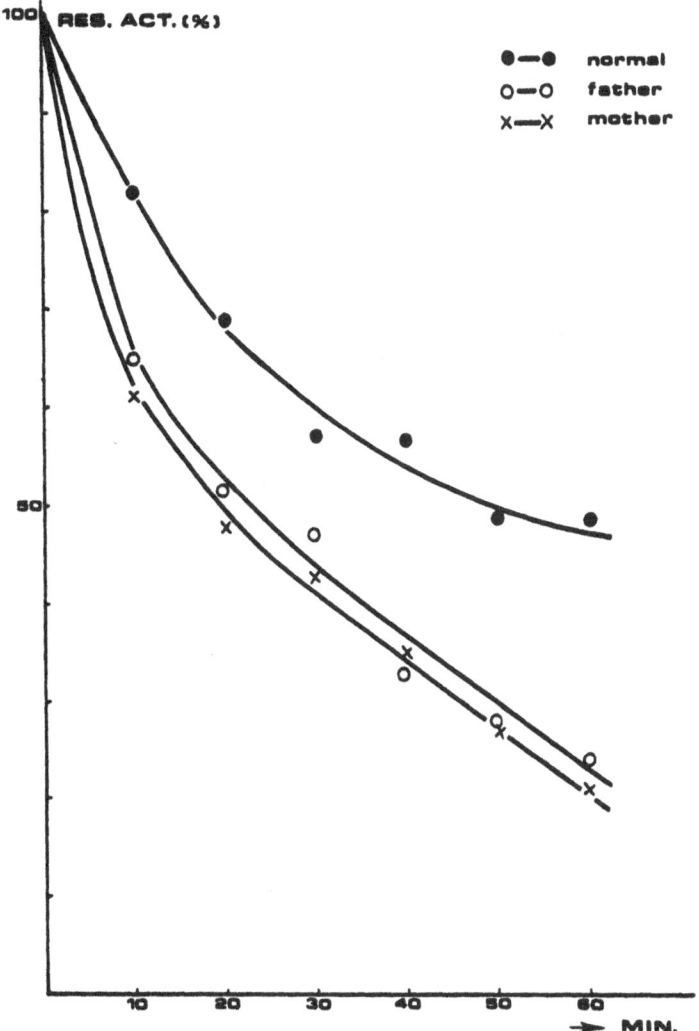

Fig. 3. Heat stability of PNP at 53°C

increased heat lability of the heterozygote form of the enzyme it
seems likely that the variant protein is unstable and that the homo-
trimer denatures rapidly and therefore could not be detected in the
patient's erythrocytes.

Purine nucleoside phosphorylase deficiency is inherited as an
autosomally recessive trait. As this trait is very rare, it is
surprising that the parents are, as far as they know, not related.
As they were born in the same area, however, some distant relationship may nevertheless exist.

In conclusion we propose that only one mutant allele is present
in this family. The Dutch variant differs from the variants reported
by Osborne[7] and Giblett[3] especially with respect to electrophoresis.
The results presented here show the genetic heterogeneity of PNP
deficiency.

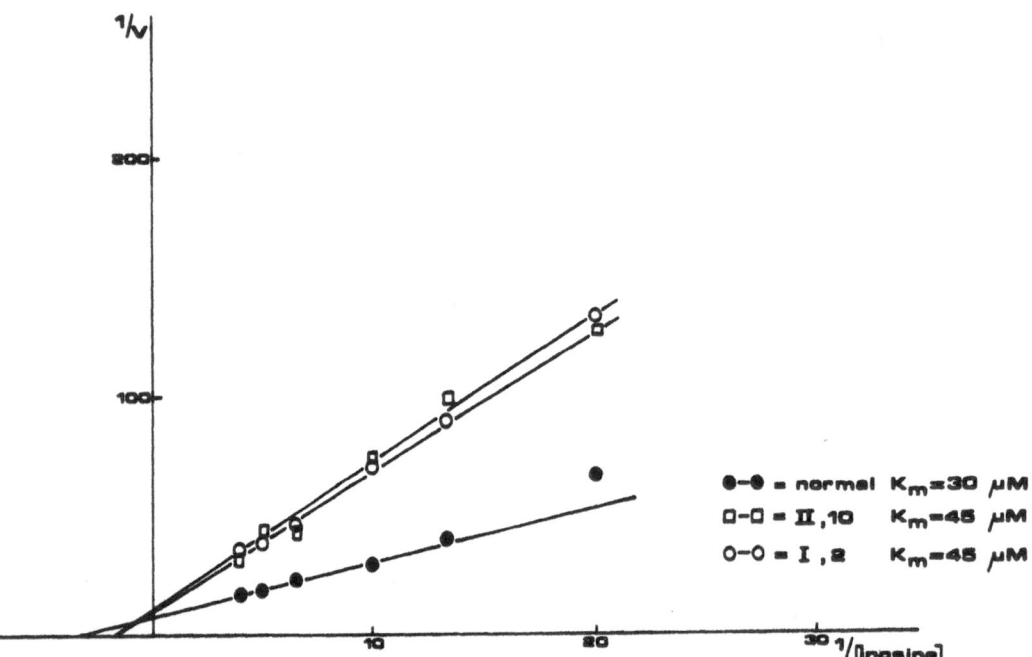

Fig. 4. Lineweaver-Burk plot for PNP of a control, II,10 and I,2
(see pedigree). The calculated K_m values are indicated.

REFERENCES

1. E.Carapella-de Luca, F.Aiuti, P.Lucarelli, L.Bruni, C.D.Baroni,
   C.Imperato, D.Roos and A.Astaldi, A patient with nucleoside
   phosphorylase deficiency, selective T-cell deficiency and auto-
   immune hemolytic anemia, J.Pediatrics 93: 1000-1003 (1978).
2. E.W.Gelfand, H.M.Dosch, W.D.Biggar and I.H.Fox, Partial purine
   nucleoside phosphorylase deficiency: studies of lymphocyte function, J.Clin.Invest. 61: 1071-1080 (1978).
3. E.L.Giblett, A.J.Ammann, D.W.Wara, R.Sandman and L.K.Diamond,
   Nucleoside phosphorylase deficiency in a child with severely defective T-cell immunity and normal B-cell immunity, Lancet i:
   1010-1013 (1975).
4. L.H.Siegenbeek van Heukelom, G.E.J.Staal, J.W.Stoop and B.J.M.
   Zegers, An abnormal form of purine nucleoside phosphorylase in a
   family with a child with severe defective T-cell and normal B-cell
   immunity. Clin.Chim.Acta 72: 117-124 (1976).
5. J.W.Stoop, B.J.M.Zegers, G.F.M.Hendrickx, L.H.Siegenbeek van
   Heukelom, G.E.J.Staal, P.K.de Bree, S.K.Wadman and R.E.Ballieux,
   Purine nucleoside phosphorylase deficiency associated with selective cellular immunodeficiency. New Engl.J.Med. 296: 651-655
   (1977).
6. L.J.Gudas, V.I.Zannis, S.M.Clift, A.J.Ammann, G.E.J.Staal and D.W.
   Martin. Characterization of mutant subunits of human purine

nucleoside phosphorylase, J.Biol.Chem. 253: 8916-8924 (1978)

7. W.R.A.Osborne, S.H.Chen, E.R.Giblett, W.D.Biggar, A.A.Ammann and C.R.Scott, Purine nucleoside phosphorylase deficiency, J.Clin. Invest 60: 741-746 (1977).

8. H.M.Kalckar, Differential spectrophotometry of purine compounds by means of specific enzymes, J.Biol.Chem. 167: 429-475 (1947).

ABNORMAL REGULATION OF PURINE METABOLISM IN A CULTURED MOUSE T-CELL LYMPHOMA MUTANT PARTIALLY DEFICIENT IN ADENYLOSUCCINATE SYNTHETASE

B. Ullman, M.A. Wormsted, B.B. Levinson, L.J. Gudas,
A. Cohen, S.M. Clift, and D.W. Martin, Jr.

Howard Hughes Medical Institute Laboratory and
Division of Medical Genetics
Department of Medicine and
Department of Biochemistry and Biophysics
University of California, San Francisco
San Francisco, California 94143

ABSTRACT

The isolation and characterization of a mutant mouse T-cell lymphoma (S49) with altered purine metabolism is described. This mutant, AU-100, was isolated from a mutagenized population of S49 cells by virtue of its resistance to 0.1 mM 6-azauridine in semi-solid agarose. The AU-100 cells are resistant to adenosine mediated cytotoxicity but are extraordinarily sensitive to killing by guanosine. High performance liquid chromatography of AU-100 cell extracts has demonstrated that intracellular levels of GTP, IMP, and GMP are all elevated about 3-fold over those levels found in wild type cells. The AU-100 cells also contain an elevated intra-cellular level of pyrophosphoribosylphosphate (PPriboseP), which accounts for its resistance to adenosine. However AU-100 cells synthesize purines de novo at a rate less than 35% of that found in wild type cells. Furthermore, the intact cells of this mutant S49 cell line cannot efficiently incorporate labeled hypoxanthine into nucleotides since the salvage enzyme HGPRTase is inhibited in situ.

The AU-100 cell line was found to be 80% deficient in adenylo-succinate synthetase, but these cells are not auxotrophic for adenosine or other purines. The significant alterations in the control of purine de novo and salvage metabolism caused by the defect in adenylosuccinate synthetase are mediated by the resulting increased levels of guanosine nucleotides.

INTRODUCTION

Alterations in the purine pathways in humans have been associated with an increased rate of purine synthesis. These alterations include: gout with purine overproduction due to increased activity (Becker et al., '73a) or to feedback resistance (Sperling et al., '72) of pyrophosphoribosylphosphate (PPriboseP) synthetase; gout with purine overproduction due to a deficiency in hypoxanthine-guanine phosphoribosyltransferase (HGPRTase) activity (Seegmiller et al., '67); and purine overproduction due to purine nucleoside phosphorylase deficiency (Cohen et al., '76).

In bacterial and mammalian systems there are two distinct sites of regulatory control over purine synthesis. The first site occurs in the portion of the pathway leading to IMP formation where the catalytic activities of PPriboseP synthetase and PPriboseP amidotransferase, the first two enzymes in the de novo pathway, are subject to regulation by purine nucleotides. The second site of regulation involves the pathways responsible for AMP and GMP synthesis from IMP. The end product of each of these branched pathways, AMP and GMP, feedback inhibit adenylosuccinate synthetase (Wyngaarden and Greenland, '63; Van der Weyden and Kelley, '74) and inosinate dehydrogenase (Magasanik et al., '57), respectively. In addition, ATP is required for the synthesis of guanosine nucleotides (Magasanik et al., '57) and GTP is required for the synthesis of adenosine nucleotides (Lieberman, '56). This unique reciprocal arrangement normally maintains a proper balance between the relative rates of production of AMP and GMP.

We have isolated from a population of mouse T-lymphoma (S49) cells a mutant clone which is prototrophic for purines but contains only 20% of the wild type activity of adenylosuccinate synthetase, the penultimate enzyme in de novo AMP synthesis. Since exogenous purines are not required for their growth, the cells of this mutant clone, AU-100, have provided considerable insight into the complex regulatory mechanisms involved in both the de novo and the salvage pathways of purine metabolism.

MATERIALS AND METHODS

The sources of all materials, chemicals, and reagents have been reported previously (Ullman et al., '76). The methods for cell maintenance, for mutant cell selection, and for the determination of growth rates, HPLC, enzyme assays, purine rates and PPriboseP levels have been described in detail by Ullman et al. ('79).

RESULTS

Growth Rate Experiments

The AU-100 cell line was selected in semisolid agarose by virtue of its resistance to 0.1 mM 6-azauridine. However, this cell line in suspension culture was as sensitive as wild type cells to 6-azauridine, which inhibited growth of both cell types by 50% (E.C.$_{50}$) at 0.3 μM (Table 1). Since Hashmi et al. ('75) have isolated a mouse cell line resistant to azauridine with concomitant adenosine resistance, we tested the AU-100 cell line for sensitivity to adenosine cytotoxicity. Interestingly, the AU-100 cell line was considerably more resistant than wild type cells to the presence of 10 μM erythro-9-(2-hydroxy-3-nonyl) adenosine (EHNA). In the presence of EHNA the E.C.$_{50}$ value for adenosine in AU-100 cells was 15-20 μM compared to an E.C.$_{50}$ value of 4 μM adenosine in wild type cells (Table 1). We have previously shown that adenosine at low concentrations is toxic to wild type S49 cells by depleting intracellular PPriboseP (Gudas et al., '78a).

Since 6-azauridine resistance and 5-fluorouracil sensitivity are both increased in S49 cell mutants with elevated levels of OPRTase and ODCase (Levinson et al., '79), we tested the AU-100 cell line for its sensitivity to 5-fluorouracil. As shown in figures 1A and 1B, the AU-100 cell line is considerably more sensitive to 5-fluorouracil than is the wild type cell line. Furthermore, since 5-fluorouracil is metabolized by OPRTase to its toxic metabolite and hypoxanthine protects against this toxic process by diminishing intracellular levels of PPriboseP (Ullman and Kirsch, '79), we attempted to reverse the supersensitivity of the AU-100 cell line to 5-fluorouracil by the addition of exogenous 0.1 mM hypoxanthine. Whereas 0.1 mM hypoxanthine increased the E.C.$_{50}$ value for 5-fluorouracil in wild type cells considerably (fig. 1A), 0.1 mM hypoxanthine had no effect on the E.C. value for 5-fluorouracil in AU-100 cells (fig. 1B). However, 0.1 mM adenine, which also protects wild type cells from the cytotoxic effects of 5-fluorouracil (fig. 1A; Ullman and Kirsch, '79), increased the E.C.$_{50}$ value for 5-fluorouracil in the AU-100 cell line from 0.2 μM to 1 μM. These observations suggested that the enzyme, HGPRTase, might be malfunctioning in the AU-100 cell mutant. Therefore, we determined in wild type and in AU-100 cells the E.C.$_{50}$ values for 6-thioguanine and for 6-mercaptopurine, both toxic analogs of HGPRTase substrates. As shown in table 1, the AU-100 cell line is approximately 4-fold more resistant than wild type cells to the toxic effects of either 6-thioguanine or 6-mercaptopurine. Interestingly the HGPRTase activity in crude extracts of AU-100 cells (260 pmol/min/mg protein) was 50-66% higher than that in extracts of wild type cells (156 pmol/min/mg protein).

Guanosine, which cannot be phosphorylated directly in S49 cells, is toxic by virtue of its ability to inhibit adenosine nucleotide

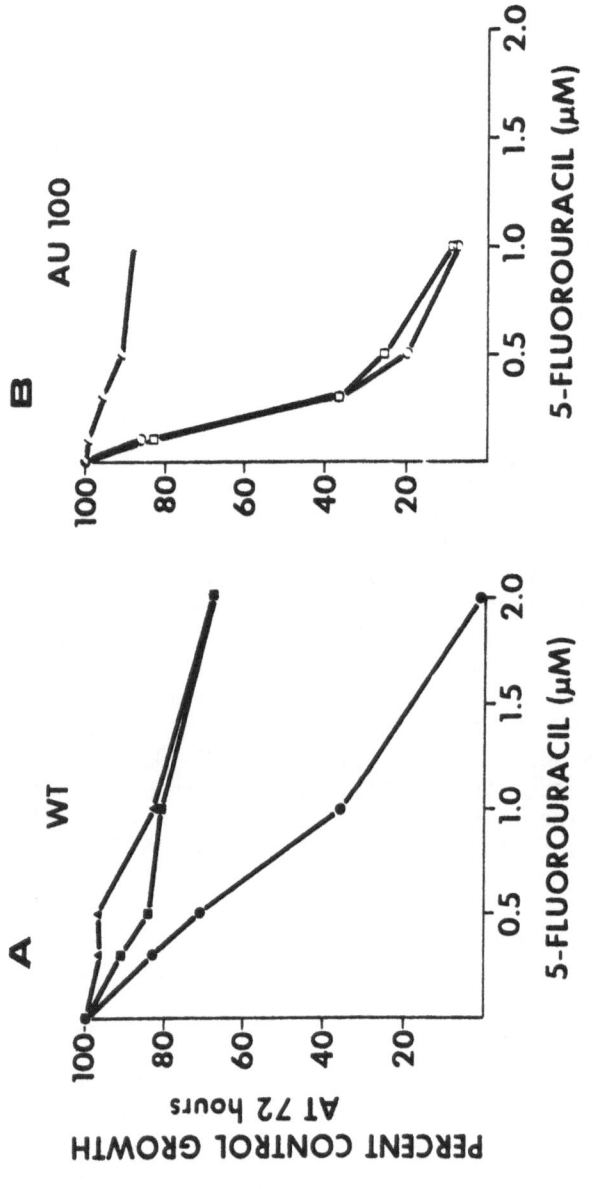

Fig. 1. Cytotoxicity of 5-fluorouracil in wild type and AU-100 cells. The toxicity of increasing concentrations of 5-fluorouracil on wild type cells is shown in Panel A. The growth inhibitory properties of 5-fluorouracil were determined in the absence (●) of added purines or in the presence of either 0.1 mM adenine (▲) or 0.5 mM hypoxanthine (■). Panel B describes identical experiments with the AU-100 cell line in the absence (○) of added purine to the medium or in the presence of either 0.1 mM adenine (△) or 0.5 mM hypoxanthine (□). The growth conditions for the two cell lines were identical and the percentages of control growth a any 5-fluorouracil concentration did not differ by more than 10% for either cell line in three separate experiments. The results depicted are those of a single experiment.

Table 1. Cytotoxicity of Purine Analogs

Additions	E.C.$_{50}$ Value (μM)	
	WT	AU-100
6-Azauridine	0.3	0.3
Adenosine	4.0	18
6-Thioguanine	1.4	5.0
6-Mercaptopurine	5.0	30
Guanosine	120	12

The growth inhibitors and cytotoxic properties of varying concentrations of freshly prepared analogs on wild type (WT) and AU-100 cells were determined as described by Ullman et al. ('79). Results depicted are the effective concentration of drug which inhibits cell growth by 50% at 72 hours (E.C.$_{50}$ value) of a single typical experiment which has been repeated at least two times with similar results.

synthesis (Gudas et al., '78b). This cytotoxic effect requires the sequential actions of purine nucleoside phosphorylase and HGPRTase (Gudas et al., '78b). We compared the sensitivity of AU-100 cells with the sensitivity of wild type cells toward guanosine. In comparable growth rate experiments the AU-100 cell line was 10-fold more sensitive to guanosine than were wild type cells. As shown in table 1, the E.C.$_{50}$ value determined for guanosine in AU-100 cells was 12 μM, whereas the E.C.$_{50}$ value for guanosine in the wild type cells is 120 μM. Addition of 0.1 mM adenine protects from guanosine mediated cytotoxicity (table 1).

PPriboseP Levels and Purine Synthetic Rates

Since the growth phenotype of the AU-100 cell line was unique, we attempted to relate the adenosine resistance and the guanosine supersensitivity to specific metabolic phenomena. Addition of exogenous adenosine to wild type S49 cell cultures causes a lethal reduction in intracellular pyrimidine nucleotides, which is mediated by severely depleted intracellular PPriboseP levels (Gudas et al.,'78a). We thought that AU-100 cells might contain elevated intracellular amounts of PPriboseP to account for their adenosine resistance. As shown in table 2, the PPriboseP levels in AU-100 cells were elevated 70% over those levels found in wild type cells. Addition of 10 μM adenosine, 10 μM EHNA to AU-100 cells caused a

Table 2. Comparison of Intracellular PPriboseP Content and Rates of Purine Synthesis in Wild Type and AU-100 Cells in the Absence and Presence of Exogenous Purines

Additions	PPriboseP Content (nmol/10^9 cells)		Purine Rates (^{14}C-glycine into purines) / (^{14}C-glycine into protein)	
	WT	AU-100	WT	AU-100
None	66	106	0.93	0.31
10 µM adenosine + 10 µM EHNA	9	24		
100 µM guanosine	69	125		
10 µM adenine			0.28	0.07
100 µM adenine			0.13	0.07

The intracellular PPriboseP content and the rates of purine synthesis were determined as described by Ullman et al. ('79). Levels of PPriboseP in wild type and AU-100 cells were also determined after a 4-hour incubation with either 10 µM adenosine and 10 µM EHNA or 100 µM guanosine. These values are the averages from two independent experiments which did not differ by more than 20%. The effects of exogenous adenine in the culture medium on purine synthetic rates were determined after a 2-hour incubation with either 10 µM or 100 µM adenine. The rates of purine synthesis are expressed as a ratio of glycine incorporation into purines versus glycine incorporation into proteins (Graf et al., '76; Cohen et al., '76) to correct for differential growth rates of the two cell lines under the various perturbing conditions. These values are also averages of two to six experiments which differed by less than 10%.

65% reduction in PPriboseP levels. However the levels of PPriboseP found in AU-100 cells after a 4-hour incubation with 10 μM adenosine-EHNA were still 60% of those levels found in unperturbed wild type cells. Thus, in AU-100 cells PPriboseP levels are elevated and adenosine does not reduce the PPriboseP to levels insufficient for de novo pyrimidine synthesis (Gudas et al., '78a). Exogenous 0.1 mM guanosine does not affect PPriboseP levels in either cell type (table 2).

PPriboseP, the product of the first reaction of de novo purine synthesis, is an important regulator of the rate of de novo purine synthesis in a variety of cell types. It has been proposed that in humans elevated intracellular PPriboseP content causes severe purine overproduction and gout. We, therefore, measured the rates of de novo purine synthesis in wild type and AU-100 cells. Unexpectedly, in five separate experiments the rates of purine synthesis in intact AU-100 cells were only 33% the rates of purine synthesis in wild type cells (table 2). The rates of purine synthesis in both cell types were inhibited equally by addition of exogenous adenine to the culture medium (table 2).

Nucleotide Levels and Uptake of Purines

The diminished purine synthesis in AU-100 cells caused us to examine the steady state levels of nucleotides in AU-100 and wild type cells. By high performance liquid chromatography it was shown (table 3) that AU-100 cells contain three times the amount of GTP as do wild type cells, although levels of UTP and ATP in both cell types are similar. Furthermore, the levels of IMP and GMP are also elevated in AU-100 cells (table 3). Since GMP (but not IMP) inhibits HGPRTase activity (with an apparent K_i for HGPRTase in S49 cells of 7.1 μM), we examined the rate of incorporation of labeled hypoxanthine into phosphorylated metabolites in intact cells. Hypoxanthine incorporation into nucleotides of AU-100 cells is 20% of that in wild type cells. Since HGPRTase activity is elevated in extracts of AU-100 cells, it is likely that their elevated GMP levels inhibit their HGPRTase catalytic activity in the intact cells.

Enzyme Assays

In AU-100 cells the elevated guanosine nucleotide levels, the underproduction of purines de novo in the face of elevated intracellular PPriboseP content, the ability of adenine but not hypoxanthine to rescue from 5-fluorouracil toxicity all suggested that the genetic defect in these cells was in the purine synthetic pathway but distal to the synthesis of IMP. As shown in table 4, levels of IMP dehydrogenase were the same in wild type and AU-100 cells. However, the enzyme, adenylosuccinate synthetase (the penultimate step in AMP synthesis), is severely but not totally deficient in

Table 3. Nucleotide Levels in Wild Type and AU-100 Cells

Nucleotide	Intracellular Content (nmol/10^9 cells)	
	WT	AU-100
ATP	1,950	2,100
UTP	448	483
GTP	126	385
IMP	7.5	22.2
GMP	14	46

Nucleotide determinations were made on cells growing asynchronously and logarithmically as described by Ullman et al. ('79). All values are expressed as nmol/10^9 cells and represent the average values of two independent experiments which differed by less than 20%.

Table 4. Intracellular Levels of Enzymes Involved in Purine Biosynthesis in Wild Type and AU-100 Cells

Enzyme	Specific Activity (nmol/min/mg protein)	
	WT	AU-100
Adenylosuccinate synthetase	0.14	0.03
IMP dehydrogenase	0.10	0.12
PPriboseP synthetase	4.20	5.80
PPriboseP amidotransferase	0.05	0.07

The specific catalytic activities of PPriboseP synthetase, PPriboseP amidotransferase, adenylosuccinate synthetase, and IMP dehydrogenase were determined as described by Ullman et al. ('79). The values expressed are in nmol/min/mg protein and are the average of two separate determinations which did not vary from the median by more than 10%.

the AU-100 cell line. Also shown in table 4, levels of the adenylosuccinate synthetase activity in AU-100 cells were only 22% the
levels found in wild type cells. The enzymes which govern the rate
of purine production, PPriboseP synthetase and PPriboseP amidophos-
phoribosyl transferase are equally active in extracts of wild type
and AU-100 cells. This observation can fully explain the phenotype
of AU-100 mutants and provides important information about the
regulation of the purine de novo and salvage pathways.

DISCUSSION

We have isolated a mutant approximately 80% deficient in
adenylosuccinate synthetase, the penultimate enzyme in adenosine
nucleotide synthesis de novo and the rate-limiting enzyme for the
conversion of IMP to AMP. It seems likely, but not certain, that
the primary genetic defect in AU-100 cells is in the gene for
adenylosuccinate synthetase. This mutant, unlike some CHO cell
mutants isolated by Patterson ('76a), does not require the addition
of any exogenous purine for normal cell growth. (The AU-100 cell
line grows in medium containing extensively dialyzed horse serum.)
Thus we were able to study the abnormal regulation of the purine de
novo and salvage pathways in AU-100 cells without any perturbation
by exogenous purines.

Since adenosine nucleotide levels are normally higher than
guanosine nucleotide levels in S49 cells (table 3), the flux of IMP
through IMP dehydrogenase and adenylosuccinate synthetase in wild
type cells must be under strict control by their respective end
products. An altered GTP to ATP ratio would reflect altered in
vivo catalytic activities of the rate-limiting enzyme or altered
regulatory patterns. The AU-100 cell line was found to contain 22%
of the wild type activity of the wild type activity of adenylosuccinate synthetase, a finding which can account for their increased
GTP to ATP ratio and their increased PPriboseP level. The increased
guanylate nucleotide levels found in AU-100 cells account for the
extraordinary sensitivity of these cells to guanosine. The elevated
GMP can also explain the observed inhibition of HGPRTase in the
intact cells. However, the functionally diminished HGPRTase activity
cannot account for the higher PPriboseP levels since S49 cells
genetically deficient in HGPRTase have near normal PPriboseP levels
(Gudas et al., '78a).

The increased intracellular levels of PPriboseP and the reduced
rate of de novo purine synthesis in AU-100 cells suggested the
existence of increased inhibition of their purine synthetic pathway
by elevated guanosine nucleotides at a site distal to the production
of PPriboseP. PPriboseP amidotransferase has been reported to be
sensitive to feedback inhibition by GMP (Wyngaarden and Ashton, '59).
In AU-100 cells the elevated guanosine nucleotides and the severely

depressed rate of purine synthesis in the presence of increased PPriboseP levels indicate that in vivo guanosine nucleotides regulate the purine de novo pathway by inhibiting the PPriboseP amidotransferase.

Since exogenous adenosine in the presence of EHNA results in the severe depletion of intracellular PPriboseP and since AMP is a potent inhibitor of PPriboseP synthetase activity in S49 cells (unpublished observation), it appears that adenosine nucleotides inhibit de novo purine and PPriboseP synthesis by inhibiting PPriboseP synthetase.

The isolation and characterization of the S49 cell mutant, AU-100, which is 78% deficient in adenylosuccinate synthtase has contributed significantly to the understanding of feedback regulatory mechanisms in the purine pathway and has enabled us to test previously proposed hypotheses concerning the purine nucleotide modulation of this multienzyme synthetic pathway. Specifically the analysis of AU-100 has provided the first direct in vivo data suggesting a physiological role for the long recognized allosteric properties of the enzyme, PPriboseP amidotransferase (Wyngaarden and Ashton, '59). Although the AU-100 cell line synthesizes purines de novo at a diminished rate, the intact mutant cells have an effectively diminished HGPRTase activity. Since diminished HGPRTase activity has been associated with purine overproduction in humans, probably by impaired salvage of exogenous purines secreted by the liver (Pritchard et al., '70), it is possible that dimished adenylosuccinate activity in humans might cause overproduction hyperuricemia and the associated clinical conditions.

REFERENCES

Cohen, A., Doyle, D., Martin, D.W., Jr., and Ammann, A.J., 1976, Abnormal purine metabolism and purine overproduction in a patient deficient in purine nucleoside phosphorylase, N. Engl. J. Med., 295:1449.

Gudas, L.J., Cohen, A., Ullman, B., and Martin, D.W., Jr., 1978a, Analysis of adenosine mediated pyrimidine starvation using cultured wild type and mutant mouse T-lymphoma cells, Somatic Cell Genet., 4:201.

Gudas, L.J., Ullman, B., Cohen, A., and Martin, D.W., Jr., 1978b, Deoxyguanosine toxicity in a mouse T-lymphoma: Relationship to purine nucleoside phosphorylase-associated immune dysfunction, Cell, 14:531.

Hashmi, S., May, S.R., Krooth, R.S., and Miller, O.J., 1975, Con-
    current development of resistance to 6-azauridine and adenosine
    in a mouse cell line, J. Cell. Physiol., 86:191.

Levinson, B., Ullman, B., and Martin, D.W., Jr., 1979, Pyrimidine
    pathway mutants of cultured mouse lymphoma cells with altered
    levels of both orotate phosphoribosyltransferase and orotidylate
    decarboxylase, J. Biol. Chem., in press.

Lieberman, I., 1956, Enzymatic synthesis of adenosine-5'-phosphate
    from inosine-5'-phosphate, J. Biol. Chem., 223:327.

Magasanik, B., Moyed, H.S., and Gehring, L.B., 1957, Enzymes essen-
    tial for the biosynthesis of nucleic acid guanine: Inosine-5'-
    phosphate dehydrogenase of Aerobacter aerogenes, J. Biol. Chem.,
    226:339.

Patterson, D., 1976, Biochemical genetics of Chinese hamster cell
    mutants with deviant purine metabolism. III. Isolation and
    characterization of a mutant unable to convert IMP to AMP,
    Somatic Cell Genet., 2:41.

Pritchard, J.B., Chavez-Peon, F., and Berlin, R.D., 1970, Purines:
    supply by liver to tissues, Am. J. Physiol., 219:1263.

Seegmiller, J.E., Rosenbloom, F.M., and Kelley, W.N., 1967, An
    enzyme defect associated with a sex-linked human neurological
    disorder and excessive purine synthesis, Science, 155:1682.

Sperling, O., Persky-Brosh, S., Boer, P., and de Vries, A., 1973,
    Human erythrocyte phosphoribosylpyrophosphate synthetase muta-
    tionally altered in regulatory properties, Biochem. Med., 7:389.

Ullman, B., Clift, S.M., Cohen, A., Gudas, L.J., Levinson, B.B.,
    Wormsted, M.A., and Martin, D.W., Jr., 1979, Abnormal regulation
    of de novo purine synthesis and purine salvage in a cultured
    mouse T-cell lymphoma mutant partially deficient in adenylo-
    succinate synthetase, J. Cell Physiol., 99:139.

Ullman, B., Cohen, A., and Martin, D.W., Jr., 1976, Characterization
    of a cell culture model for the study of adenosine deaminase-
    and purine nucleoside phosphorylase-deficient immunologic
    disease, Cell, 9:205.

Ullman, B. and Kirsch, Jr., 1979, Metabolism of 5-fluorouracil in
    cultured cells: protection from 5-fluorouracil cytotoxicity
    by purines, Molecular Pharmacology, 15:357.

Van der Weyden, M.B. and Kelley, W.N., 1974, Human adenylosuccinate synthetase. Partial purification, kinetic and regulatory properties of the enzyme from placenta, J. Biol. Chem., 249:7282.

Wyngaarden, J.B. and Ashton, D.M., 1959, The regulation of activity of phosphoribosylpyrophosphate amidotransferase by purine ribonucleotides: A potential feedback control of purine biosynthesis, J. Biol. Chem., 234:1492.

Wyngaarden, J.B. and Greenland, R.A., 1963, The inhibition of succinoadenylate kinosynthetase of Escherichia coli by adenosine and guanosine 5'-monophosphates, J. Biol. Chem., 238:1054.

SUPERACTIVE PHOSPHORIBOSYLPYROPHOSPHATE SYNTHETASE WITH ALTERED

REGULATORY AND CATALYTIC PROPERTIES

Michael A. Becker, Kari O. Raivio, Bohdan Bakay,
William B. Adams and William L. Nyhan
San Diego Veterans Administration Hospital and
Departments of Medicine and Pediatrics, University
of California, San Diego, La Jolla, California 92161

One line of evidence supporting a role for 5-phosphoribosyl-
1-pyrophosphate (PP-Ribose-P) in the regulation of the rate of
purine synthesis de novo has come from the study of inherited
abnormalities of the enzyme PP-ribose-P synthetase (E.C. 2.7.6.1)
(1-7). PP-ribose-P synthetase catalyzes synthesis of PP-ribose-P
from ATP and ribose-5-phosphate (Rib-5-P) in a reaction requiring
Mg^{2+} and inorganic phosphate (Pi). The activity of the enzyme is
inhibited by a variety of compounds including: purine, pyrimidine,
and pyridine nucleotides (6,8-10); 2, 3-diphosphoglycerate (6,8);
and the products of the reaction, AMP and PP-ribose-P (6,8).

Several families have been identified in which superactive
forms of PP-ribose-P synthetase result in excessive PP-ribose-P
production in association with increased purine nucleotide and uric
acid synthesis and clinical gout (1-7). Superactive PP-ribose-P
synthetases due either to diminished sensitivity to feedback inhibitors (3,4) or to excessive maximal reaction velocity per molecule
of enzyme (6) have been described and extensively characterized.
To date the clinical manifestations of all the males bearing mutant
forms of this X-linked (11-13) enzyme have been restricted to gouty
arthritis, uric acid urolithiasis or renal insufficiency first noted
in early adulthood (1,5,7).

We have recently characterized a mutant form of PP-ribose-P
synthetase identified in fibroblasts from a child with uric acid
overproduction apparent since the first year of life and from his
clinically affected mother. Our studies provide evidence for both
purine nucleotide feedback-resistance and excessive maximal reaction
velocity in the aberrant enzyme and indicate that this variant form
of PP-ribose-P synthetase is abnormally labile in erythrocytes
in vivo.

Patient SM is a 14-year-old male first studied in detail at age 3 by Dr. Nyhan and his colleagues (14) who demonstrated marked uric acid overproduction and clinical features which included dysplastic teeth, failure to cry tears, presumed mental retardation, absence of speech and bizarre and autistic behavior. Recent clinical reevaluation of the patient by Dr. Nyhan has revealed a severe hearing loss suggesting milder behavioral abnormalities than originally suspected. The patient's hyperuricemic mother, SuM, has had attacks of gouty arthritis and uric acid urolithiasis and, in addition, has a significant hearing loss.

Comparison of the PP-ribose-P metabolism and purine nucleotide synthesis in fibroblasts cultured from patient SM and normal individuals is shown in Table 1 in which values for the individual measurements are expressed relative to an arbitrary mean value of 1.0 for 4 normal fibroblast strains. The rate of the early steps of purine synthesis de novo in SM fibroblasts was 2.7-fold greater than normal, and fibroblast PP-ribose-P concentration and rate of generation were increased 4.0 and 2.7-fold, respectively. Rates of incorporation of the labeled purine bases adenine,hypoxanthine, and guanine into intracellular purine compounds were increased in SM cells, while the rate of incorporation of [^{14}C] adenosine was normal. These studies, which demonstrated increased rates of PP-ribose-P-dependent purine biosynthetic processes in SM fibroblasts, suggested that an increased rate of intracellular PP-ribose-P synthesis underlay the excessive rate of purine synthesis de novo in these cells. Subsequent measurement of PP-ribose-P synthetase activity in dialyzed extracts of patient SM's fibroblasts showed a greater than 2-fold increased activity of this enzyme when measured at saturating substrate and Pi concentrations.

Table 1. PP-Ribose-P Metabolism and Purine Nucleotide Synthesis in SM Fibroblasts.

Study	Value relative to mean value in 4 normal strains (1.0)
Purine synthesis (FGAR accumulation)	2.7
PP-Ribose-P concentration	4.0
PP-Ribose-P generation	2.7
Incorporation of [^{14}C]:	
adenine	2.7
hypoxanthine	2.0
guanine	2.6
adenosine	0.9
Activity of:	
PP-Ribose-P synthetase	2.2
HGPRT	1.0
APRT	1.1

In addition to increased PP-ribose-P synthetase activity in crude or dialyzed fibroblast extracts from patient SM, distinct differences in the shapes of the curves relating enzyme activity and Pi concentration were observed (Fig. 1A). In contrast to the sigmoidal curve of Pi activation shown by the normal enzyme in dialyzed extracts, hyperbolic activation of SM PP-ribose-P synthetase was apparent. While at 32 mM Pi, SM PP-ribose-P synthetase specific activity was 2-fold greater than normal, at more physiologic Pi concentrations, activities of the variant form of the enzyme were nearly 10-fold greater than normal. The sigmoidal shape of the normal Pi activation curve results from greater efficiency of feed-back inhibitors of enzyme activity at low Pi concentration (3,15). Thus, these studies suggested diminished responsiveness of SM's enzyme to feedback inhibitors. Accordingly, procedures more effective than dialysis in removing small molecule inhibitors from the milieu of the enzyme were adopted. After chromatography of crude fibroblast extracts on Sephadex G-25, hyperbolic Pi activation curves were obtained for both normal and SM PP-ribose-P synthetases, and under these circumstances the variant form of the enzyme ex-hibited a 2-fold greater activity at all Pi concentrations (Fig. 1A). Partially purified normal and SM PP-ribose-P synthetases also showed hyperbolic Pi activation.

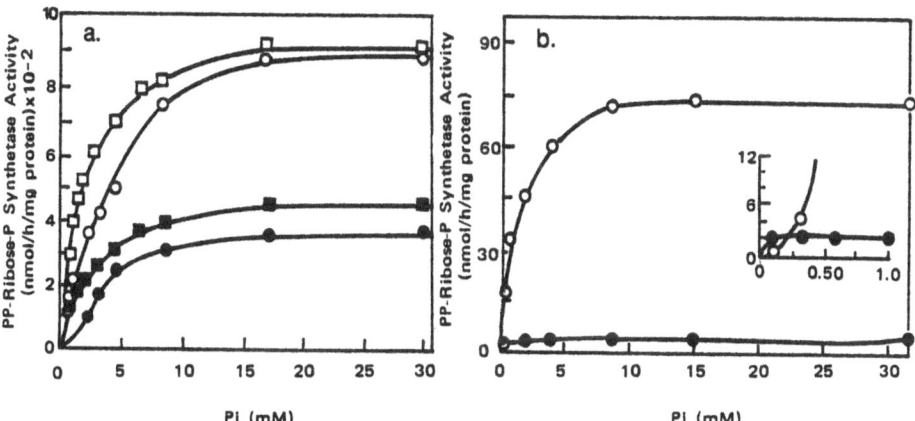

Figure 1. Activation of normal and SM fibroblast (a) and erythrocyte (b) PP-ribose-P synthetases by Pi. In a, enzyme activities were measured at the appropriate final Pi concentrations in fibroblast extracts prepared either by dialysis (circles) for 2 h against a buffer at pH 7.4 containing 2 mM sodium phosphate, 10 mM reduced glutathione, and 1 mM EDTA or by chromatography (squares) on Sephadex G-25 equilibrated with this buffer. SM, open symbols; normal, closed symbols. In b, enzyme activities measured in dialyzed hemolysates are shown with the inset expanding the range of values obtained at Pi concentrations below 1.0 mM. SM, closed symbols; normal, open symbols.

In order to determine whether the increased enzyme activity observed in SM fibroblast extracts even after chromatography resulted from an increased number of enzyme molecules or from increased catalytic activity per molecule of enzyme, immunochemical studies were carried out. Chromatographed normal and SM extracts, equated for protein concentration, were titrated with increasing amounts of the IgG fraction of monospecific antiserum to normal PP-ribose-P synthetase (16). Inactivation of SM PP-ribose-P synthetase was 2-fold greater than that of normal enzyme for any given amount of antiserum in the range of the titration curve, although, for any given ratio of antiserum to extract, protein in the resulting immunoprecipitate was identical for mutant and normal extracts. These studies indicated that fibroblast PP-ribose-P synthetase from patient SM had 2-fold greater catalytic activity per enzyme molecule.

Additional experiments were undertaken to delineate the substrate binding and inhibitor responsiveness of PP-ribose-P synthetase from the fibroblasts of patient SM. Table 2 provides a summary of the results obtained when partially (30-fold) purified preparations of normal and SM PP-ribose-P synthetase were studied. Apparent binding constants (Km's) for MgATP, Rib-5-P and Mg^{2+} were similar for normal and SM enzymes tested at either 1 mM or 32 mM Pi. In contrast, the concentrations of ADP and GDP at which 50% inactivation of PP-ribose-P synthetase was observed differed for SM and normal enzymes at both Pi concentrations. PP-ribose-P synthetase from patient SM was 4- to 5-fold more resistant to ADP and GDP than was the normal enzyme. These findings indicate that despite normal affinity for substrates, PP-ribose-P synthetase from patient SM is relatively more resistant to the inhibitory effects of purine nucleotide compounds. Thus, mutant PP-ribose-P synthetase had both increased maximal reaction velocity and diminished sensitivity to certain inhibitors of enzyme activity.

Table 2. Kinetic Constants of Partially Purified SM and Normal PP-Ribose-P Synthetases

Study	Pi Concentration	PP-Ribose-P synthetase	
		Normal	SM
	mM	μM	
Km MgATP	1	16	15
	32	24	22
Km Rib-5-P	1	12	15
	32	27	33
Km Mg^{2+}	1	36	32
	32	44	49
$I_{0.5}$	1	22	85
	32	135	280
Ki slope ADP	32	22	80
$I_{0.5}$ GDP	1	34	153
	32	320	>750

An apparent paradox was encountered when PP-ribose-P synthetase activity was measured in erythrocyte lysates from patient SM (Fig. 1B). Regardless of whether lysates were dialyzed or chromatographed, enzyme activity in lysates from patient SM was markedly diminished when measured at Pi concentrations greater than 1 mM. Nevertheless, PP-ribose-P synthetase activity in SM hemolysates showed hyperbolic Pi activation as observed in the fibroblast studies. The explanation for these findings was found to lie in the diminished stability of SM PP-ribose-P synthetase. Thermal inactivation of SM fibroblast PP-ribose-P synthetase at 54° was substantially more rapid than that of the normal form of the fibroblast enzyme. This finding led us to perform density fractionation of normal and SM erythrocytes. Fresh normal and SM erythrocytes were separated into four density (and thus age) classes on a phthalate gradient (17). Lysates of erythrocytes contained in each of these density classes were then prepared, and activities of PP-ribose-P synthetase, HGPRT, APRT and, glucose-6-phosphate dehydrogenase were determined. While SM and normal HGPRT, APRT and glucose-6-phosphate dehydrogenase activities showed similar patterns of activity relative to increasing red cell age, PP-ribose-P synthetase activity in SM erythrocytes was much less stable than this activity in normal cells and was virtually absent in all but the youngest erythrocytes. Despite the apparently diminished amount of PP-ribose-P synthetase in the overall SM erythrocyte population, however, SM cells showed a greater than normal rate of incorporation of [^{14}C] adenine into nucleotides when incubated in buffer containing 1 mM Pi. During incubation in buffer containing 32 mM Pi, SM erythrocytes showed a lower than normal rate of incorporation of the same label into nucleotides. We interpret these findings to reflect the major extent to which activity of normal PP-ribose-P synthetase is ordinarily attenuated by feedback inhibitors under physiological conditions of Pi concentration and conversely, the extent to which the regulatory and catalytic alterations in SM PP-ribose-P synthetase permit even a diminished amount of enzyme to produce excessive PP-ribose-P.

Finally, we have recently had the opportunity to confirm the hereditary nature of the defect in SM PP-ribose-P synthetase by studying fibroblasts cultured from his mother, SuM. Although PP-ribose-P synthetase activity in erythrocyte lysates from this woman was normal, enzyme activities in crude and dialyzed fibroblast extracts were intermediate to normal and SM enzyme activities and showed hyperbolic Pi activation. In addition, SuM fibroblast PP-ribose-P synthetase showed patterns of responsiveness to ADP and GDP which were intermediate to those found for normal and SM enzymes. These findings suggest that SuM is a heterozygous carrier of the mutant enzyme expressed in hemizygous form by her son, SM.

In summary, our studies of PP-ribose-P synthetase from the

fibroblasts of a child with purine overproduction have revealed a
mutant enzyme in which both regulatory and catalytic properties
are abnormal and favor excessive intracellular PP-ribose-P synthesis.
In addition, we have provided evidence suggesting that instability
of the mutant enzyme accounts for an apparent deficiency in
erythrocyte PP-ribose-P synthetase activity under usual assay
conditions.

REFERENCES

1. O. Sperling, G. Eilam, S. Persky-Brosh, and A. de Vries,
    Biochem.Med. 6:310 (1972)
2. O. Sperling, P. Boer, S. Persky-Brosh, E. Kanarek, and
    A. de Vries, Rev.Eur.Etud.Clin.Biol. 17: 703 (1972).
3. O. Sperling, S. Persky-Brosh, P. Boer, and A. de Vries,
    Biochem. Med. 7:389 (1973).
4. E. Zoref, A. de Vries, and O. Sperling, J.Clin.Invest. 56:1093
    (1975).
5. M.A. Becker, L.J. Meyer, and J.E. Seegmiller, Am.J.Med. 55:
    232 (1973).
6. M.A. Becker, P.J. Kostel, and L.J. Meyer, J.Biol.Chem. 250:
    6822 (1975).
7. M.A. Becker, J.Clin.Invest. 57:308 (1976).
8. I.H. Fox and W.N. Kelley, J.Biol.Chem. 247:2126 (1972).
9. P.C.L. Wong and A.W. Murray, Biochemistry 8:1608 (1969).
10. D.G. Roth and T.F. Deuel, J.Biol.Chem. 249:297 (1974).
11. E. Zoref, A. de Vries, and O. Sperling, Adv.Exp.Med.Biol. 76A:
    287 (1977).
12. R.C.K. Yen, W.B. Adams, C. Lazar, and M.A. Becker, Proc.Natl.
    Acad.Sci. USA 75:482 (1978).
13. M.A. Becker, R.C.K. Yen, P. Itkin, S.J. Goss, J.E. Seegmiller,
    and B. Bakay, Science 203:1016 (1979)
14. W.L. Nyhan, J.A. James, A.J. Teberg, L. Sweetman and L.G.Nelson,
    J.Pediat. 74: 20 (1969).
15. A. Hershko, A. Razin and J. Mager, Biochim.Biophys.Acta 184:
    64 (1969).
16. M.A. Becker, P.J. Kostel, L.J. Meyer and J.E. Seegmiller,
    Proc.Natl.Acad.Sci. USA 70:2749 (1973).
17. T.D. Beardmore, J.S. Cashman and W.N. Kelley, J.Clin.Invest.
    51:1823 (1972).

ACKNOWLEDGMENTS

    This work was supported in part by: the Medical Research
Service of the Veterans Administration; Grants AM-18197 and GM-17702
from the National Institutes of Health; National Foundation - March
of Dimes Grant 1-377 (W.L.N.) and grants to Dr.J.Edwin Seegmiller
from the Kroc Foundation and to Dr. Raivio from the Sigrid Juselius
Foundation.

AMP PHOSPHATASE ACTIVITY IN HUMAN TERM PLACENTA: STUDIES ON PLACENTAL 5'-NUCLEOTIDASE

M. Helen Maguire and T.P. Krishnakantha

Department of Pharmacology
University of Kansas Medical Center
Kansas City, Kansas U.S.A.

Measurement of the levels of AMP-metabolizing enzymes in homogenates of human term placenta showed the presence of strong AMP phosphatase activity at pH 7.2 (1). At this pH it was not possible to detect AMP deaminase, even in the presence of ATP to inhibit concomitant dephosphorylation of AMP by 5'-nucleotidase. Our recent studies used conditions selected to optimize AMP deaminase activity, i.e. increased K^+ concentration, lower assay pH and inclusion of the potent 5'-nucleotidase inhibitor adenosine 5'-methylenediphosphonate, and resulted in measurement of 0.475 units of AMP deaminase per g placenta (Table 1). Even under these conditions the rate of adenosine formation was greater than the rate of formation of IMP. These findings substantiated the conclusion from our earlier study (1), that the preferred pathway of catabolism of AMP in human term placenta is via adenosine rather than via IMP. The levels of adenosine deaminase and adenosine kinase in the placenta are low compared to those AMP phosphatase and AMP deaminase (cf Table 1), indicating that under conditions which lead to accelerated AMP formation and breakdown, such as anoxia, adenosine may accumulate, and indeed 230 nmoles of adenosine per g tissue was isolated from post-labor term placentas (1). The dephosphorylation of AMP at pH 7.2 is catalyzed in part by heat-stable enzyme activity, presumably due to alkaline phosphatase, but more than 80% of the total phosphatase activity is mediated by heat-labile 5'-nucleotidase (Table 1). In order to study the regulation and role of human placental 5'-nucleotidase in the catabolism of AMP, the enzyme was purified from term placentas.

5'-Nucleotidase was assayed at pH 7.25 in 0.05 M Tris-HCl, either spectrophotometrically by coupling the dephosphorylation of AMP with adenosine deaminase-catalyzed hydrolysis of adenosine to

Table 1.   Activities of AMP-Metabolizing Enzymes in
Placental Homogenates

Enzyme	Assay pH	Activity (units[a]/g)	n[b]
Total AMP phosphatase	7.25	2.151 ± 0.169	(11)
5'-Nucleotidase (total AMP phosphatase - heat-stable activity)	7.25	1.828 ± 0.146	(11)
AMP deaminase[c]	6.8	0.475 ± 0.04	(4)
Adenosine deaminase[d]	7.0	0.171 ± 0.015	(4)
Adenosine kinase[e]	5.6	0.042	(2)

[a] μMoles substrate converted/min at $30°$ $\pm$ S.E.

[b] Number of placentas.

[c] Radiometric assay following conversion of 8^{14}C-AMP to 8^{14}C-IMP in the presence of 0.5 M KCl and adenosine 5'-methylenediphosphonate.

[d] Spectrophotometric assay following oxidation of hypoxanthine at 293 nm using xanthine oxidase and purine nucleoside phosphorylase as coupline enzymes.

[e] Radiometric assay following conversion of 8^{14}C-adenosine to 8^{14}C-AMP in the presence of an ATP-regenerating system (pyruvate-PEP).

inosine (1), or radiometrically using ^{14}C-labeled 5'-mononucleotides and measuring the conversion to labeled nucleosides after paper chromatographic separation of reaction products. The contribution on non-specific phosphatase to the observed total dephosphorylation preparation at 65°C for 30 minutes.

Term placentas were obtained immediately after delivery and 5'-nucleotidase was purified from aqueous homogenates of the placentas. The insoluble fraction of the homogenates was extracted with a sodium deoxycholate-Triton X-100 mixture to give a solubilized enzyme preparation which was fractionated by salt precipitation and chromatography on DEAD-cellusose. Filtration through Sephadex

G-100 and G-200 and further ion exchange fractionation yielded a
500-800-fold purified 5'-nucleotidase which hydrolyzed 11-20
μmoles AMP per min per mg at $30^{\circ}C$, and which had no non-specific
phosphatase activity. Polyacrylamide gel electrophoresis of the
purified 5'-nucleotidase showed one major and two minor protein
bonds. The former was positive for both glycoprotein and AMP
phosphatase activity. The enzyme exhibited a discontinuous concave
Arrhenius plot with a break at $38^{\circ}C$, and a pH optimum of 7.2-7.3.
In the neutral pH range there was no significant activation by
Mg^{2+}, but Mg^{2+} induced a second pH optimum between pH 9 and 10.
Comparison of K_m and V_{max} values of the 5'-mononucleotides, AMP,
GMP, IMP, UMP, CMP and IMP, indicated that AMP was the preferred
substrate. Dephosphorylation of AMP by 5'-nucleotidase was inhibited
by nucleoside di- and triphosphates in a linear competitive fashion.
ADP was the most potent inhibitor with a K_i of 90 nM.

It is concluded that 5'-nucleotidase is an intrinsic membrane
protein which is predominantly responsible for the dephosphorylation
of AMP to adenosine in the human tern placenta. The kinetic
constants of AMP as substrate and ADP and ATP as inhibitors suggest
that the activity of the enzyme can be modulated by large changes
in the adenine nucleotide pool, such as result from anoxia.
Definition of the role of membrane-bound 5'-nucleotidase in AMP
catabolism in the placenta awaits determination of the cellular
location of the enzyme in the placental trophoblast.

These findings will be reported more extensively in a paper
to be published separately.

ACKNOWLEDGMENT

This study was supported by a grant from the National Institutes
of Health (HD 08653).

REFERENCE

1. Sim, M.K. and Maguire, M.H.: Presence of adenosine in
    the human term placenta: determination of adenosine
    content and pathways of adenosine metabolism. Circulation
    Res. 31:779-788, 1972.

ADENOSINE AND DEOXYADENOSINE METABOLISM IN THE ERYTHROCYTES OF A

PATIENT WITH ADENOSINE DEAMINASE DEFICIENCY

A. Sahota, H. A. Simmonds, C. F. Potter, J. G. Watson,
K. Hugh-Jones and D. Perrett
Clinical Science Laboratories, Guy's Hospital, London;
Westminster Hospital, London; and St. Bartholomew's
Hospital, London

The excretion of deoxyadenosine in the urine[1,2] and the
accumulation of adenine deoxynucleotides in erythrocytes[3,4] and
lymphocytes[4] have now been reported in several cases of adenosine
deaminase (ADA) deficiency in children with severe combined immunodeficiency (SCID).

We have recently[5] studied a child with ADA deficiency and
found dAMP, in addition to dADP and dATP, in the red cells.
Significant amounts of deoxyadenosine were found in the urine, as
in the previous case studied by us[1]. Deoxyadenosine and deoxynucleotides disappeared from the urine and red cells respectively
after bone marrow transplant, coincident with apparent clinical
improvement and restoration of immune function[5].

This paper describes studies of adenosine and deoxyadenosine
metabolism in the patient's erythrocytes on two separate occasions,
each about ten weeks after red cell transfusion. The effect on ADA
activity of storage of lysates at different temperatures was also
studied.

METHODS

The clinical history of the patient (L.S.) has been described[5].
She was given packed red cell transfusions after the marrow graft
following extensive immunological investigations. The metabolism
of 8-C^{14} adenosine and deoxyadenosine was studied in intact cells
by the method of Dean et al[6] on two occasions, each about ten weeks
after the last transfusion. The metabolites were separated by
high-voltage electrophoresis[7], or in the case of adenine deoxynucleotides by HPLC[5], and the radioactivity measured. The effect

of storage on ADA activity was examined after storing erythrocyte
lysates from normal subjects at room temperature, +4°C, -20°C and
-70°C, for several weeks.

RESULTS AND DISCUSSION

     Several interesting results have been obtained in these
studies of adenosine and deoxyadenosine metabolism in the red cells
of the ADA deficient patient, L.S.

     We have confirmed and extended the results of Korber et al[8]
that reliable measurements of ADA activity are dependent on the
length and temperature of storage of red cells. Storage at -20°C
resulted in 80% loss in activity in five days, whereas the enzyme
was stable for at least eleven days at room temperature, +4°C or
-70°C (Fig. 1). Consequently, ADA measurements should be done on
fresh cells only or on samples which have been properly stored.
All results in this paper refer to studies in fresh cells.

     Pre-treatment adenosine deaminase was almost negligible in
the patient's red cells (0.6 nmol/mgHb/h). The normal range for
our laboratory is 69.5±17.4 (n = 31). Packed cell transfusions
restored ADA activity to just below normal levels, which then fell

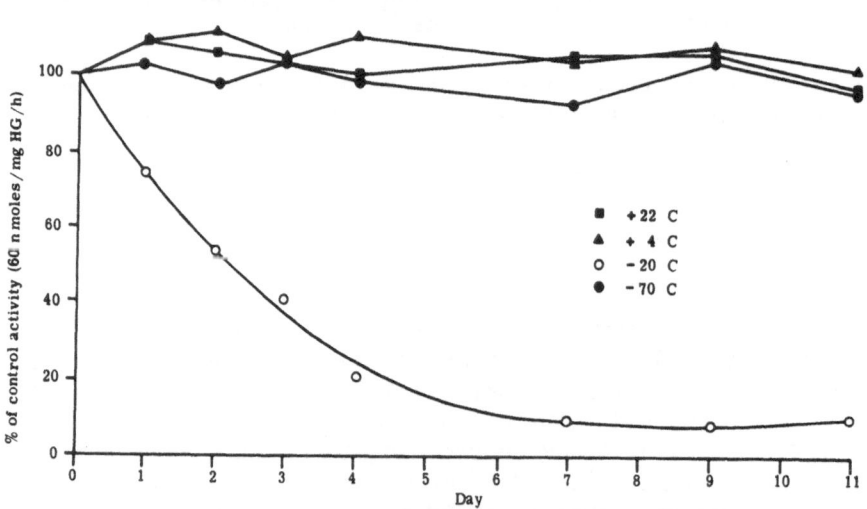

Fig. 1.   Adenosine deaminase activity during storage at different
          temperatures.
              +22°C;   +4°C;   -20°C;   -70°C.

Table 1. Erythrocyte adenosine deaminase activity on the days indicated following red cell transfusion after bone marrow transplant in patient L.S. All four deoxy compounds disappeared post marrow graft, coincident with restoration of immune function[5].

PATIENT (L.S.)			LYSATE ADA ACTIVITY (nmol/mgHb/h)	ERYTHROCYTE NUCLEOTIDE LEVELS (nmol/ml packed cells)						URINARY DEOXYADENOSINE (mmol/mmol creatinine)
				ATP	ADP	AMP	dATP	dADP	dAMP	
PRE-TREATMENT LEVELS			0.6	760	135	30	750	116	15	0.19
BONE MARROW TRANSPLANT										
DAYS AFTER RED CELL TRANSFUSION	1st	15	42.2	832	-	30	NOT DETECTABLE			
		29	36.9	1020	214	52				
		48	22.0	1133	133	16				
		67	9.2	1340	137	15				
	2nd	66	11.4	1260	131	10				
NORMAL VALUES			69.5 ±17.4 (n = 31)	1278 ±127	114 ±24	10 ±3				
				(n = 9)						

progressively with time (Table 1). However, 15% of normal activity was still present nearly ten weeks after transfusion on two separate occasions. This was somewhat surprising as erythrocyte ADA is reported to fall to almost undetectable levels four weeks after transfusion[2] or ten weeks after bone marrow transplantation[3]. This may indicate red cell chimerism; the percentage of donor cells, however, could not be quantified because of the close serological match between the patient and donor cells. Fibroblast ADA activity (≈5% control) was unchanged after marrow graft.

The most interesting finding was that, despite the low lysate activity, intact erythrocytes from the patient metabolised adenosine and deoxyadenosine, at concentrations up to 75 µM in a manner very similar to controls (Fig. 2). Adenosine was metabolised principally by deamination at high concentrations (above 10 µM) and by phosphorylation at the lower more physiological concentrations (1-10 µM). Deoxyadenosine, on the other hand, was poorly phosphorylated (1 mM Pi) at any concentration and was metabolised mainly by deamination (not shown), confirming that it is a very weak substrate for adenosine kinase. A kinase specific for deoxynucleosides has been purified from calf thymus[9] but it has not been reported in red cells. In

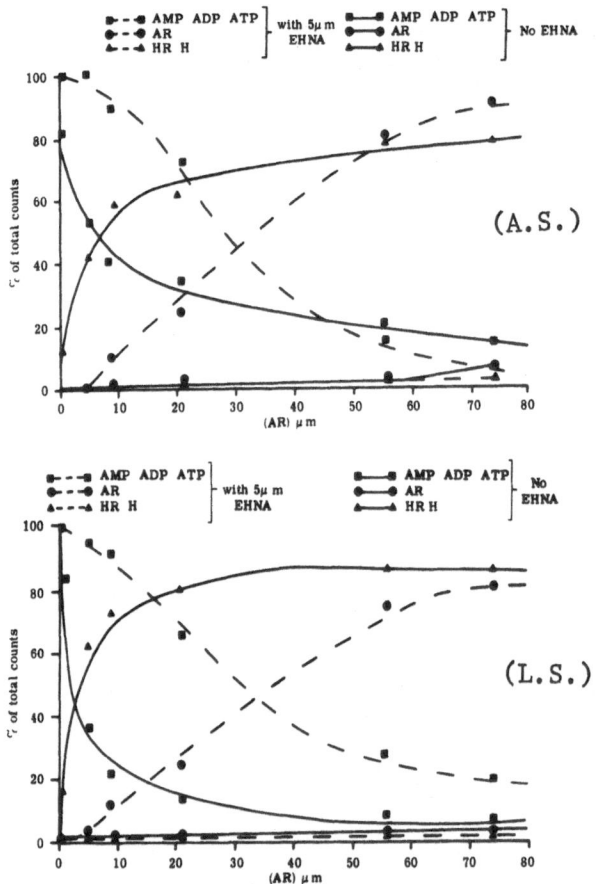

Fig. 2.   Adenosine metabolism in red cells from an ADA⁻ child
          (L.S.) post marrow graft and a control (A.S.), in the
          presence and absence of EHNA, an ADA inhibitor. A 30%
          cell suspension was incubated for 5 minutes in 1 mM Pi in
          Earl's medium at the adenosine concentration stated.

the presence of 5 μM EHNA, which completely inhibited ADA activity,
there was increased salvage of adenosine to adenine nucleotides at
the lower concentrations, but a large percentage of the substrate
remained unmetabolised at the higher concentrations (Fig. 2).
IMP synthesis was negligible under these conditions (1 mM Pi). No
significant formation of adenine from either nucleoside occurred
(with or without EHNA), confirming earlier studies in erythrocytes
deficient in APRT and the absence of any appreciable activity of
adenosine phosphorylase in intact red cells[6].

     These results show that studies on intact cells may give a
better indication of the in vivo capacity of ADA than cell extracts.
They may also explain why some patients with low but detectable

levels of ADA in red cell lysates are not immunodeficient[10]. In view of these findings, and the variability in ADA activity with storage[11] low ADA levels should initially be interpreted with reservation.

REFERENCES

1. H. A. Simmonds, A. Sahota, C. F. Potter and J. S. Cameron, Purine metabolism and immunodeficiency, Clin. Sci. Mol. Med., 54:579 (1978).

2. F. C. Schmalstieg, G. C. Mills, J. A. Nelson, L. T. May, A. S. Goldman and R. M. Goldblum, Limited effect of erythrocyte and plasma infusions in adenosine deaminase deficiency, J. Pediat., 93:597 (1978).

3. S-H. CHen, H. D. Ochs, C. R. Scott, E. R. Giblett and A. J. Tingle, Adenosine deaminase deficiency, J. Clin. Invest., 62:1386 (1978).

4. J. Donofrio, M. S. Coleman, J. J. Hutton, A. Daoud, B. Lampkin and J. Dyminski, Overproduction of adenine deoxynucleosides and deoxynucleotides in adenosine deaminase deficiency with severe combined immunodeficiency, J. Clin . Invest., 62:884 (1978).

5. H. A. Simmonds, A. Sahota, C. F. Potter, D. Perrett, K. Hugh-Jones and J. D. Watson, Purine metabolism in adenosine deaminase deficiency, in: "Enzyme Defects and Immune Dysfunction", Ciba Foundation 68 (New Series), Elsevier, Amsterdam, pp252 (1979).

6. B. M. Dean, D. Perrett, H. A. Simmonds, A. Sahota and K. J. Van Acker, Adenine and adenosine metabolism in intact erythrocytes deficient in adenosine monophosphate pyrophosphate phosphoribosyltransferase: a study of two families, Clin. Sci. Mol. Med., 55:407 (1978).

7. H. A. Simmonds, Two-dimensional thin-layer high voltage electrophoresis and chromatography for the separation of urinary purines, pyrimidines and pyrazolopyrimidines, Clin. Chim. Acta, 23:319 (1969).

8. W. Korber, E. B. Meisterernst and G. Hermann, Quantitative measurement of adenosine deaminase from human erythrocytes, Clin. Chim. Acta, 63:323 (1975).

9. V. Krygier and R. L. Momparler, Mammalian deoxynucleoside kinases. II. J. Biol. Chem., 246:2752 (1971).

10. T. Jenkins, Red blood cell adenosine deaminase activity in a 'healthy' ! Kung individual, Lancet, 2:736 (1973).

11. P. P. Trotta, E. M. Smithwick and M. E. Balis, A normal level of adenosine deaminase activity in the red cell lysates of carriers and patients with severe combined immunodeficiency disease, Proc. Nat. Acad. Sci. U.S.A., 73:104 (1976).

COMPLETE ADENOSINE DEAMINASE (ADA) DEFICIENCY WITHOUT IMMUNODEFI-

CIENCY, AND PRIMARY HYPEROXALURIA, IN A 12-YEAR-OLD BOY

J.L. Perignon, M. Hamet, P. Cartier, C. Griscelli

Laboratoire de Biochimie et Département de Pédiatrie

CHU Necker-Enfants Malades, Paris

Adenosine deaminase (ADA) deficiency was the first described biochemical defect giving rise to an immune deficiency (1) ; ADA deficiency is present in about 25 to 30 % of patients with severe combined immunodeficiency. Conversely, three patients have been reported to present an ADA deficiency without detectable immunodeficiency, but detailed studies of the immune status have been published for two of them only (2, 3). We had the opportunity to study an observation of ADA deficiency not associated with immunodeficiency, in a child who presented a primary hyperoxaluria. This association, not previously described, of two rare metabolic disorders, raise both the problem of an eventual relation between the two defects, and that of the correlation between ADA deficiency and immunodeficiency.

CASE REPORT

K.S. is a twelve-year-old male Caucasian Algerian child born on November 1965, without any known pathological history during the pregnancy nor the first years of life. He was submitted to legal vaccinations including BCG and vaccine without any adverse reaction, and presented several eruptive diseases with simple evolution. From the age of 6, he was repeatedly hospitalized in Algiers because of urinary tract infections and urolithiasis, and finally referred to us at the age of 12. Physical examination showed a short-statured boy (third percentile after French standards), apparently well but for a right otitis media. There was no hepatosplenomegaly ; peripheral lymph nodes were normaly present and tonsils were large. On four occasions, the child emitted calculi, which were analysed (cf. infra).

Laboratory investigations showed an alteration of glomerular

(Inulin clearance 58 ml/mn/1.73 m^2) and tubular functions (DDAVP concentration test, acidification test). The weight of the stools and the fecal fat were normal.

X-ray examination showed nephrolithiasis and nephrocalcinosis ; there was no osseous alterations imputable to oxalosis, whereas minor metaphyseal abnormalities, i.e. little spicules perpendicular to the bone axis, were seen on the femora and humeri ; ribs, pelvis, and vertebrae were normal.

METABOLIC STUDY

Calculi consisted in 50 to 100 % monohydrated calcium oxalate. Serum oxalate, urinary oxalate and glycolate were measured by the technique of Charransol et al. (4). On 3 determinations, serum oxalate level was 1.30, 5.50 and 5.50 mg/l (normal 3,00) ; oxaluria was 150 mg/24 hours (normal $\leqslant$ 50), and urinary glycolic acid 145 mg/24 hours (normal 15 to 60).

ENZYMATIC STUDY

Enzyme activities were assayed by radio-isotopic methods (5). ADA activity in erythrocytes was 2 to 3 nmol/minute/ml RBC (normal 494 $\pm$ 61) ; in lymphocytes 0.27 nmol/minute/10^6cells (normal 1.9 $\pm$ 0.6) ; in jejunum 16 nmol/minute/mg of protein (normal 63), and in fibroblasts 3.1 nmol/minute/mg protein (normal 14 to 118, mean 50). Other RBC enzymes activities tested (hypoxanthine-guanine-phosphoribosyltransferase, adenine phosphoribosyltransferase, purine nucleoside phosphorylase (PNP), adenosine kinase, uridine kinase, 5'-pyrimidine nucleotidase) were normal.

IMMUNOLOGICAL STUDY

Lymphocytes counts were within normal values (1800 to 2900/mm^3). Percentage of T lymphocytes detected by rosetting with sheep erythrocytes was normal (82 %) and percentage of B lymphocytes, detected by rosetting with sensitized sheep erythrocytes in the presence of complement and by surface immunoglobulin studies using a polyvalent fluorescent F(ab) $_2$ antiserum, was normal, respectively 16 and 11 %. Surface immunoglobin studies with monospecific antisera showed a normal distribution of B lymphocytes bearing IgM (5 %), IgD (4 %), IgG (1 %) and IgA (1 %). Serum immunoglobulin levels determined by a radial immuno-diffusion method using monospecific antisera anti α, γ and μ -heavy chains, or, for IgE, by radioimmunoassay (Pharmacia), were normal or increased for all classes : IgG 19 g/l (normal 11 $\pm$ 2) ; IgM 1.5 g/l (normal 0.8 $\pm$ 0.3) ; IgA 2 g/l (normal 1.2 $\pm$ 0.3) ; IgD 0.08 g/l (normal $\leqslant$ 0.1) and IgE 274 U/ml (normal 60 $\pm$ 40). Antibody synthesis was normal : anti-A and anti-B allohemagglutinin levels were 1:128 ; antibodies after a booster injection increased for Tetanus (1 to 4 IU), Diphteria (1:32 to 1:256), H. Pertussis (1:64 to 1:128) and Polioviruses (1:64 to 1:128 for the three viruses).

Skin-delayed reactivity to antigens (streptokinase - streptodornase, candidine, PPD) and mitogens (PHA) was normally positive. Proliferative responses to mitogens or to antigens were comparable to controls (non stimulated 300 cpm ; PHA 94.10^3, PWM 22.10^3, Con A 71.10^3, PPD 22.10^3, CD 13.10^3 cpm). Mixed leukocyte reaction (MLR) was positive (non stimulated cells 600 cpm ; patients-irradiated stimulant cells 23.10^3 cpm). Generation of cytotoxic cells during MLR (6) was also normal, comparable to controls. The patient's leukocytes cultured in the presence of PWM (7) were capable to mature into immunoglobulin containing cells (IgM 116, IgG 178, IgA 110 %o) comparably to normal leukocytes (IgM 90, IgG 90, IgA 140 %o).

FAMILY STUDY

The father and the mother are first-cousins ; the patient is the 11th of 12 children. Two children died one from diarrhea at one year of age, the second from dehydration when 3 months old. A maternal aunt has recurrent nephrolithiasis. None of the other members of the family has clinical features of immune dysfunction nor symptomatology of nephrolithiasis.

RBC ADA activities and serum oxalate levels are indicated on Fig. 1 : III.4 (father), III.11 (maternal aunt) are heterozygotes for ADA deficiency, whereas III.7 (mother) has low-level activity, but more than 50 % of normal activity, an opportunity already mentioned (8) ; IV.9 (sister) is homozygote for ADA deficiency. PNP and adenosine kinase activities were normal in all the hemolysates. Serum oxalate was abnormally high in six members of the family.

DISCUSSION

We report an observation associating a primary hyperoxaluria and an ADA deficiency without immunological disorder.

The diagnosis of primary hyperoxaluria was based upon coexistence of calcium oxalate nephrolithiasis and nephrocalcinosis, hyperoxaluria and hyperoxalemia, without any clinical or biological evidence of intestinal malabsorption. The high excretion rate of glycolate is consistent with type I primary hyperoxaluria (9). A treatment with pyridoxine (250 mg/24 hours) during two months failed to produce any reduction of oxalate excretion.

The association between primary hyperoxaluria and ADA deficiency appears to be fortuitous since : a) the genetic study disclosed subjects heterozygotes for ADA deficiency with normal serum oxalate level, and, conversely, subjects with normal ADA activity and elevated serum oxalate level (serum oxalate however, is not an accurate means of detection of heterozygotes for primary hyperoxaluria) ; b) we found normal ADA activity in RBC of three patients with oxauria and, conversely, normal serum oxalate and oxaluria in two patients with ADA deficiency and immune disorder (data not shown) ; c) ADA activity in our patient was the same when measured in dialysed hemolysate, and addition of oxalate did not affect ADA activity

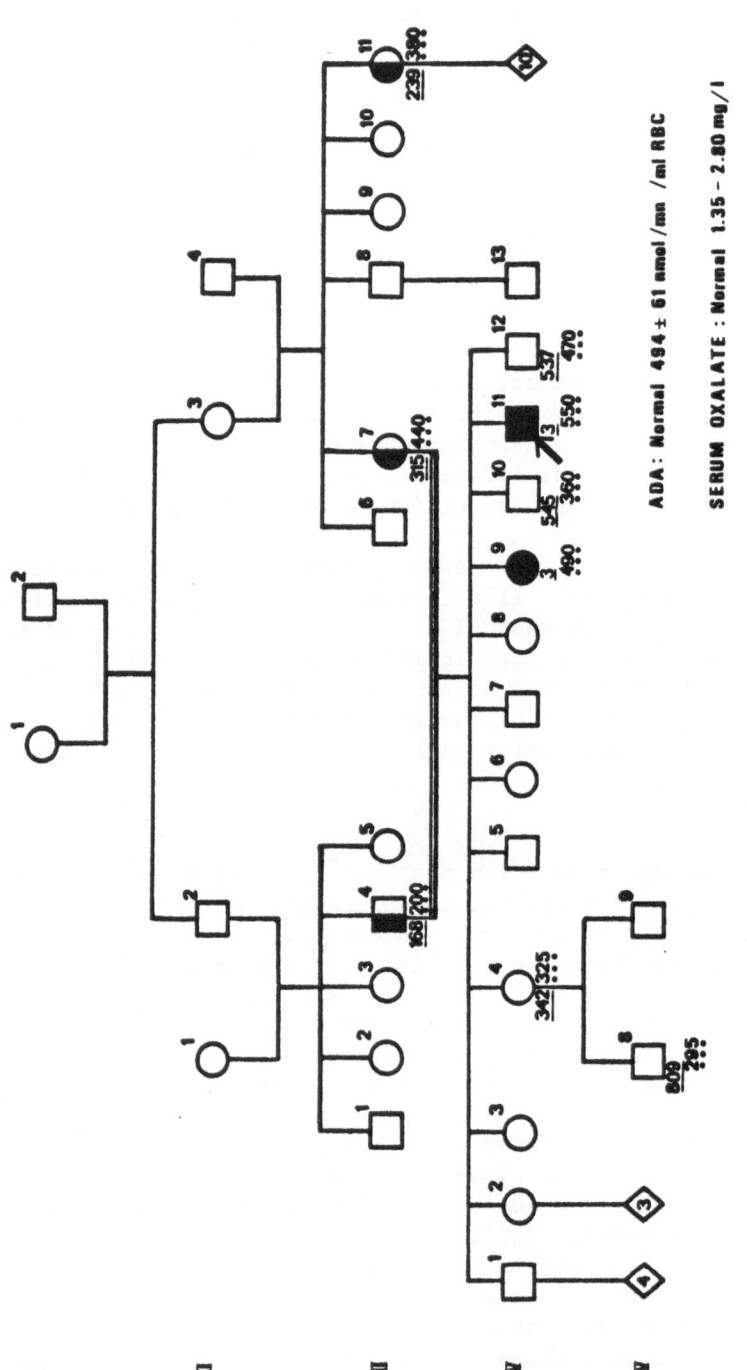

ADA : Normal 494 ± 61 nmol / mn / ml RBC

SERUM OXALATE : Normal 1.35 - 2.80 mg / l

Fig. 1   Pedigree of affected family

in control hemolysates.

This observation raises the problem of the correlation between ADA activity and immune function. Recent data published by Hirschhorn et al. (10) about an ADA-deficient child with SCID prove that a non negligible tissular residual ADA activity (liver:31 % of normal) is not sufficient to prevent immunodeficiency. Conversely, this observation and those of Jenkins (2) and Hirschhorn (3) show that normal RBC ADA activity is not absolutely necessary to insure normal immunocompetence. ADA activity in lymphocytes of immunodeficient patients is difficult to determine because of the lymphopenia presented by such patients. Reported residual activities in lymphocytes range from 0 to 6 % of normal values (11) ; in a post-mortem study of an affected child, Hirschhorn et al. (10) found ADA activities in the thymus, the spleen and a lymph node which were 0.2, 1.1 and 3.6 % of normal values, respectively. Finally, lymphocyte ADA activity appears to be the critical point, and the finding, in K.S., of a lymphocyte ADA reaching 15 % of normal activity, might reasonably account for the normal immune status of this patient.

REFERENCES

1 E.R. Giblett, J.E. Anderson, F. Cohen, B. Pollara and H.J. Meuwissen : Adenosine deaminase deficiency in two patients with severely impaired cellular immunity. Lancet 2:1067 (1972).
2 T. Jenkins, A.R. Rabson, G.T. Nurse, A.B. Lane and D.A. Hopkinson : Deficiency of adenosine deaminase not associated with severe combined immunodeficiency. J. Pediat. 89:732 (1976).
3 R. Hirschhorn, W. Borkowsky, S. Bajaj and A. Gershon : Adenosine deaminase deficiency in a normal child : immunologic and biochemical studies. Clin. Res. 26:552 A (1978).
4 G. Charransol, C. Barthelemy and P. Desgrez : Rapid determination of urinary oxalic acid by gas-liquid chromatography without extraction. J. Chromatogr. 145:452 (1978).
5 P. Cartier and M. Hamet : Dosage de l'activité adénosine désaminasique dans les érythrocytes et les lymphocytes humains. Clin. Chim. Acta 71:429 (1976).
6 Griscelli, C., A. Durandy, D. Guy-Grand, F. Daguillard, C. Herzog. and M. Prunieras : A syndrome associating partial albinism and immunodeficiency. Am. J. Med. 65:691 (1978).
7. L.Y. Wu, A.R. Lawton and M.D. Cooper : Differentiation capacity of cultured B lymphocyte from immunodeficient patient. J. Clin. Invest. 52:3180 (1973).
8 R. Hirschhorn : Adenosine deaminase deficiency and immunodeficiency. Fed. Proc. 36:2166 (1977).
9 H.E. Williams and L.H. Smith, Jr. : Primary hyperoxaluria, in Stanbury J.B., Wyngaarden, J.B. and Fredrickson, D.S. editors : The metabolic basis of inherited disease, 4th edition, New York, 1978, McGraw-Hill, pp 182-204.

10 R. Hirschhorn, F. Martiniuk and F.S. Rosen : Adenosine deaminase
   activity in normal tissues and tissues from a child with severe
   combined immunodeficiency and adenosine deaminase deficiency.
   Clin. Immunol. Immunopathol. 9:287 (1978).

11 S.H. Polmar, R.C. Stern, A.L. Schwartz, E.M. Wetzler, P.A.
   Chase and R. Hirschhorn : Enzyme replacement therapy for adeno-
   sine deaminase deficiency and severe combined immunodeficiency.
   New Engl. J. Med. 295:1337 (1976).

12 A. Cohen, R. Hirschhorn, S.D. Horowitz, A. Rubinstein, S.H.
   Polmar, R. Hong and D.W. Martin Jr. : Deoxyadenosine triphosphate
   as a potentially toxic metabolite in adenosine deaminase deficiency. Proc. Natl. Acad. Sci. USA 75:472 (1978).

METABOLISM OF ADENOSINE AND DEOXYADENOSINE

BY STORED HUMAN RED CELLS

Grant R. Bartlett

Laboratory for Comparative Biochemistry
San Diego, California 92109

INTRODUCTION

A striking feature of human adenosine deaminase deficiency is the accumulation in the circulating erythrocyte of a large amount of deoxyATP,[1-3] a change which does not appear to interfere with the viability or oxygen transport function of the cell. Even if. the red cell can not serve to give direct information on the pathogenesis of this disease, it does provide a convenient tool for its assay and for studies on cellular mechanisms involved in the metabolism of purine nucleosides.

Several years ago my laboratory carried out studies which demonstrated that fresh and stored human red cells could readily metabolize adenosine to lactate with a simultaneous conversion of part of the intact nucleoside into ATP.[4] We also found that metabolism of deoxyadenosine by the human red cell led to a considerable synthesis of deoxyATP.[5] We have recently returned to this study, which has important implications to red cell hematology as well as to disorders of purine metabolism in man, and the following is a progress report of the results.

MATERIALS AND METHODS

Red cells from human blood which had been stored for eight weeks under the usual blood-banking conditions were used for these studies for three reasons:[6] 1) It appeared that this was the longest time of storage compatible with significant rejuvenation of viability associated with regeneration of ATP. 2) The eight-week stored red cells, although unable to use glucose, still

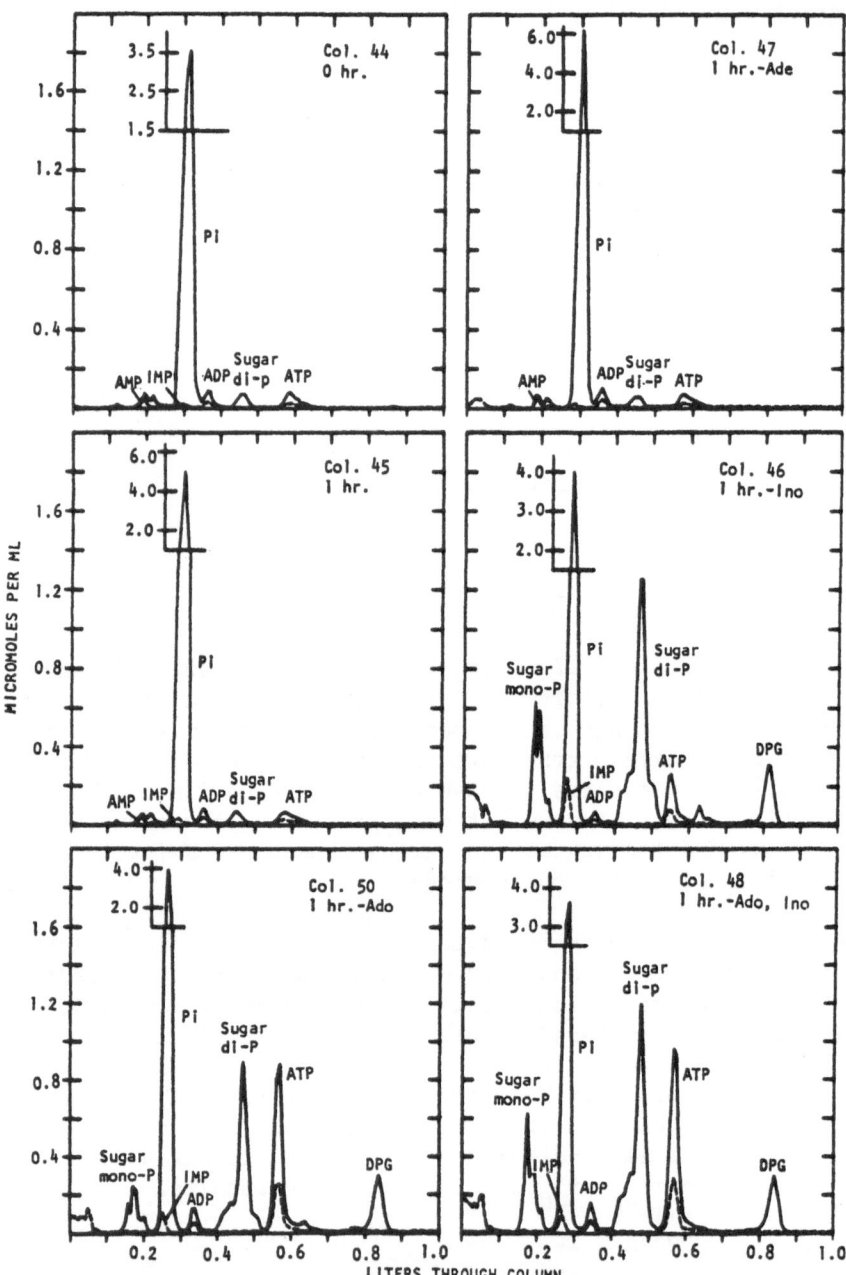

Figure 1. Phosphate compounds in stored red cells after incubation
at 37° C in isotonic pH 7.4 saline-phosphate with various combina-
tions of inosine (Ino), adenine (Ade), adenosine (Ado) and inorganic
phosphate (Pi). Dowex 1-formate column chromatography. Phosphate,
20 mM; adenine, 1.0 mM; inosine, 5 mM; adenosine, 5 mM. Total
phosphorus, solid line; A at 260 nm, dash line.

showed a high capacity for the metabolism of nucleosides. 3) Most
of the original pools of organic phosphate metabolites had disappeared during the storage making it easier to follow net synthesis
of metabolites from added substrates.

Phosphate metabolites were isolated by ion-exchange column
chromatography from trichloroacetic or perchloric acid extracts
of red cells separated from blood which had been stored for eight
weeks and then incubated under different conditions with various
mixtures of inorganic phosphate, adenine, inosine, adenosine,
2-deoxyinosine and 2-deoxyadenosine.[7] The most useful protocol
for demonstration of details of the metabolic activity was by
incubation of a 30% suspension of red cells in 20 mM pH 7.4
phosphate buffer for one hour at 37° with 5 mM nucleoside with
and without 1 mM adenine.

RESULTS

Results of the metabolic studies with the purine ribonucleosides are summarized in Figure 1. Red cells incubated in phosphate
without additions, or with adenine alone, showed no changes from
the original content of metabolites which consisted of traces of
adenylates and glycolytic intermediates. Added inosine was removed
completely with the purine portion accumulating as free hypoxanthine
and with a variety of phosphorylated intermediates and lactic acid
formed from the ribose part. There was little or no increase in
total adenylate. The incubation with a combination of inosine and
adenine brought about a synthesis of ATP to a level equal to or
above that of the fresh red cell. This process was extremely
efficient in that more than 90% of the added adenine was incorporated into ATP. Non-nucleotide intermediates were much the same
as found with inosine alone with the exception of a marked reduction
in the pool of PRPP in the presence of adenine. In the adenosine
experiment all of the nucleoside disappeared with the base appearing in hypoxanthine, a very small amount of inosine, and adenylate.
Most of the adenylate was ATP which reached a concentration in the
range of that of the fresh red cell. The ribose part of adenosine
which did not move into inosine and adenylate was changed to a
variety of phosphate intermediates and lactate. Intracellular
intermediates which accumulated during the metabolism of inosine
and adenosine included relatively large pools of fructose diphosphate and ribose 5-phosphate and lesser amounts of other pentose
phosphates, triose phosphates, phosphoglycerates, and sedoheptulose and octulose phosphates.

The stored red cells readily metabolized deoxyinosine and
deoxyadenosine under the same conditions as described for the
ribonucleosides. Figure 2 shows a chromatograph of phosphate
compounds found in the stored human red cells following incubation

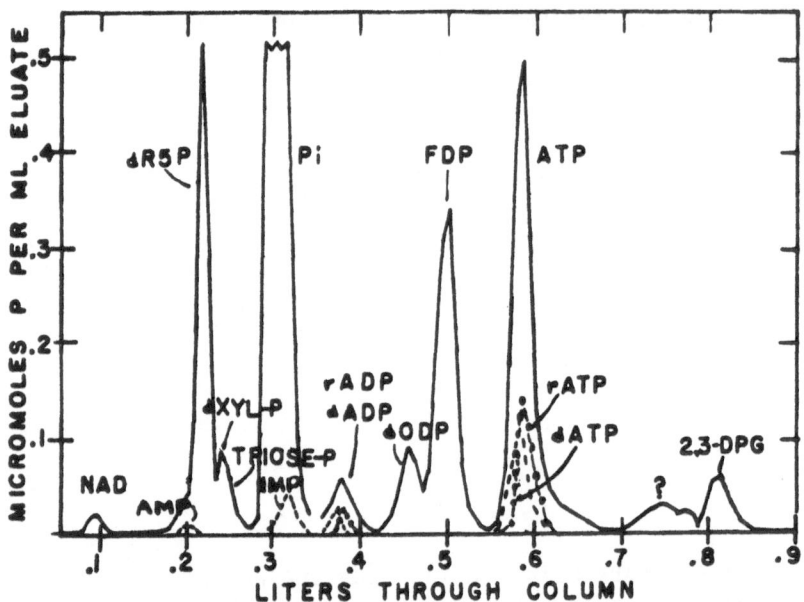

Figure 2. Phosphate compounds in stored red cells after incubation
for one hour at 37°C in pH 7.4 saline-phosphate (20 mM) with
deoxyadenosine (5mM). Dowex 1-formate column chromatography.
Solid line = phosphorus.

with deoxyadenosine. By the end of one hour of incubation most
of the nucleoside had disappeared with the sugar part metabolized
to a variety of phosphorylated intermediates, acetaldehyde and
lactate. Deoxyinosine gave no synthesis of adenylate with or
without the addition of adenine. The incubation with deoxyadenosine however led to the formation of a large pool of adenine
nucleotide triphosphate which was found to contain roughly equal
amounts of ATP and deoxyATP. Non-nucleotide intermediates which
accumulated in the red cell during the incubations with deoxyinosine or deoxyadenosine included fructose diphosphate, deoxyribose 5-phosphate, triose phosphate, phosphoglycerates, deoxy-
xylulose 1-phosphate and deoxyoctulose diphosphate.

DISCUSSION

The non-nucleotide metabolites found in the red cell following
an incubation with inosine or with adenosine were those expected
for the metabolism of ribose phosphate, produced by phosphorolysis
of the nucleoside, through pentose shunt and glycolytic pathways.
A relatively slow oxidation of triose phosphate by NAD, as compared
to earlier steps, would account for the accumulation of fructose,

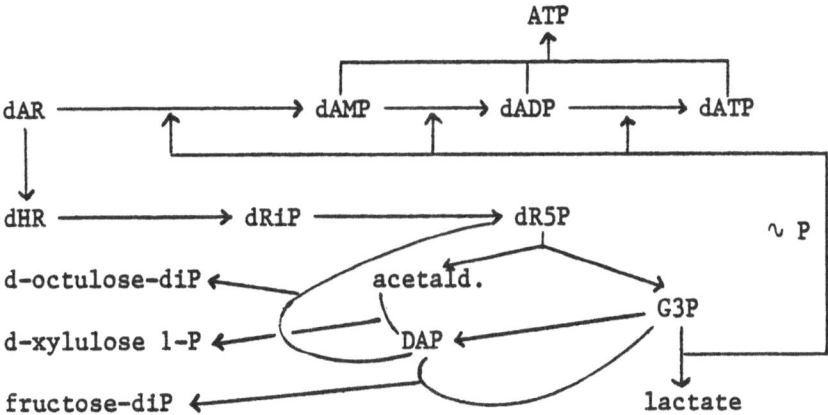

sedoheptulose and octulose diphosphates. This was supported by
the fact that the addition of pyruvate led to a reduction in the
amounts of these intermediates and an increase in lactate and
2,3-diphosphoglycerate.

     The non-nucleotide intermediates which were found in the
incubations with the deoxyribonucleosides could be explained by
reaction of deoxyribose 5-phosphate with the known dR5P-aldolase
to give acetaldehyde and triose phosphate. A slow reaction at the
glyceraldehyde 3-P dehydrogenase step would account for the
accumulation of fructose diphosphate, xylulose-phosphate and
deoxyoctulose diphosphate by an aldolase reaction of dihydroxy-
acetone phosphate with glyceraldehyde 3-P, acetaldehyde and
deoxyribose 5-P. The amount of acetaldehyde which was found was
considerably less than expected indicating the presence of an as
yet unidentified additional metabolic pathway for this compound.

     The ATP formed during the incubation of the stored red cells
with adenosine and the deoxyATP from deoxyadenosine was a result,
most probably, of initial phosphorylation of the nucleoside by the
trace of ATP present, catalyzed by adenosine kinase, with subsequent conversion of the resulting AMP and deoxyAMP to the triphosphates by phosphorylations derived from the metabolism of ribose
and deoxyribose phosphates to lactate. For these reactions to
have taken place required a remarkably close synchronization
between the sugar catabolic and the nucleotide synthetic pathways.
Another interesting aspect of the metabolic control system was
that as adenylate was synthesized an equilibrium was maintained
to give a very high ratio of ATP to ADP and of ADP to AMP so that,
although the nucleotide monophosphate was the first product, the
final mixture contained an overwhelming preponderance of the
triphosphate.

A remarkable finding of this study was that the metabolism of deoxyadenosine not only produced a large amount of deoxyATP but also gave an almost equivalent amount of ATP. We have little information yet on the nature of this hitherto unknown reaction. It is well established that deoxyribonucleotides are formed by a reductase enzyme system from ribomononucleotides, a process not considered to be reversible. Formation of ribose phosphate from any of the non-nucleotide intermediates did not seem possible and it seemed likely that ribose-adenylate was formed by direct oxidation of deoxyribose-adenylate.

ACKNOWLEDGMENT

Supported by Grant HLB-6931 from the National Heart, Lung, and Blood Institute.

REFERENCES

1. M.S. Coleman, J. Donofrio, J.J. Hutton, L. Hahn, A. Daoud, B. Lampkin and J. Dyminski (1978) Identification and quantitation of adenine deoxynucleotides in erythrocytes of a patient with adenosine deaminase deficiency and severe combined immunodeficiency. J. Biol. Chem. 253:1619-1626.
2. J. Donofrio, M.S. Coleman, J.J. Hutton, A. Daoud, B. Lampkin and J. Dyminski (1978) Overproduction of adenine deoxynucleosides and deoxynucleotides in adenosine deaminase deficiency with severe combined immunodeficiency disease. J. Clin. Invest. 62:884-887.
3. A. Cohen, R. Hirschhorn, S.D. Horowitz, A. Rubinstein, S. Polmar, R. Hong and D.W. Martin, Jr. (1978) Deoxyadenosine triphosphate as a potentially toxic metabolite in adenosine deaminase deficiency. Proc. Natl. Acad. Sci. USA 75:472-476.
4. G.R. Bartlett and G. Bucolo (1968) The metabolism of ribonucleoside by the human erythrocyte. Biochim. Biophys. Acta 156:240-253.
5. G.R. Bartlett (1968) The metabolism of deoxyribonucleoside by the human erythrocyte. Biochim. Biophys. Acta 156:254-265.
6. G.R. Bartlett (1974) Red cell metabolism: Review highlighting changes during storage. In The Human Red Cell In Vitro, T.J. Greenwalt and G.A. Jamieson, Eds. Grune and Stratton, New York, pp. 5-29.
7. G.R. Bartlett (1968) Phosphorus compounds in the human erythrocyte. Biochim. Biophys. Acta 156:221-230.

ADENOSINE DEAMINASE AND PURINE NUCLEOSIDE PHOSPHORYLASE ACTIVITIES

DURING CULTURING OF FIBROLASTS

M.P. Uitendaal[+], F.T.J.J. Oerlemans[o], C.H.M.M.
de Bruyn[o], T.L. Oei[o] and P. Hösli[*]
[+]Cardiochemical Lab., Thoraxcentre, Erasmus University,
P.O.B. 1738, Rotterdam, The Netherlands
[o]Department of Human Genetics, Faculty of Medicine,
University of Nijmegen, Nijmegen, The Netherlands
[*]Department of Molecular Biology, Institut Pasteur,
25, Rue du Dr. Roux, Paris 15, France

INTRODUCTION

Fibroblast cultures can be used as a valuable model for the
study of inborn errors and as an important diagnostic tool, e.g.
amniotic fluid derived fibroblasts in prenatal diagnosis (1-3).
In order to develop reliable enzyme assays of fibroblasts the effect
of cell culture conditions on enzyme activities should be known.
In addition, trypsinisation of fibroblasts for harvesting and subsequent enzymatic analysis causes uncontrolled protein loss.
Therefore, expressing specific activities of trypsinised cells on
a protein basis might have disturbing effects and thus broaden
the range of specific enzyme activities.

When small fluctuations in gene expression are to be studied
other ways of expressing the enzyme activity should be chosen
(e.g. DNA content, cell number). In order to reduce errors caused
by protein loss during trypsinisation in the present study the
fibroblasts were lyophilised in situ. Activities of some purine
enzymes were determined using conventional "macro"-methods
(endvol. 30 μl) with a suspension of the lyophilised cells and
using ultramicro-methods with small numbers (five cells per
incubation in 0.3 μl) of visually selected lyophilised fibroblasts
(4,5,6). The activities of adenosine deaminase (E.C. 3.5.4.4, ADA)
purine nucleoside phosphorylase (E.C. 2.4.2.1, NP) and hypoxanthine
phosphoribosyltransferase (E.C. 2.4.2.8, HPRT) were measured at
different days after seeding of normal and HPRT deficient cells.

In addition, ADA and NP activity measurements were done in different
passages of normal fibroblast cultures.

MATERIALS AND METHODS

Fibroblast cultures were established either from forearm skin
biopsies from healthy volunteers or from the skin of a HPRT deficient fetus (2). Stored under liquid nitrogen in early passages
(pass. 5-7), cells were thawed and cultured in HAM F 10 medium
(Difco) supplemented with 15% fetal calf serum, (Gibco) and 100
I.U. pencillin and 100 mg/ml streptomycin (both from Difco).
Cells were seeded at an initial density of 3.10^5 cells/25 cm^2. The
specific culture conditions (4,5) and ultramicro-enzyme assays
have been described previously (7,8). For the "macro"-assays the
medium was removed from the growing cultures, followed by three
rinses with physiological saline. The cells, still attached to
the bottom of the culture flasks (Falcon; 25 cm^2), were quickly
frozen at -20 $^\circ$C and subsequently lyophilised. The lyophilisate
was suspended in distilled water (250 - 750 μl, depending on the
cell density). Final incubation volume was 30 μl; the same concentrations as in the ultramicro radiochemical assays were used
(1,7-9). Protein determinations were performed with the Folin-
Ciocalteu reagent using bovine serum albumine as a standard (10).

RESULTS AND DISCUSSION

Cells with the same passage number from a control strain
were seeded into twenty culture flasks with equal densities. At
different times after seeding cells from two flasks were taken
for analysis of protein content, ADA, NP and HPRT activities.
Protein content gradually increased (Fig. 1), even after
confluency was reached between 72 hr and 92 hr after seeding.
In the latter period a dramatic increase in ADA and NP activity
but not in HPRT activity was observed. The ADA and NP activity
decreased during further culturing in confluency, whereas
HPRT activity rose gradually. The NP activity shown in fig. 1 is
the activity with inosine as a substrate. The reverse reaction, with
hypoxanthine and ribose-1-phosphate as substrates, showed essentially
the same pattern.

In a similar experiment the activities of these three enzymes
with HPRT deficienct fibroblasts are shown in table 1. The residual
HPRT activity was approximately 5% of normal. The increase in ADA
activity at confluency was more pronounced than in the case of
control cells, whereas the increase in NP activity was less
pronounced. It should be noted that the degree of confluency can
not be controlled completely. Therefore, the differences between

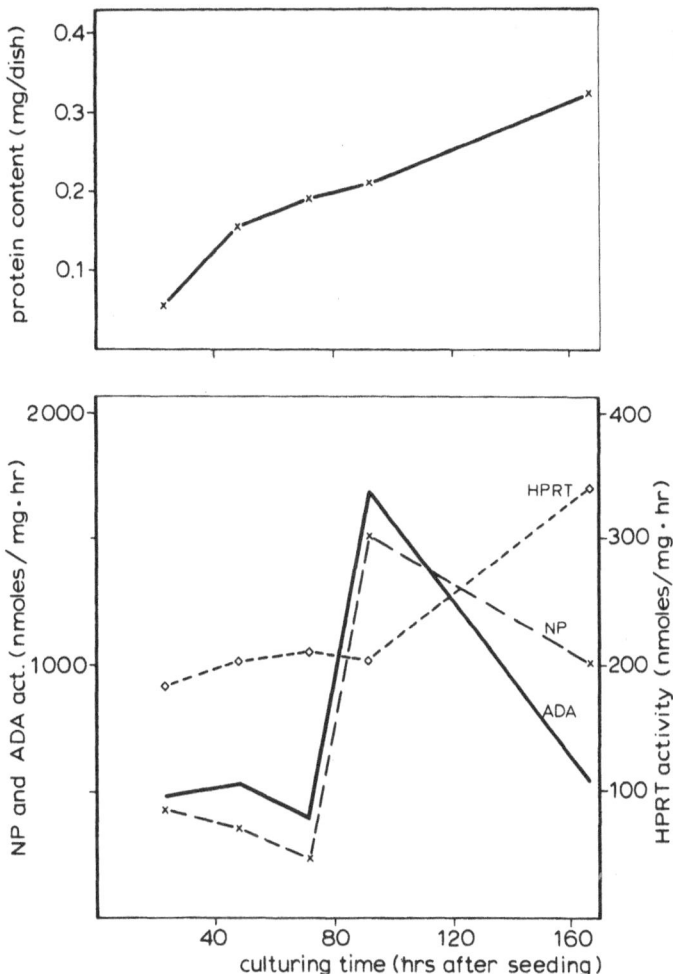

Fig 1. Activity of three enzymes of purine metabolism (lower part) and protein content (upper part) of control fibroblasts at different times after seeding. Initial cell concentration was 5 x 10^5 cells/25 cm^2. Confluency was reached after 72-92 hr.

control and HPRT deficient cells can only be interpreted in qualitative terms. More detailed studies, using smaller intervals and more cell strains, are needed before conclusions can be drawn regarding the differing ADA and NP activity patterns. However, it can already be stated that there is a clear tendency of a several fold rise in ADA and PNP activity when confluency is reached.

Table 1. Activity of Three Enzymes of Purine Metabolism in HPRT
         Deficient Fibroblasts at Different Times after Seeding.

time after seeding	H-PRT	ADA	NP-Hx
23 hr	14.0	735	245
48 hr[*]	9.7	560	225
72 hr	8.9	1229	224
92 hr	8.7	4445	644
167 hr	11.3	1581	574

[*] For routine determinations cells are lyopholised 48 hr. after
seeding.
Value of a normal control strain for HPRT is 203 nmoles/mg prot.
hr; for ADA and NP-Hx 526 and 353 nmoles/mg prot. hr, respectively.

Table 2. Ratios of Highest vs. Lowest Enzyme Activities Measured
         in Different Subseedings of Two Control Fibroblast
         Strains.

cell strain	passage no.	number of dishes tested	ADA H/L ratio	NP-Hx[*] H/L ratio
I	7,8,12 and 13	5	3.7	1.4
II	7,9,12,13 and 17	8	2.6	1.5

Enzyme determinations on each dish were carried out in fivefold
with five cells per incubation. For each dish the mean value was
calculated. This mean value was compared with that of the other
dishes of the same cell strain.
[*] NP activity determined with hypoxanthine and ribose-1-phosphate as
substrates.

When using conventional "macro"assays it seems preferable to lyophilise rather than trypsinise cells before use. Not only uncontrolled protein loss is prevented but lyophilisation also yields higher specific enzyme activities. For instance HPRT, a relatively stable enzyme, displays a mean activity of approximately 200 nmoles/mg protein.hr. under our conditions, whereas values of 120 - 160 nmoles/mg prot.hr. are reported from laboratories with similar analytical procedures except for the trypsinisation (11-13). Moreover measurements should be done on standard times after seeding, preferably before confluency is reached: only insignificant fluctuations are observed up to 72 hr.

The present procedure settles with two disturbing factors: the cells are lyophilised in situ 48 hrs after seeding, before confluency. Moreover the enzyme activity is expressed on a per cell basis. The advantages of expressing enzyme activities in this way have been discussed elsewhere (4,6,14). Using the ultramicro-assay, ADA and NP activities were measured in different passages of two control cell strains (always 48 hrs after seeding). In contrast to NP, a considerable difference of ADA activity was found between the various passages. This is illustrated by table 2, where the ratio of the highest and lowest ADA and NP activities are listed. From cell strain I four different passages were tested. Here the highest/lowest ADA ratio was 3.7. For NP this ratio was 1.4. From cell strain II five different passages were tested (total 8 dishes). The ratio of the highest and lowest ADA activities measured was 2.6; the corresponding value for NP was 1.5. Although culture conditions were kept as constant as possible with respect to medium, serum, pCO_2, cell density, time of harvesting, etc., it was not possible to eliminate the factors that are responsible for the fluctuations in ADA activity in a genetically homogeneous cell population.

As has been discussed previously (8) the fluctuating ADA activities in cultured fibroblasts constitute a possible source of error in both experimental and diagnostic procedures, e.g. in prenatal diagnosis where normal, heterozygous and mutant cells must be clearly discrimated.

REFERENCES

1. M.P. Uitendaal, T.L; Oei, C.H.M.M. de Bruyn and P. Hösli, Biochem. Biophys. Res. Commun. 71:574 (1976).
2. P. Hösli, C.H.M.M; de Bruyn, F.T.J.J. Oerlemans, M. Verjaal and R.E; Nobrega. Hum. Genet. 37:195 (1977).
3. C.H.M.M. de Bruyn, M.P. Uitendaal, T.L. Oei, S.J. Geerts and P. Hösli. In: Models for the Study of Inborn Errors of Metabolism (F. Hommes, Ed.). Elsevier/North Holland. Biomedical Press, Amsterdam, 179 (1979).

4. P. Hösli. In: Birth Defects (A.G; Motulsky and W. Lenz, Eds.).
    Excerpta Medica, Amsterdam p. 226 (1974).

5. P. Hösli. Clin. Chem. 23:1476 (1977).

6. P. Hösli and C.H.M.M. de Bruyn. Adv. Exp. Med. Biol. 76A:
    591 (1977).

7. M.P. Uitendaal, C.H.M.M. de Bruyn, T.L. Oei, P. Hösli and
    C. Griscelli. Analyt. Biochem. 84:147 (1978).

8. M.P. Uitendaal, C.H.M.M. de Bruyn, T.L. Oei, S.J. Geerts and
    P. Hösli. Biochem. Med. 20:54 (1978).

9. C.H.M.M. de Bruyn, T.L; Oei, and P. Hösli. Biochem. Biophys.
    Res. Commun. 68:483 (1976).

10. O.H. Lowry, N.J. Rosebrough, A.L. Farr and R.J. Randall. J.
    Biol. Chem. 193:265 (1951).

11. E. Zoref, O. Sperling and A. de Vries. Adv. Exp. Med. Biol.
    41A:15 (1974).

12. R.B. Gordon, L. Thompson and B.T. Emmerson. Adv. Exp. Med.
    Biol. 76A:314 (1977).

13. L. Sweetman, M. Borden, P. Lesh, B. Bakay and M.A. Becker.
    Adv. Exp. Med. Biol. 76A:319 (1977).

14. M.P. Uitendaal. Studies on purine metabolism in cultured
    fibroblasts. Thesis, Nijmegen University, Studentenpers,
    Nijmegen, 1978.

S-ADENOSYLHOMOCYSTEINE METABOLISM IN ADENOSINE DEAMINASE

DEFICIENT CELLS

Michael S. Hershfield and Nicholas M. Kredich

Duke University Medical Center

Durham, North Carolina USA 27710

At the Second International Symposium on Purine Metabolism (Baden, 1976) there was much discussion of the potential role of adenosine cytotoxicity as a cause for the immune defect associated with adenosine deaminase (ADA) deficiency. Since then attention has shifted to the role of deoxyadenosine toxicity with the discovery of elevated dATP pools in blood cells of affected patients (1). I wish to briefly review the studies which have led us to propose related mechanisms of toxicity for both adenosine and deoxyadenosine to ADA deficient cells. Each mechanism involves the compound S-adenosylhomocysteine (AdoHcy) and the enzyme AdoHcy hydrolase, which catalyzes its reversible cleavage to adenosine and L-homocysteine.

S-Adenosylhomocysteine Metabolism and Transmethylation

AdoHcy is the demethylated derivative of S-adenosylmethionine (AdoMet) which is formed as a by-product of all AdoMet-dependent transmethylation reactions. About 100 such reactions, most catalyzed by specific methyltransferases, occur in mammalian cells. Residues in DNA, RNA, proteins, phospholipids, and biogenic amines are a few of the compounds which are modified by transmethylation, or whose biosynthesis involves this process. Methylation of RNA may be necessary for the maturation or function of messenger, transfer, and ribosomal RNA, while phospholipid methylation is responsible for many properties of membranes. These and other methylation reactions are required in many aspects of cell differentiation and function. AdoHcy is not only a by-product of transmethylation, but is also a potent competetive (with AdoMet) inhibitor of many methylases. Therefore AdoHcy must be (and normally is) rapidly eliminated via AdoHcy hydrolase catalyzed cleavage. However, because the thermodynamic equilibrium of this reaction greatly favors condensation of adenosine and L-homocysteine to form AdoHcy (2), further metabolism of adenosine and L-homocysteine is also necessary for efficient catabolism of AdoHcy.

AdoHcy Accumulation and Nucleotide-Independent, Uridine-Resistant Adenosine Toxicity

Until recently there was general agreement that adenosine toxicity resulted from expansion of some intracellular adenine nucleotide pool, and depletion of pyrimidine nucleotides was considered a likely mechanism (3). However, Hershfield, Snyder, and Seegmiller found that the toxicity of adenosine to the WI-L2 human B lymphoblastoid cell line was not altered by mutational loss of the enzymes necessary to convert adenosine to nucleotides (4). Furthermore, uridine did not reverse the toxicity of adenosine to adenosine kinase deficient cells. The cause of this nucleotide-independent, uridine-resistant form of adenosine toxicity was suggested by studies with mouse lymphoma cells by Kredich and Martin (5). They showed that when ADA was inhibited, exogenous adenosine was rapidly converted to intracellular AdoHcy by reversal of the AdoHcy hydrolase reaction, and that AdoHcy accumulation caused the inhibition of DNA methylation. Addition of L-homocysteine (not toxic by itself) greatly augmented both the accumulation of AdoHcy and the toxicity of adenosine, and uridine did not alter these effects.

We have extended these observations to human peripheral blood lymphocytes (6) and WI-L2 lymphoblasts (7), documenting the inhibition of RNA as well as DNA methylation. In addition we found that L-homocysteine is selectively toxic to ADA-inhibited cells which lack adenosine kinase, even in the absence of added adenosine. AdoHcy accumulated in these cells owing to the availability of endogenously produced adenosine whose alternate routes of metabolism had been eliminated (7). This result not only confirms the toxic effects of AdoHcy accumulation, but also suggests that conversion of adenosine to AMP by adenosine kinase might be important in limiting the toxicity of adenosine in the ADA-deficient patient by preventing its utilization by AdoHcy hydrolase.

Deoxyadenosine Toxicity Via Inactivation of AdoHcy Hydrolase

It had been suggested (4) that the nucleotide-independent form of adenosine toxicity might be mediated by a high affinity adenosine binding protein. Subsequently we showed that the major cytoplasmic protein which could form a stable complex with adenosine ($K_D \sim 2$-$5 \times 10^{-7} M$) copurified with AdoHcy hydrolase (8). In addition to providing independent evidence that this enzyme was the postulated "target" for adenosine, this observation led to

further studies of the interaction of AdoHcy hydrolase with adenosine analogs, and to the finding that binding of deoxyadenosine
and adenine arabinoside (Ara-A) caused the irreversible inactivation of AdoHcy hydrolase by an active site directed, "suicide-like"
mechanism (9). As a result of this observation we measured AdoHcy
hydrolase activity in the erythrocytes of 3 unrelated ADA deficient
children, and in each found <2% of control activity (10). This
study showed that irreversible inactivation of AdoHcy hydrolase by
undegraded deoxyadenosine occurred in vivo, and suggested an entirely new mechanism of deoxyadenosine toxicity which might contribute to the immune defect associated with ADA deficiency: inactivation of AdoHcy hydrolase should lead to accumulation of AdoHcy
and inhibition of methylation.

    Our current efforts are aimed at defining the consequences of
inactivation of AdoHcy hydrolase in cultured human lymphoblastoid
cell lines and in peripheral blood lymphocytes. In preliminary
studies we have documented the inactivation of AdoHcy hydrolase,
accumulation of AdoHcy, and inhibition of nucleic acid methylation
in ADA-inhibited cells exposed to deoxyadenosine. We find that
the degree to which methylation is impaired is dependent upon the
ratio of AdoMet-to-AdoHcy concentration in a cell. This is expected since AdoHcy is a competetive (with AdoMet) inhibitor of
methylases. Normally the concentration of AdoMet in cells is at
least 10 times that of AdoHcy. Inhibition of a specific methylase
appears to occur when this ratio decreases to a characteristic
value.

    These initial results predict that the greatest impairment in
methylation should occur in those cell types with lowest AdoMet
concentrations, since the AdoMet-AdoHcy ratio will be lowered to
a greater extent by a given increase in AdoHcy concentration when
AdoHcy hydrolase is inactivated. Preliminary studies suggest that
the normal AdoMet concentration in peripheral blood lymphocytes
is relatively low, and we achieved more appreciable decreases in
the AdoMet-AdoHcy ratio in these cells, and in a cultured T lymphoblastoid cell line, than in a B cell line with much higher Ado-
Met concentration.

    Several factors govern the extent of inactivation of adenosyl-
homocysteine hydrolase in the intact cell: 1) the concentration
of deoxyadenosine to which the cell is exposed; 2) the length of
time of exposure; 3) the capacity of a cell to replace inactive
enzymes via new protein synthesis; 4) the intracellular concentrations of adenosine and adenosylhomocysteine which block binding
of deoxyadenosine to the enzyme. Concentrations of deoxyadenosine
in the range of 1-2µM have been measured in the plasma of adenosine deaminase deficient patients. Whether or not there is any
fluctuation in deoxyadenosine concentrations at different times of

day or on different types of diet is not known. In addition, higher concentrations of deoxyadenosine may arise in tissues with high cell turnover in which there is considerable degradation of DNA. Assuming, however, a more or less equal exposure of all tissues to deoxyadenosine in the ADA deficient patient, and also assuming low steady state concentrations of both adenosine and adenosylhomocysteine, then the major factor which may determine the extent of inactivation of AdoHcy hydrolase in a given cell may be the rapidity with which the cell synthesizes new enzyme.

The red blood cell has a long life span and is incapable of new protein synthesis, and hence is the cell type most susceptible to inactivation of AdoHcy hydrolase. Some lymphoid cells also have long life spans, from weeks to years. Furthermore, these cells are normally in a dormant state with relatively low rates of new protein synthesis, which could further predispose their adenosylhomocysteine hydrolase to inactivation. Because of the difficulty in obtaining sufficient homogeneous preparations of peripheral blood lymphocytes from severely lymphopenic ADA deficient children, we currently do not know whether there is in fact any AdoHcy hydrolase deficiency in ADA deficient lymphocytes in vivo. Studies of AdoHcy hydrolase activity and of AdoMet and AdoHcy pool sizes in various tissues will be helpful in determining whether lymphoid cells may be selectively susceptible to AdoHcy toxicity in ADA deficient patients.

In summary we have defined two mechanisms involving AdoHcy hydrolase and AdoHcy metabolism which might operate in the ADA deficient patient. The first mechanism involves conversion of adenosine and homocysteine to AdoHcy via reversal of the normal physiologic direction of the AdoHcy hydrolase catalyzed reaction. The determinants of this mechanism are the high affinity of AdoHcy hydrolase for adenosine, and the thermodynamic equilibrium for the hydrolase catalyzed reaction which greatly favors AdoHcy formation. Evidence for operation of this mechanism in the intact ADA deficient patient has not yet been obtained. The second mechanism we have studied involves the irreversible 'suicide-like' inactivation of AdoHcy hydrolase by deoxyadenosine. This inactivation occurs in the intact cell and we have documented >98% inactivation of erythrocyte AdoHcy hydrolase activity in three ADA deficient patients. This represents a unique mechanism for producing secondary enzyme deficiency in an inborn error of metabolism. Our results suggest the possibility that deficiency of AdoHcy hydrolase in lymphocytes may contribute to the immune defect in ADA deficient patients by causing accumulation of AdoHcy to levels sufficient to inhibit critical methyl transfer reactions.

1. M. S. Coleman, J. Donofrio, J. J. Hutton, A. Daoud, B. Lampkin, and J. Dyminski, Abnormal concentrations of

deoxynucleotides in adenosine deaminase (ADA) deficiency and severe combined immunodeficiency disease (SCID). Blood 50 (Suppl. 1): 292 (Abstr.) (1977).

2. G. DeLa Haba and G. L. Cantoni, The enzymatic synthesis of S-adenosyl-L-homocysteine from adenosine and homocysteine. J. Biol. Chem. 234: 603-608 (1959).

3. H. Green and T-S. Chan, Pyrimidine starvation induced by adenosine in fibroblasts and lymphoid cells: Role of adenosine deaminase. Science 182: 836-837 (1973).

4. M. S. Hershfield, F. F. Snyder, and J. E. Seegmiller, Adenine and adenosine are toxic to human lymphoblast mutants defective in purine salvage enzymes. Science 197: 1284-1287 (1977).

5. N. M. Kredich and D. W. Martin, Jr., Role of S-adenosylhomocysteine in adenosine mediated toxicity in cultured mouse T lymphoma cells. Cell 12: 931-938 (1977).

6. J. Johnston and N. M. Kredich, Inhibition of methylation by adenosine in adenosine deaminase-inhibited, phytohemaglutinin-stimulated human lymphocytes. J. Immunol., in press (1979).

7. N. M. Kredich and M. S. Hershfield, S-adenosylhomocysteine toxicity in normal and adenosine kinase-deficient lymphoblasts of human origin. Proc. Nat. Acad. Sci. 76: 2450-2454 (1979).

8. M. S. Hershfield and N. M. Kredich, S-adenosylhomocysteine hydrolase is an adenosine-binding protein: A target for adenosine toxicity. Science 202: 757-760 (1978).

9. M. S. Hershfield, Apparent suicide inactivation of human lymphoblast S-adenosylhomocysteine hydrolase by 2'-deoxyadenosine and adenine arabinoside. J. Biol. Chem. 254: 22-25 (1979).

10. M. S. Hershfield, N. M. Kredich, D. R. Ownby, H. Ownby, and R. Buckley, In Vivo inactivation of erythrocyte S-adenosylhomocysteine hydrolase by 2'-deoxyadenosine in adenosine deaminase-deficient patients. J. Clin. Invest. 63: 807-811 (1979).

ALTERED DEOXYNUCLEOSIDE TRIPHOSPHATE LEVELS PARALLELING DEOXYADENOSINE TOXICITY IN ADENOSINE DEAMINASE INHIBITED HUMAN LYMPHOCYTES

Harry G. Bluestein, Linda F. Thompson, Daniel A. Albert, and J. Edwin Seegmiller

Department of Medicine, University of California, San Diego
San Diego, California   92103

Deficiency of the enzyme adenosine deaminase (ADA) has been associated with an inherited form of severe combined immunodeficiency disease.[1] The accumulation of deoxyadenosine-containing ribonucleotides (dATP and dADP) in the cells of these patients[2,3] has been implicated as the primary cause for the immune dysfunction. Deoxy-ATP is a potent inhibitor of ribonucleotide reductase whose activity is responsible for the formation of deoxyribonucleotides from the corresponding ribonucleoside diphosphates. Thus the accumulation of dATP could deprive the cells of ADA deficient individuals of the deoxyribonucleotides needed for DNA synthesis and thereby prevent the expansion of the lymphocyte clones needed for immune responses.

We have been examining the effects of deoxyadenosine toxicity on ADA deficient lymphoid cells in an in vitro model system. Since enzyme deficient patients are generally profoundly lymphopenic, we have been using normal lymphocytes which have been made functionally ADA deficient using 2'-deoxycoformycin, a potent ADA inhibitor. We found that the proliferation of those cells induced by the mitogen phytohemagglutinin (PHA) is inhibited by deoxyadenosine (AdR) and that the inhibition is accompanied by the intracellular accumulation of dATP. Significantly, the inhibition of PHA responsiveness by deoxyadenosine can be at least partially overcome by adding deoxycytidine (CdR), deoxyguanosine (GdR) or thymidine (TdR) to the cultures. The response to PHA can be completely reconstituted by a mixture containing all three of those deoxynucleosides.[4] Those results are compatible with the hypothesis that immunodeficiency results from the inhibition of ribonucleotide reductase

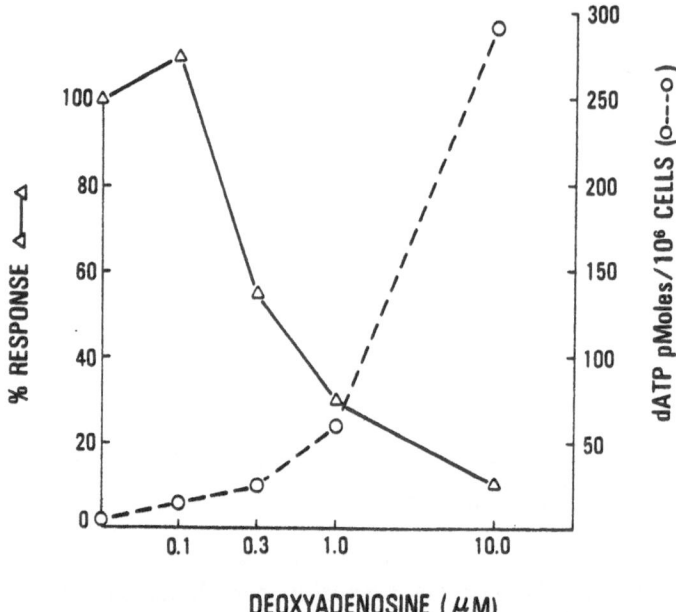

Fig. 1. Relationship between the intracellular accumulation of dATP and
the inhibition of PHA stimulation in ADA inhibited lymphocytes
cultured with AdR. Normal human PBL (10^6/ml) were cultured
with 2'-deoxycoformycin (1 µM) PHA (1 µg/ml) and AdR at the
concentrations indicated for 48 hours. ^{3}H-leucine was added to
an aliquot of the cultures for a 4 hour pulse prior to harvesting
and the response to PHA determined as:

$$\% \text{ PHA response} = \frac{^3\text{H-CPM with AdR}}{^3\text{H-CPM without AdR}}$$

The remainder of the cultured cells were extracted with 0.4N
perchloric acid and the extract neutralized with alamine-Freon.
The concentration of dATP in the extracts (pmoles/10^6 cells) was
determined by a DNA polymerase assay.[5]

by elevated intracellular dATP. That hypothesis predicts that the deoxyadenosine toxicity to ADA inhibited cells can be overcome by providing
the cells with an alternate source of the deoxynucleotides needed for DNA
synthesis. However, deoxyadenosine toxicity in this model system is
dependent upon the uptake of AdR from the culture medium, and it must
be demonstrated that the deoxynucleoside reversal of the toxicity is not
due to the blockade of the uptake of AdR. In addition, the ribonucleotide reductase inhibition hypothesis predicts that deoxyadenosine toxicity
will be accompanied by depleted pools of deoxyribonucleoside triphosphates
other than dATP. The current studies are directed at examining those
predictions.

    The response of ADA inhibited human lymphocytes to PHA is increas-
ingly suppressed as the deoxyadenosine concentration is increased above
0.1 μM (Fig. 1).   Normal human lymphocytes obtained by Ficoll-Hypaque
density gradient centrifugation of venous blood, cultured with 2'-dCf
respond normally to mitogenic stimulation by PHA.   When the PHA response
is measured as the incorporation of ^{3}H-leucine into acid-precipitable
material, the addition of 1 μM AdR suppressed the PHA response more than
60%.   The intracellular concentration of dATP in those cells increases to
60 picomoles/10^6 cells, a fivefold increase over the dATP concentration in
ADA inhibited cells cultured without AdR.   The PHA response with
10 μM AdR in the cultures is less than 10% of normal and is accompanied
by the accumulation of dATP to 290 picomoles/10^6 cells.

    The suppression of PHA stimulation of the ADA inhibited lympho-
cytes by AdR is reversed by adding a mixture of CdR, GdR and TdR to
the cultures.   The response of peripheral blood lymphocytes (PBL) cultured
with 1 μM 2'-dCf and 10 μM AdR increases from less than 10% to greater
than 50% of normal by adding a mixture of the deoxynucleosides at
concentrations of 3 μM each, and the response increases to greater than

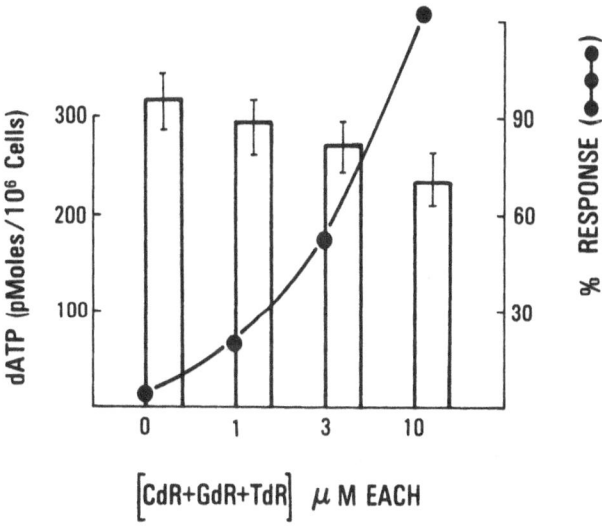

Fig. 2. The effect of deoxynucleoside reversal of AdR toxicity in ADA
         inhibited lymphocytes on the intracellular accumulation of dATP.
         Normal human PBL (10^6 cells/ml) were cultured with 2'-deoxyco-
         formycin (1 μM), PHA (1 μg/ml), AdR (10 μM) and a mixture of
         CdR, GdR, and TdR at the concentrations indicated.   After 48
         hours incubation, the % PHA response (•-•-•) and the intracellular
         dATP pools (vertical bars) were measured as described in Fig. 1.

100% with a mixture containing 10 μM concentrations of each deoxynucleoside. The return of PHA responsiveness was not due to an inhibition of
AdR uptake or phosphorylation by the lymphocytes since the dATP accumulation
decreased only slightly from 315 to 230 picomoles/10^6 cells as the concentrations of deoxynucleosides was increased from 0 to 10 μM of each (Fig. 2).

The addition of deoxyadenosine to proliferating ADA inhibited PBL leads
to a fall in the deoxycytidine triphosphate (dCTP) concentration concomi-
tantly with an accumulation of dATP (Table 1). PBL maximally stimulated by
PHA were rendered ADA deficient with 2'-dCf and then given 10 μM AdR.
After an additional two hours in culture, the intracellular pools of dATP, dCTP
and TTP were measured. The addition of AdR to the Cells led to a 3-4
fold increase in dATP concentration. Thymidine triphosphate levels were not
significantly altered by the AdR pulse.

The formation of TTP is dependent on ribonucleotide reductase
conversion of UDP to dUDP and yet TTP does not fall as dATP increases.
It has been demonstrated that the conversion of UDP to dUDP is more
resistant to the inhibitory effect of dATP than is the conversion of CDP
to dCDP.[6] In addition it has been demonstrated in mouse embryo cells
that the inhibition of ribonucleotide reductase by hydroxyurea causes a drop
in the dATP, dCTP and dGTP pool sizes, but TTP concentrations remain
elevated.[7] The less efficient inhibition of the formation of thymidylates
by dATP may actually lead to the accumulation of TTP when DNA synthesis
is completely inhibited.

Table I

Effect of Deoxyadenosine on Deoxynucleoside Triphosphates
in Proliferating ADA Inhibited Lymphocytes*

AdR (μM)	Nucleotide Concentration (pmoles/10^6 cells)		
	dATP	dCTP	TTP
0	11.1 ± 3.1	37.8 ± 1.0	16.6 ± 2.4
10	41.4 ± 9.1	14.8 ± 3.7	22.0 ± 3.7

*Normal human PBL (10^6/ml) were cultured with PHA (1 μg/ml). After
48 hours 2'-dCf (1 μM) was added to the cultures, followed by the addition
of AdR (10 μM) to one of the groups. The cells were harvested at 72 hours
and extracted with 0.4NPCA followed by neutralization with alamine-Freon.
Deoxynucleotides were assayed by a DNA polymerase assay.[5]

The in vitro response to PHA requires the triggering of lymphocytes from a resting state into active proliferation. The triggering and explosive expansion of lymphocyte clones by antigen is a unique characteristic of the immune system which may render it particularly susceptible to the biochemical defects accompanying ADA deficiency.

REFERENCES

1. E. L. Giblett, J. E. Anderson, F. Cohen, B. Pollara and H. J. Meuwissen, Adenosine deaminase deficiency in two patients with severely impaired cellular immunity, Lancet ii:1067 (1972).
2. M. S. Coleman, J. Donofrio, J. J. Hutton, L. Hahn, A. Daoud, B. Lampkin and J. Dyminski, Identification and quantitiation of adenine deoxynucleotides in erythrocytes of a patient with adenosine deaminase deficiency and severe combined immunodeficiency, J. Biol. Chem. 253:1619 (1978).
3. A. Cohen, R. Hirschhorn, S. D. Horowitz, A. Rubinstein, S. H. Polmar, R. Hong and D. W. Martin, Jr., Deoxyadenosine triphosphate as a potentially toxic metabolite in adenosine deaminase deficiency, Proc. Natl. Acad. Sci. USA 75:472 (1978).
4. H. G. Bluestein, R. C. Willis, L. F. Thompson, S. Matsumoto and J. E. Seegmiller. Accumulation of deoxyribonucleotides as a possible mediator of immunosuppression in hereditary deficiency of adenosine deaminase, Trans. Assoc. Amer. Phys. 91:394 (1978).
5. J. K. Lowe and G. B. Grindley, Inhibition of growth rate and deoxyribonucleoside triphosphate concentrations in cultured leukemia L1210 cells, Mol. Pharmacol. 12:177 (1976).
6. E. C. Moore and R. B. Hurlbert, Regulation of mammalian deoxyribonucleotide biosynthesis by nucleotides as activators and inhibitors. J. Biol. Chem. 241:4802 (1966).
7. L. Skoog and B. Nordenskjold, Effects of hydroxyurea and 1-B-D-arabinofuranosylcytosine on deoxyribonucleotide pools in mouse embryo cells, Eur. J. Biochem. 19:81 (1971).

ACKNOWLEDGMENT

This work was supported in part by grants A1-10931, GM-17702, AM-07062 and AM-13622 from the National Institutes of Health.

AUTHOR INDEX